# Earthquake-Resistant Structures:
## Design and Analytical Aspects

# Earthquake-Resistant Structures: Design and Analytical Aspects

Edited by **Bruno Crump**

NY RESEARCH PRESS

New York

Published by NY Research Press,
23 West, 55th Street, Suite 816,
New York, NY 10019, USA
www.nyresearchpress.com

Earthquake-Resistant Structures: Design and Analytical Aspects
Edited by Bruno Crump

International Standard Book Number: 978-1-63238-119-4 (Hardback)

Printed in the United States of America.

# Contents

Preface                                                                                                                 VII

Chapter 1    **Design Principles of Seismic Isolation**                                          1
             George C. Lee and Zach Liang

Chapter 2    **Seismic Design Forces and Risks**                                               37
             Junichi Abe, Hiroyuki Sugimoto and
             Tadatomo Watanabe

Chapter 3    **Seismic Bearing Capacity of Shallow Foundations**                 57
             Francesco Castelli and Ernesto Motta

Chapter 4    **Advanced Base Isolation
             Systems for Light Weight Equipments**                                         79
             Chong-Shien Tsai

Chapter 5    **Seismic Damage Estimation in
             Buried Pipelines Due to Future Earthquakes –
             The Case of the Mexico City Water System**                              131
             Omar A. Pineda-Porras and Mario Ordaz

Chapter 6    **Masonry and Earthquakes: Material Properties,
             Experimental Testing and Design Approaches**                        151
             Thomas Zimmermann and Alfred Strauss

Chapter 7    **The Equivalent Non-Linear Single Degree of
             Freedom System of Asymmetric Multi-Storey
             Buildings in Seismic Static Pushover Analysis**                       179
             Triantafyllos K. Makarios

Chapter 8    **Seismic Performance of Masonry Building**                          205
             Xiaosong Ren, Pang Li,
             Chuang Liu and Bin Zhou

Chapter 9   **Seismic Response of Reinforced Concrete Columns**       **227**
Halil Sezen and Muhammad S. Lodhi

Chapter 10  **Seismic Behavior and Retrofit of Infilled Frames**        **251**
Mohammad Reza Tabeshpour,
Amir Azad and Ali Akbar Golafshani

Chapter 11  **Seismic Vulnerability Analysis of
RC Buildings in Western China**       **279**
Zhu Jian

**Permissions**

**List of Contributors**

# Preface

This book has been an outcome of determined endeavour from a group of educationists in the field. The primary objective was to involve a broad spectrum of professionals from diverse cultural background involved in the field for developing new researches. The book not only targets students but also scholars pursuing higher research for further enhancement of the theoretical and practical applications of the subject.

The book provides an overview of the latest developments and advances related to earthquake-resistant structures. It comprises of research works contributed by various experts and researchers in the field of earthquake engineering. The book discusses seismic-resistance design of masonry and reinforcement of concrete structures with safety measurements of strengthening and rehabilitation of existing structures against earthquake loads. It also covers topics dedicated to assessment and rehabilitation of jacket platforms, electromagnetic sensing mechanisms for health assessment of structures, post-earthquake examining of steel buildings in fire environment and response of underground pipes to blast loads. This book will be of help to graduate students, researchers and practicing structural engineers.

It was an honour to edit such a profound book and also a challenging task to compile and examine all the relevant data for accuracy and originality. I wish to acknowledge the efforts of the contributors for submitting such brilliant and diverse chapters in the field and for endlessly working for the completion of the book. Last, but not the least; I thank my family for being a constant source of support in all my research endeavours.

**Editor**

# 1

# Design Principles of Seismic Isolation

George C. Lee and Zach Liang

*Multidisciplinary Center for Earthquake Engineering Research,*
*University at Buffalo, State University of New York*
*USA*

## 1. Introduction

In earthquake resistance design of structures, two general concepts have been used. The first is to increase the capacity of the structures to resist the earthquake load effects (mostly horizontal forces) or to increase the dynamic stiffness such as the seismic energy dissipation ability by adding damping systems (both devices and/or structural fuses). The second concept includes seismic isolation systems to reduce the input load effects on structures. Obviously, both concepts can be integrated to achieve an optimal design of earthquake resilient structures. This chapter is focused on the principles of seismic isolation.

It should be pointed out that from the perspective of the structural response control community, earthquake protective systems are generally classified as passive, active and semi-active systems. The passive control area consists of many different categories such as energy dissipation systems, toned-mass systems and vibration isolation systems. This chapter addresses only the passive, seismic isolation systems [Soong and Dargush, 1997; Takewaki, 2009; Liang et al, 2011]

Using seismic isolation devices/systems to control earthquake induced vibration of bridges and buildings is considered to be a relatively matured technology and such devices have been installed in many structures world-wide in recent decades. Design guidelines have been established and they are periodically improved as new information based on research and/or field observations become available during the past 20-30 years [ATC 1995; SEAONC 1986; FEMA 1997; IBC 2000; ECS 2000; AASHTO 2010, ASCE 2007, 2010].

Besides the United States, base isolation technologies are also used in Japan, Italy, New Zealand, China, as well as many other countries and regions. [Naiem and Kelly, 1999; Komodromos, 2000; Christopoulos, C. and Filiatrault 2006]

Affiliated with the increased use of seismic isolation systems, there is an increased demand of various isolation devices manufactured by different vendors. This growth of installing seismic isolation devices in earthquake engineering has been following the typical pattern experienced in structural engineering development, which begins from a "statics" platform by gradually modifying the design approach to include the seismic effects based on structural dynamics principles as they develop and new field observations on the responses of real-world structures. The process is typically slow because most studies and laboratory observations have been concentrated on the performances of the devices with scaled-down experiments. Results could not be readily scaled-up for design purposes. At the same time, there were very limited field data on the actual performances of seismically isolated structures. In recent years, some limited successful stories were reported in the literature on the seismic performance of base-isolated bridges and buildings during real earthquakes, as well as reports of unsuccessful cases including the failure of isolation bearings and falling spans of bridges and magnification (rather than reduction) of vibration levels of buildings. These structural failures have not been systematically examined for their contributing factors. Some of them include: construction quality, improper choice of the type of isolation bearings, incomplete design principles and methods, unknown ground motion and soil characteristics, etc. In summary, current practice is mainly based on past research and observations on the performance of the isolation devices themselves, with minimum information on the dynamic performance of the structure-device as a system.

The working principle of seismic isolation may be explained in several ways. It is a general understanding that isolation devices/systems are used to reduce the seismic force introduced base shear. Designers often understand the working principles from the viewpoint of design spectrum in that, when the vibration period of a structure is longer than a certain level, continue to increase the period will reduce the magnitude of the spectral value and thus reduce the base shear accordingly. To qualitatively explain the working principle of seismic isolation in this manner is reasonable, but it is insufficient to use it in actual design. Refinements and additional design principles are necessary. Another commonly used explanation of seismic isolation is the "decoupling" of the superstructure vibration from the ground motions excitations to reduce the vibration of structures. This statement again requires quantitative elaborations from the viewpoint of isolation design. In general, an isolation device/system can be viewed as a low-pass mechanical filter of the structure being isolated, to filter out excitations with the undesirable high frequencies to reduce the level of acceleration. In order to establish the cut-off frequency the period of the isolation system must be carefully addressed, and this requires a basic design principle to guide the design.

In order to reduce the base shear, an isolation system must be allowed to deform. This relative displacement cannot be filtered like the absolute acceleration. In general working range, the longer the relative displacement associated with longer periods, the more reduction in base shear can be achieved, except for the fact that the latter will introduce certain negative effects. The most significant issue of large relative displacement is the large P-delta effect and for falling spans in bridges. In this regard, a design principle is needed to achieve the best compromise in seismic isolation design. In reality, the only approach to effectively reduce the relative displacement is to increase damping, which, in turn, will result in higher level of acceleration. This conflicting demand of controlling acceleration and acceptable displacement in essence defines the limiting range of the effectiveness of seismic isolation systems. Quantitatively this issue can be addressed, and this is an important design principle to be conceptually discussed in this chapter.

In Section 2 of this chapter, several important design issues (e.g. P-delta effect, vertical motions, etc.) will be discussed and seismic isolation design principles will be described. In Section 3, the quantitative basis of treating seismically-isolated structures will be briefly reviewed and simplified models will be established for the dynamic analysis and design of the structure-device system. Design methods will be briefly discussed in Section 4, and a newly developed seismic isolation device to address some of the issues facing today's practice is briefly introduced in Section 5. Finally, the key issues and parameters in seismic isolation is summarized and future research needs are briefly noted.

## 2. Some issues and principles of seismic isolation

In this section, the theories, design and practical considerations of seismic isolation are briefly discussed.

### 2.1 State-of-practice on seismic isolation

The principle of base isolation is typically conceptually explained by using figure 2.1.

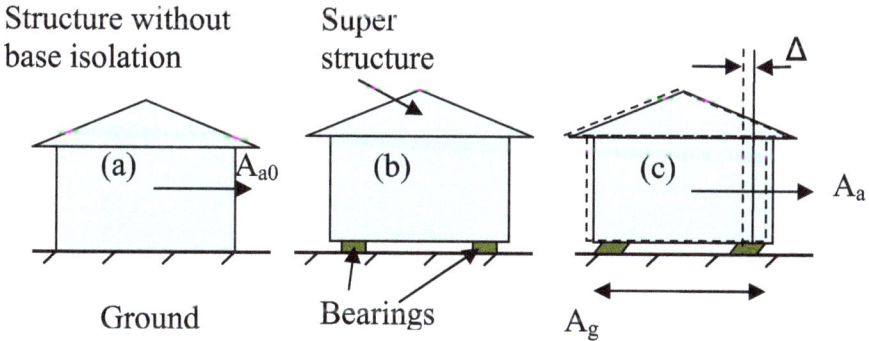

Fig. 2.1. Concept of base isolation

Figure 2.1 (a) and (b) show a structure without and with base isolation, respectively. It is seen that the major difference between (a) and (b) is that in (b), the structure is set on top of isolation several bearings. When the ground moves with acceleration $A_g$, the superstructure will move accordingly with a displacement $\Delta$, and a reduced level of absolute acceleration, denoted by $A_a$, rather than the original level $A_{a0}$.

This principle has been the basis for many research and development efforts in design guidelines and devices for seismic isolation. These have been many important contributions during the past 25 years led by Kelly and his associates at Berkeley and Constantinou and his associates at Buffalo as well as many others. (See references listed at the end of this chapter.)

### 2.2 Basic concept

The major purpose of using the seismic isolation is to reduce the base-shear of the structure. Physically, large base shear is one of the main reasons of structural damages due to strong horizontal ground accelerations. Thus, to reduce the lateral acceleration is a basic principle.

From the viewpoint of design, many aseismic codes use the base shear as a control parameter. For example, if the base shear of a building is reduced, then the upper story lateral forces floor drifts are also reduced. In the case of a bridge, base shear reduction will minimize damage to the piers.

### 2.2.1 Base shear

Base shear V can be calculated through various approaches. The following are several examples, first

$$V = C_s W \tag{2.1}$$

where $C_s$ is the seismic response factor and $W$ is the total weight of a structure. Base isolation is intended for reducing $C_s$, second

$$V = \Sigma f_{Lj} \; (kN) \qquad (2.2)$$

where $f_{Lj}$ is the later force of the $j^{th}$ story of the structure. Base isolation is intended to reduce $f_{Lj}$ simultaneously, so that the base shear will be reduced, in addition

$$V = K_b \Delta \; (kN) \qquad (2.3)$$

where $K_b$ is the lateral stiffness of the bearing system; $\Delta$ is the nominal relative displacement of the bearing. The stiffness $K_b$ of the bearing system will be much smaller than the structure without the bearing, so that the base shear is reduced.

### 2.2.2 Lateral acceleration
In equation (2.1), $C_s$ is in fact the normalized lateral absolute acceleration $A_a$, which is in general zero-valued unless earthquake occurs.

$$C_s = A_a / g \qquad (2.4)$$

Also note that

$$W = Mg \; (kN) \qquad (2.5)$$

where M is the total mass and $g = 9.8 \, m/s^2$ is the gravity.
In equation (2.2), $f_{Lj}$ is also caused by lateral absolute acceleration $a_{aj}$ of the $j^{th}$ story, that is

$$f_{Lj} = m_j a_{aj} (kN) \qquad (2.6)$$

where $m_j$ is the mass of the $j^{th}$ floor.
From (2.5) and (2.6), it is seen that, it is difficult to change or reduce the mass M or $m_j$ in a design; however, if the acceleration can be reduced, the lateral forces will be reduced. Therefore, we will focus the discussion on the acceleration.

### 2.3 Issues of base isolation
Seismic isolation is considered as a relatively matured technology as evidenced by the many practical applications. These applications have been designed based on codes and provisions that have been established incrementally over time. In the following, seismic isolation principles are examined from a structural dynamics perspective with an objective to suggest additional future research needs.

### 2.3.1 Absolute acceleration vs. relative displacement
Seen in figure 2.1, to achieve the goal of acceleration reduction, in between the ground and the super structure, there will be installed in a group of bearings, which have much soft stiffness so that the period of the total system will be elongated.

Thus, to achieve acceleration reduction, a major sacrifice is the relative displacement between base and structure must be significantly large. Due to nonlinearities of isolation system, the dynamic displacement can be multiple-centered, which can further notably enlarge the displacement. In addition, permanent residual deformation of bearing may worsen the situation.

Generally speaking, the simplest model of a base isolation system can be expressed as

$$M\,a_a(t) + C\,v(t) + K_b d(t) = 0 \tag{2.7}$$

where $C\,v(t)$ is the viscous damping force and $C$ is the damping coefficient; $v(t)$ is relative velocity.

In most civil engineering structures, the damping force is very small, that is

$$C\,v(t) \approx 0 \tag{2.8}$$

Thus, (2.7) can be re-written as

$$M\,A_a \approx K_b \Delta \tag{2.9}$$

where $A_a$ and $\Delta$ are amplitudes of the absolute acceleration $a_a(t)$ and relative displacement $d(t)$. Equation (2.9) describes the relationship between acceleration and displacement of a single-degree-of-freedom (SDOF) system, which can be used to generate the design spectra. Since the damping force is omitted, the generated acceleration is not exactly real, which is referred to as pseudo acceleration, denoted by $A_s$. Thus, (2.9) is rewritten as

$$M\,A_s = K_b \Delta \tag{2.10}$$

Furthermore, we have

$$A_s = \omega_b^2 \Delta \tag{2.11}$$

where $\omega_b$ is the angular natural frequency of the isolation system

$$\omega_b = \sqrt{K_b/M} \ \ (\text{rad/s}) \tag{2.12}$$

Since the natural period $T_b$ is

$$T_b = 2\,\pi\,/\,\omega_b \ \ (\text{s}) \tag{2.13}$$

Equation (2.11) can be rewritten as

$$A_s = 4\pi^2 / T_b^2 \ \Delta \tag{2.14}$$

From (2.14), the acceleration $A_s$ and displacement $\Delta$ are proportional, that is

$$A_s \propto \Delta \tag{2.15}$$

Since $A_s$ and $\Delta$ are deterministic functions, (2.15) indicates that between $A_s$ and $\Delta$ , only one parameter is needed, usually, the acceleration is considered.

Combine (2.12), (2.13) and (2.14), it is also seen that $A_s$ is proportional to the stiffness $K_b$, that is

$$A_s \propto K_b \tag{2.16}$$

That is, the weaker the stiffness is chosen, the smaller value of $A_s$ can be achieved, which is the basis for current design practice of seismic isolation.

From the above discussion, it seems that, as long as $K_b$ is smaller than a certain level, the base isolation would be successful.

However, the above mentioned design principle may have several problems if the assumptions and limitations are not examined. First, from (2.10), it is seen that, only if $\Delta$ is fixed, (2.16) holds. On the other hand, if V, that is $A_a$, is fixed, one can have

$$\Delta \propto 1/K_b \tag{2.17}$$

That is, the weaker the stiffness is chosen, the larger value of $\Delta$ can result.

A more accurate model will unveil that, to realize the base isolation, only one parameter, say $A_a$ or $\Delta$, is not sufficient. This is because $A_a$ and $\Delta$ are actually independent. In fact, both of them are needed. That is, whenever a claim of the displacement being considered in an isolation design, as long as they are not treated as two independent parameters, the design is questionable. Later, why they must be independent will be explained. Here, let us first use certain group of seismic ground motions as excitations applied on a SDOF system, the ground motion are suggested by Naiem and Kelly (1999) and normalized to have peak ground acceleration (PGA) to be 0.4 (g). The responses are mean plus one standard deviation values, plotted in figure 2.2.

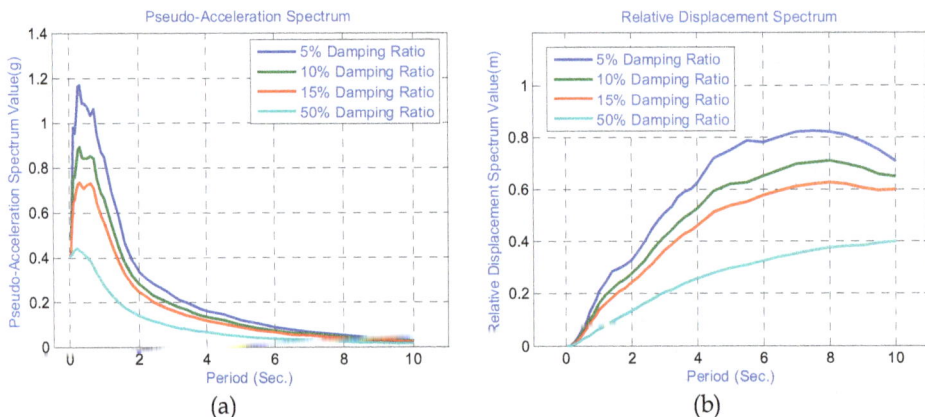

Fig. 2.2. Statistical seismic responses of SDOF systems

Figure 2.2 (a) shows the accelerations of SDOF systems as functions of period for selected damping ratios. When the period becomes larger, the accelerations do reduce, especially when $T_b > 2$ (s). The responses are all smaller than 0.4 (g) and namely, the acceleration is reduced.

Figure 2.2 (b) shows the displacements. It is seen that, when the periods increase, the displacements can become rather large. When $T_b > 2$ (s), the responses can be larger than 0.1 (m), especially if the damping ratio is small, say, 5%.

In figure 2.2 the parameter, damping ratio denoted by $\xi$ is defined by

$$\xi = C / 2\sqrt{MK_b} \qquad (2.18)$$

## 2.3.2 Displacement and center position

Another seismic isolation design issue is the self-centering capacity. Because the SDOF system used to generate the response is linear, and many commercially available bearings are nonlinear systems, the displacement time history can be multiple centered. Figure 2.3 shows examples of a bi-linear (nonlinear) system under Northridge earthquake excitations. The plot in Fig. 2.3 (a) is the displacement time history of the bi-linear system with load and unload stiffness ratio = 0.1 and damping ratio = 0.01. Here the damping ratio is calculated when the system is linear. The plot in Fig 2.3 (b) is the same system with damping ratio = 0.2. It is seen that, the biased deformations exist in both cases. This example illustrates that center shifting can enlarge the displacement significantly, even with heavy damping.

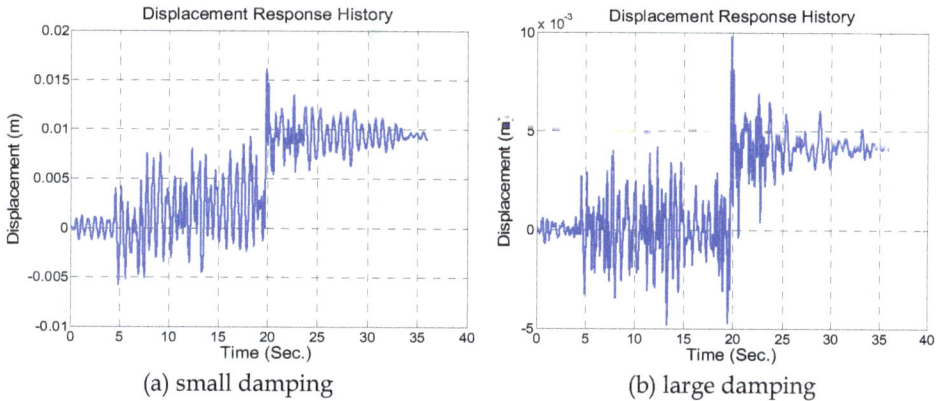

(a) small damping                  (b) large damping

Fig. 2.3. Multiple centers of displacement responses

Briefly speaking, the current isolation design practice can be based on the spectral analysis, dealing with linear systems with the seismic response coefficient $C_s$ given by

$$C_s = A_s / g = \frac{AS}{T_b B} \qquad (2.19)$$

and the spectral displacement $d_D$ given by

$$d_D = D = C_s\, T_b^2 / 4\pi^2\, g = \frac{AST_b}{4\,B}\ (m) \qquad (2.20)$$

In the above equations, A is the input level of ground acceleration; S is the site factor; D, instead of $\Delta$, is used to denote the dynamic amplitude and B is called the numerical damping coefficient. Approximately

$$B = 3\xi + 0.9 \qquad (2.21)$$

Equations (2.19) and (2.20) will work for most (but not all) situations but cannot handle the position shifting which is a nonlinear response. Some of the nonlinear modeling issues are discussed in Section 3.

### 2.3.3 SDOF and MDOF models

Many analyses and design of seismic isolations are based on SDOF model, whereas realistic structures are mostly MDOF systems. The acceleration of a higher story of an MDOF system can be much more difficult to reduce.

Equation (2.15) is based on a SDOF system, when the superstructure can be treated as a lumped mass. Namely, the relative deformation among different stories of the structure is negligible. Realistically, such a case is rare; therefore, the acceleration $a_{aj}$ at different stories can be different.

A conventional idea is to decouple a MDOF system into several vibration modes. Each mode is treated as a SDOF system. The total response of the MDOF can then obtained through certain method of modal combinations, such as SRSS method. That is, (2.19) can be rewritten as

$$C_{si} = \frac{AS}{T_{bi} B_i} \tag{2.22}$$

where the subscript i stands for the $i^{th}$ mode. And the $i^{th}$ spectral displacement $d_{iD}$, (2.20), is rewritten as:

$$d_{iD} = C_{si} \, T_{bi}^2 / 4\pi^2 \, g = \frac{AST_{bi}}{4 \, B_i} \; (m) \tag{2.23}$$

where

$$B_i = 3\xi_i + 0.9 \tag{2.24}$$

Furthermore, for multi-story structures, an additional parameter, mode shape, is needed to distribute the acceleration and displacement at different levels. By denoting the mode shape by $\mathbf{P}_i$, which is a vector with the $j^{th}$ element representing the model displacement $p_{ji}$. The acceleration vector $\mathbf{A}_{si}$ is given as

$$\mathbf{A}_{si} = \Gamma_i C_{si} \mathbf{P}_i \, g = \left\{ a_{ji} \right\} \; (m/s^2) \tag{2.25}$$

where $a_{ji}$ is the acceleration of the $i^{th}$ mode at the $j^{th}$ story and $\Gamma_i$ is the $i^{th}$ modal participation factor.

The displacement vector is given by

$$\mathbf{d}_{si} = \Gamma_i d_{iD} \mathbf{P}_i = \left\{ d_{ji} \right\} \; (m) \tag{2.26}$$

where $d_{ji}$ is the displacement of the $i^{th}$ mode at the $j^{th}$ story.

### 2.3.3.1 Acceleration of higher stories

One of the shortcomings of approximately a MDOF system by a SDOF model is the inability to estimate the responses at the higher levels of a MDOF structure. Typically, through SRSS, the acceleration of the $j^{th}$ story of MDOF system can be calculated as

$$a_{sj} = \sqrt{\sum_i a_{ij}^2} \ (m/s^2) \qquad (2.27)$$

And the displacement of the $j^{th}$ story is

$$d_j = \sqrt{\sum_i d_{ji}^2} \ (m) \qquad (2.28)$$

The reduction of MDOF accelerations is discussed in the literature. The general conclusion is that, (2.25) and (2.27) may not work well so that the acceleration of the higher story $a_{sj}$ may not be calculated correctly. In fact, $a_{sj}$ can be significantly larger.

The reasons of this inaccuracy mainly come from several factors. The first is the damping effect. The second is the error introduced by using only the first mode for design simplicity (the triangular shape function) as illustrated in figure 2.4

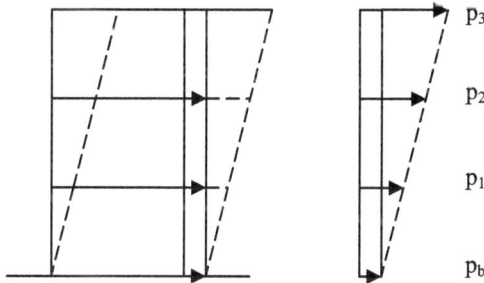

Fig. 2.4. Mode shape function of the $1^{st}$ mode

In addition, seen from figure 2.4, even through the acceleration of the base, denoted by $p_b$ is rather small by using isolators, the top story will have a rather large acceleration.

### 2.2.3.2 Cross effects

Another shortcoming of using SDOF models is the inability to estimate the cross effect. Typical MDOF structures have "cross effects" in their dynamic responses. Different from a single member of a structure, which has principal axes, a three-dimensional structure often does not have principal axes. This is conceptually shown in figure 2.5, which is a two story structure. Suppose the first story does have its own principal axes, marked as $x_1$-$y_1$, and the second story also has its own principal axes, marked as $x_2$-$y_2$. However, from the top view, if $x_1$ and $x_2$ are not pointing exactly the same direction, say, there exists an angle $\theta$, then the entire structure will not have principal axes in general. In this case, the inputs from any two perpendicular directions will cause mutual responses. The resulted displacement will be further magnified. This third reason of large displacement. At present, there are no available methods to quantify cross effects associated with seismic responses, although in general this effect in base isolation design may be small for regularly shaped structures.

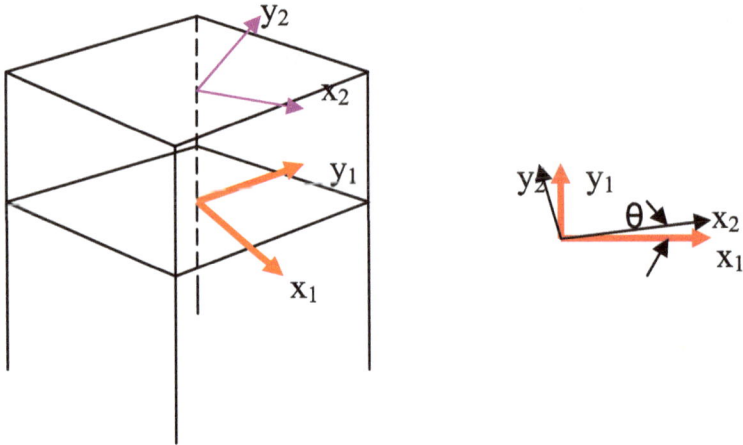

Fig. 2.5. Cross effect

### 2.3.4 Overturning moment

The product of large vertical load and large bearing displacement forms a large P-$\Delta$ effect, conceptually shown in figure 2.6.

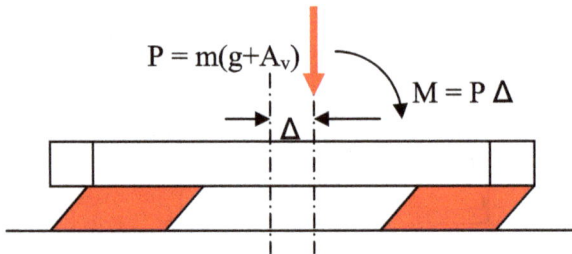

$$P = m(g+A_v)$$ $$M = P\,\Delta$$

Fig. 2.6. Isolation P-$\Delta$ effect

This large displacement will result in an overturning moment, given by

$$M = P\Delta = m\left(g + A_v\right)\Delta \tag{2.29}$$

where, the vertical load P is a product of the total weight and the vertical acceleration, the vertical acceleration is the sum of gravity g and earthquake induced acceleration $A_v$. For example, suppose the total mass is 1000 (ton), the displacement is 0.5 (m), and the additional vertical acceleration is 0.4 (g), the total overturning moment will be 6.86 (MN-m). This is a large magnitude, which requires special consideration in the design of foundations, structure base as well as bearings.

### 2.3.5 Horizontal and vertical vibrations

As mentioned above, the primary purpose of base isolation is to reduce the horizontal load and/or acceleration. By installing bearings, the lateral stiffness will be significantly reduced so the horizontal vibration can be suppressed. Moreover, by using bearings, the vertical stiffness can also be reduced to a certain level and the vertical vibration can be magnified. Since the earthquake induced vertical load is often not significantly large, the vertical vibration is often ignored in design. However, due to the magnification of vertical acceleration as well as the above-mentioned large overturning moment, care must be taken to check the vertical load. In the worst scenario, there can be an uplift force acting on bearings with many of them not manufactured to take the uplift load (e.g. rubber bearings).

## 3. Dynamics of seismically Isolated MDOF systems

In this section, base isolation is examined from a different perspective, as a second order mechanical filter of a dynamic system. The working principle of base isolation system is to increase the dynamic stiffness of acceleration without sacrifice too much "dynamic stiffness." Dynamic stiffness is a function of effective period and damping.

### 3.1 Models of isolation systems
### 3.1.1 Linear SDOF model

The linear SDOF model is used here to provide a platform to explain the essence of isolation systems under sinusoidal excitation.

A more detailed base isolation model can be rewritten as

$$M\,a(t) + C\,v(t) + K_h d(t) = -M\,a_g(t) \tag{3.1}$$

where $a(t)$ and $a_g(t)$ are respectively the relative and ground accelerations. Note that

$$a_a(t) = a(t) + a_g(t) \tag{3.2}$$

$$a(t) = \dot{v}(t) = \ddot{d}(t) \tag{3.3}$$

where the overhead dot and double dot stand for the first and the second derivatives with respect to time t.

Let the ground displacement $d_g(t)$ be sinusoidal with driving frequency $\omega_f$,

$$d_g(t) = D\cos(\omega_f t) \tag{3.4}$$

The ground acceleration is

$$a_g(t) = A_g\cos(\omega_f t) = -D_g\,\omega_f^2\cos(\omega_f t) \tag{3.5}$$

Here, $D_g$ and $A_g$ are respectively the amplitudes of the displacement and acceleration. The amplitude of steady state responses of the relative displacement D can be written as

$$D = \frac{D_g\dfrac{\omega_f^2}{\omega_b^2}}{\sqrt{\left(1-\dfrac{\omega_f^2}{\omega_b^2}\right)^2 + \left(2\xi\dfrac{\omega_f}{\omega_b}\right)^2}} = \frac{r^2}{\sqrt{\left(1-r^2\right)^2 + \left(2\xi r\right)^2}}D_g = \beta_d D_g \tag{3.6}$$

Here r is the frequency ratio

$$r = \frac{\omega_f}{\omega_b} \tag{3.7}$$

and $\beta_d$ is the dynamic magnification factor for the relative displacement

$$\beta_d = \frac{r^2}{\sqrt{\left(1-r^2\right)^2 + \left(2\xi r\right)^2}} \tag{3.8}$$

Note that r = 0 means the driving frequency $\omega_f$ =0 and the system is excited by static force only. In this case, D(r =0) = 0.

The amplitude of steady state responses of the absolute acceleration $A_a$ can be written as

$$A_a = \sqrt{\frac{1+\left(2\xi r\right)^2}{\left(1-r^2\right)^2 + \left(2\xi r\right)^2}} \ A_g = \beta_A A_g \tag{3.9}$$

and $\beta_A$ is the dynamic magnification factor for the relative displacement

$$\beta_A = \sqrt{\frac{1+\left(2\xi r\right)^2}{\left(1-r^2\right)^2 + \left(2\xi r\right)^2}} \tag{3.10}$$

Note that when the driving frequency $\omega_f$ = 0 and the system is excited by static force only. $A_a(r =0) = A_g$. If there exist a static force $F_{SA}$ so that

$$A_g = \frac{F_{SA}}{M} \tag{3.11}$$

Then, when r ≠ 0 or $\omega_f$ ≠ 0, the dynamic response of the acceleration can be seen as

$$A_a = \frac{F_{SA}}{M}\beta_A = \frac{F_{SA}}{M/\beta_A} = \frac{F_{SA}}{M_\beta} \tag{3.12}$$

where $M_\beta$ is called apparent mass or dynamic mass and

$$M_\beta = M/\beta_A \tag{3.13}$$

That is, the value of the dynamic response $A_a$ can be seen as a static force divided by a dynamic mass. From (3.12),

$$M_\beta = \frac{F_{SA}}{A_a} \tag{3.14}$$

The essence described by (3.14) is that the "dynamic mass" equals to a force divided by response. Generally speaking, this term is stiffness. Since the response is dynamic, it can be called a dynamic stiffness, $K_{dA}$. In this particular case, the dynamic stiffness is the apparent mass.

$$F_{SA} = K_{dA}A_a \qquad (3.15)$$

Similarly, the relationship between a static force $F_{Sd}$ and a dynamic displacement D can be written as

$$F_{Sd} = K_{dD}D \qquad (3.16)$$

It can be seen that

$$K_{dA} \propto \frac{1}{\beta_A} \qquad (3.17)$$

And

$$K_{dD} \propto \frac{1}{\beta_b} \qquad (3.18)$$

Both the dynamic stiffness and the dynamic factors are functions of frequency ratio r. It can be proven that, the proportional coefficients for (3.17) and (3.18) are constant with respect to r. That is, we can use the plots of dynamic magnification factors to shown the variations of the dynamic stiffness with respect to r. Figure 3.1(a) and (b) show examples of $\beta_A$ and $\beta_b$. It can be seen from figure 3.1(a), the plot of $\beta_A$, when the frequency ratio varies, the amplitude will vary accordingly. Note that,

$$r = \frac{\omega_f}{\omega_b} = \frac{\omega_f}{2\pi} T_b \qquad (3.19)$$

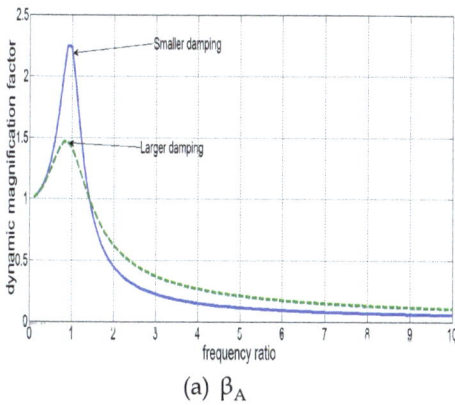

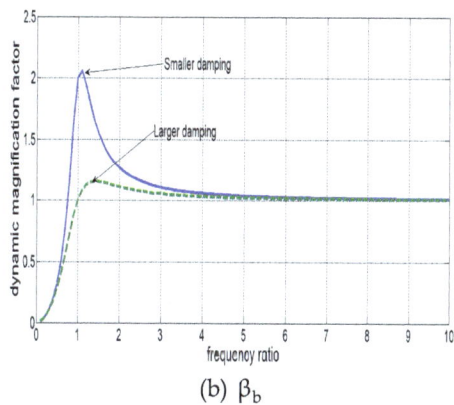

(a) $\beta_A$

(b) $\beta_b$

Fig. 3.1. Dynamic magnification factors

That is, if the driving frequency is given, then the frequency ratio r is proportional to the natural period $T_b$ . Therefore, the horizontal plot can be regarded as the varying period. Recall figure 2.2(a) where x-axis is also period. We can realize the similarity of these two plots. First, when the period increases, the amplitudes in both plot increases. After a certain level, these amplitudes start to decrease. It is understandable that, the target of acceleration reduction should be in the region when the accelerations being smaller than a certain level. In figure 3.1 (a), we can clearly realize that this region starts at r = 1.414, which is called *rule 1.4*. To let the isolation system start to work, the natural period should be at least 1.4 times larger than the driving period $T_f = 2\pi/\omega_f$. In the case of earthquake excitation, due to the randomness of ground motions, there is no clear-cut number. However, from figure 2.2 (a), it seems that this number should be even larger than 1.4. In fact, with both numerical simulations and shaking table tests, for linear and nonlinear as well as SDOF and MDOF systems, much higher values of the frequency ratio have been observed by the authors. In general, this limiting ratio should be indeed greater than 1.4.

$$r > 1.4 \qquad\qquad (3.20)$$

Additionally, different damping will result in different response level. In figure 3.1(a), it is seen that, when r > 1.4, larger damping makes the reduction less effective. In figure 2.2 (a), it can be seen that when the period becomes longer, increase damping still further reduce the responses, but the effectiveness is greatly decreased. Note that, this plot is based on statistical results. For many ground excitations, this phenomenon is not always true. That is, increasing damping can indeed reduce the effectiveness of reduction.

Now, consider the displacement. From figure 3.1(b), when the period is near zero, the relative displacement is rather small. Then, as the period increases, the displacement will reach the peak level and then slowly decreases. However, it will never be lower than 1, which means that, the displacement will never below the level of ground displacement.

Increasing the damping, however, will help to reduce the large displacement. This may be seen from Fig. 2.2(b) and it is true for both sinusoidal and earthquake excitations.

### 3.1.2 Nonlinear SDOF model

The above discussion is based on a linear model. Nonlinear system will have certain differences and can be much more complex to analyze. Generally speaking, many nonlinear systems considerably worsen the problem of large displacement. In the working region, increasing the period will decrease the acceleration but the displacement will remain at a high level. Increasing the damping will help to reduce the displacement but does not help the acceleration.

### 3.1.2.1 Effective system

It is well understood that, in general, a nonlinear dynamic system does not have a fixed period or damping ratio. However, in most engineering applications, the effective period $T_{eff}$ and damping ratio $\xi_{eff}$ are needed, and are calculated from

$$T_{eff} = \frac{2\pi}{\sqrt{M/K_{eff}}} \qquad\qquad (3.21)$$

And

$$\xi_{eff} = \frac{E_d}{2\pi K_{eff} D^2}$$

(3.22)

where $E_d$ is the energy dissipation of the nonlinear system during a full vibration cycle.
From (3.21) and (3.22), it is seen that, to have the effective values, the key issue is to establish the effective stiffness $K_{eff}$.
In the following, the bi-linear system is used to discuss the effective stiffness, for the sake of simplicity. Although this system is only a portion of the entire nonlinear systems, the basic idea of conservative and dissipative forces seen in the bilinear system can be extended to general nonlinear systems. And, in many cases, nonlinear isolation is indeed modeled as a bi-linear system. Figure 3.2 shows (a) the elastic-perfectly-plastic (EEP) system and (b) the general bi-linear system. Conventionally, the secant 'stiffness" is taken to be the effective stiffness. That is

$$K_{eff} = K_{sec} = \frac{f_N}{D}$$

(3.23)

where D is the maximum displacement and $f_N$ is the maximum nonlinear force. In the following, since the system is nonlinear and the .response is random, we use lower case letters to denote the responses in general situations, unless these responses will indeed reach their peak value.

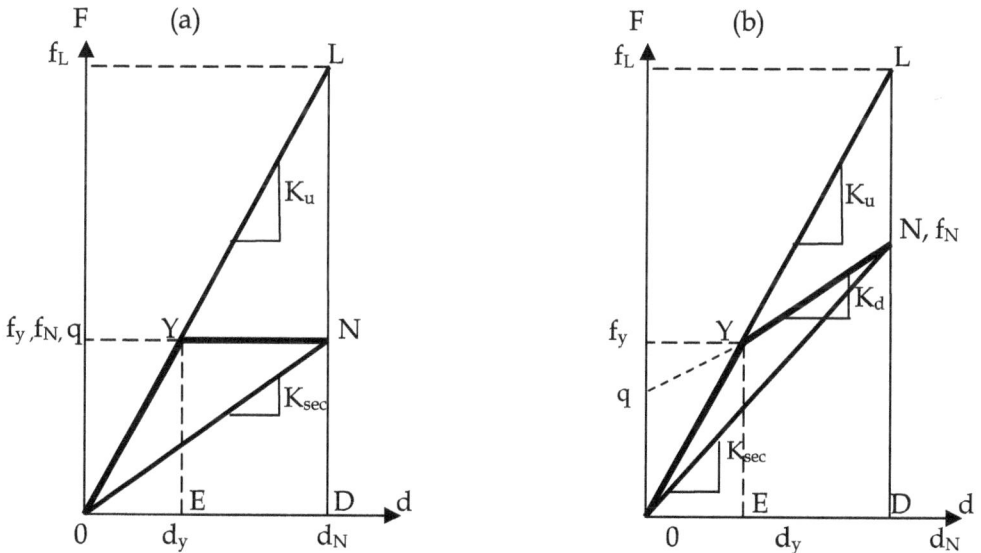

Fig. 3.2. Secant stiffness

In figure 3.2, $d_y$ and $d_N$ are the yielding and the maximum displacements. $f_y$ and $f_N$ are the yielding and the maximum forces. $K_d$ and $K_u$ are the loading and unloading stiffness. If the system remains elastic, then we will have a linear system when the displacement reaches the maximum value D, the force will be $f_L$. However, since the system is nonlinear, the maximum force $f_N$ will be smaller. Thus the corresponding effective stiffness will be affected.

For seismic isolation, the measurement of the effective stiffness can be considerably overestimated. To see this point, let us consider the definition of stiffness in a linear system. It is well known that the stiffness denotes capability of how a linear system can resist external force. Suppose under a force $f_L$, the system has a deformation d, and then the rate defines the stiffness. That is

$$K = \frac{f_L}{D} \tag{3.24}$$

When the load $f_L$ is released, the linear system will return to its original position. Thus, the stiffness also denotes the capacity for how a system can bounce back after the force $f_L$ is removed, for at displacement d, the linear system will have a potential energy $E_p$

$$E_p = \frac{K\,d^2}{2} \tag{3.25}$$

Therefore, we can have alternate expression for the stiffness, which is

$$K = \frac{2E_p}{d^2} \tag{3.26}$$

Apparently, in a linear system, the above two expressions of the stiffness, described by (3.24) and (3.26), are identical. This is because the potential energy can be written as

$$E_p = \frac{f_L d}{2} \tag{3.27}$$

However, in a nonlinear system, this equation will no longer hold, because the maximum force $f_N$ can contain two components: the dissipative force $f_d$ and the conservative force, $f_c$,

$$f_N = f_c + f_d \tag{3.28}$$

For example, in Figure 3.2(a), when $f_N$ is reached,

$$f_c = 0$$

$$f_d = q$$

Thus,

$$f_N = f_d = q$$

Namely, the EPP system does not have a conservative force. Its restoring force is zero. From another example of the general bilinear system shown in figure 3.2 (b),

$$f_d = q$$

and

$$f_c = f_N - f_d$$

Note that, only the conservative force contributes to the potential energy $E_p$. That is,

$$E_p = \frac{f_c d}{2} < \frac{f_N d}{2} \tag{3.29}$$

In this case, using (3.24) and (3.26) will contradict each other. In order to choose the right formulae to estimate the effective period and damping, the estimation of effective stiffness must be considered more precisely. Generally speaking, a nonlinear system will have the same problem as the above-mentioned bi-linear system, that is, the restoring force is smaller than the maximum force, as long as the nonlinear system has softening springs.

### 3.1.2.2 Estimation of effective period

From the above-discussion, when we use an effective linear system to represent a nonlinear system, the effective stiffness should satisfy the following:

$$K_{eff} = \frac{2E_p}{d^2} \tag{3.30}$$

$$K_{eff} = \frac{f_c}{d} \tag{3.31}$$

By using (3.30) as well as (3.31), the effective stiffness $k_{eff}$ will be smaller than the secant stiffness $k_{sec}$.

Furthermore, it can be seen that, vibration is caused by the energy exchange between potential and kinetic energies. The natural frequency of a linear system can be obtained by letting the maximum potential energy equal the maximum kinetic energy, that is, through the relation

$$\frac{k\, d^2}{2} = \frac{M\, v^2}{2} = \frac{M\, \omega_b^2\, d^2}{2} \tag{3.32}$$

In nonlinear systems, we should modify the above equation as

$$\frac{k_{eff}\, d^2}{2} = \frac{m\, \omega_{eff}^2\, d^2}{2} \tag{3.33}$$

Or,

$$\omega_{eff} = \sqrt{\frac{k_{eff}}{M}} = \sqrt{\frac{f_c/d}{M}} < \sqrt{\frac{f_N/d}{M}} \tag{3.34}$$

In other words, considering the dynamic property of a nonlinear system, we should not use the secant stiffness as the effective stiffness. Following this logic, the effective stiffness should be defined in a nonlinear system by considering the restoring potential energy as follows.

In the bi-linear case (see the shaded areas in Figure 3.3), when the system moves from 0 to d, the potential energy is

$$E_p = \frac{1}{2}\left[K_u d_y^2 + K_d\left(d - d_y\right)^2\right] \tag{3.35}$$

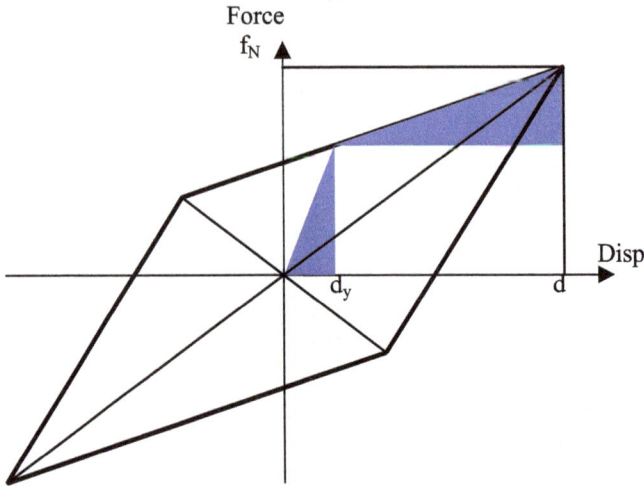

Fig. 3.3. Maximum potential energy of a bi-linear system

Therefore,

$$K_{eff} = \frac{2E_p}{d^2} = \frac{K_u d_y^2 + K_d\left(d - d_y\right)^2}{d^2} \tag{3.36}$$

By using the displacement ductility $\mu$, we can write

$$K_{eff} = \frac{K_u + K_d\left(\mu - 1\right)^2}{\mu^2} \tag{3.37}$$

where $\mu$ is the displacement ductility and

$$\mu = \frac{d}{d_y} \tag{3.38}$$

The relationship between $K_d$ and $K_u$ is given as

$$K_d = a\, K_u \tag{3.39}$$

Using this notation, we have

$$K_{eff} = \frac{1 + a\left(\mu - 1\right)^2}{\mu^2} K_u \tag{3.40}$$

Therefore, the corresponding effective period is

$$T_{eff} = \sqrt{\frac{\mu^2}{1 + a(\mu - 1)^2}} \, T_1 = 2\pi \, \mu \sqrt{\frac{M}{\left[1 + a(\mu - 1)^2\right] K_u}} \qquad (3.41)$$

From (3.41) and (3.23), the effective stiffness estimated by the secant method can overestimate the period by the factor

$$\frac{\frac{1 + a(\mu - 1)}{\mu}}{\frac{1 + a(\mu - 1)^2}{\mu^2}} = \frac{1 + a(\mu - 1)}{1 + a(\mu - 1)^2} \mu \qquad (3.42)$$

Note that, the term a is smaller than unity and μ is larger than unity. Therefore, from (3.42), it is seen that the period can be notably underestimated by using the secant stiffness. For example, suppose a = 0.2 and μ = 4, the factor will be greater than 2.

Similar to the estimation of effective period, the term $K_{eff}$ will also affect the measurement of the effective damping ratio. Using the same logic, the damping ratio can also be underestimated. These double underestimations may cancel each other to a certain degree for acceleration computation, but will certainly underestimate the displacement. For a detailed discussion on structural damping, reference is made to Liang et al (2012).

### 3.1.3 MDOF models

Now, let use consider a linear n-DOF systems along one direction, say the X-direction. The governing equation can be written as

$$M\ddot{X}(t) + C\dot{X}(t) + KX(t) = -M J a_g(t) \qquad (3.43)$$

Here M, C and K are nxn mass, damping and stiffness matrices, J = $\{1\}_{nx1}$ is the input column vector and

$$X(t) = \begin{Bmatrix} d_1(t) \\ d_2(t) \\ ... \\ d_n(t) \end{Bmatrix}_{nx1} \qquad (3.44)$$

where $d_j(t)$ is the displacement at the jth location.

If the following Caughey criterion hold

$$CM^{-1} K = K M^{-1} C \qquad (3.45)$$

The system is proportionally damped, which can be decoupled into n-SDOF systems, which are referred to as n-normal modes, so that the analysis of SDOF system can be used.

In base isolation, to regulate the displacement, large damping must be used. When the damping force is larger, isolation system is often not proportionally damped. That is, (3.45) will no longer hold and the system cannot be decoupled into n SDOF normal modes. In this case, it should be decoupled in 2n state space by using the state equations.

Configuration of MDOF structures and non-proportional damping will both affect the magnitude of the accelerations of higher stories. However, the acceleration of the base, as well as the relative displacement between the base and the ground will not be significantly affected by using the normal mode approach. Practically speaking, more accurately estimate the displacement is important then the computation of the acceleration of higher stories.

Furthermore, when the isolation system is non-linear, its first "effective" mode will dominate. Thus, we can use its first "effective" mode only to design the system.

However, care must be taken when the cross effect occurs. It is seen that, if both the motion of both X and Y directions, which are perpendicular, are considered, we can have

$$\begin{bmatrix} \mathbf{M}_X & \\ & \mathbf{M}_Y \end{bmatrix}\begin{Bmatrix} \ddot{X} \\ \ddot{Y} \end{Bmatrix}+\begin{bmatrix} \mathbf{C}_{XX} & \mathbf{C}_{XY} \\ \mathbf{C}_{YX} & \mathbf{C}_{YY} \end{bmatrix}\begin{Bmatrix} \dot{X} \\ \dot{Y} \end{Bmatrix}+\begin{bmatrix} \mathbf{K}_{XX} & \mathbf{K}_{XY} \\ \mathbf{K}_{YX} & \mathbf{K}_{YY} \end{bmatrix}\begin{Bmatrix} X \\ Y \end{Bmatrix}=-\begin{bmatrix} \mathbf{M}_X & \\ & \mathbf{M}_Y \end{bmatrix}\begin{Bmatrix} J\,a_{xg} \\ J\,a_{yg} \end{Bmatrix} \qquad (3.46)$$

where $a_{xg}$ and $a_{yg}$ are ground accelerations along X and Y directions respectively, which are time variables and for the sake of simplicity, we omit the symbol (t) in equation (3.46). The subscript X and Y denote the directions. The subscript XX means the input is in x direction and the response is also in X direction. The subscript XY means the input is in X direction and the response is in Y direction, and so on.

In (3.46), $\mathbf{M}_X = \mathbf{M}_Y$ are the mass matrices. It is seen that, both $\mathbf{C}_{XY} = \mathbf{C}_{YX}^T$ and $\mathbf{K}_{XY} = \mathbf{K}_{YX}^T$ are the cross terms. Generally speaking, if the system is rotated with the help of rotation matrix we can minimize the cross terms $\mathbf{C}_{XY}$ and $\mathbf{K}_{XY}$.

$$\Theta=\begin{bmatrix} \cos(\theta)\mathbf{I} & -\sin(\theta)\mathbf{I} \\ \sin(\theta)\mathbf{I} & \cos(\theta)\mathbf{I} \end{bmatrix}_{2n\times2n} \qquad (3.47)$$

The angle $\theta$ can be chosen from 0 to $90^0$. However, in this case no matter how the angle $\theta$ is chosen, at least one of $\mathbf{C}_{XY}$ and $\mathbf{K}_{XY}$ is not null and hence, the input in X direction will cause the response in Y direction. That is defined here as the *cross effect*, which implies energies of ground motions from one direction is transferred to another. There are many reasons to generate the cross effect. In certain cases, the cross effect can considerably magnify the displacement.

## 3.2 Bearings and effect of damping
### 3.2.1 Role of damping
From the above discussion, it is seen that there are several possible inaccuracies that may exist in estimating the displacement. The results can often be underestimation of displacement in isolation design.

As pointed out earlier, the only way to reduce the displacement in the working range of isolation systems is to increase the damping. For example, from figure 2.2(b), it is seen that, at 3 second period, if the damping ratio is taken to be 50%, the displacement is about 0.2 (m) whereas using 5% only can cause about 0.5 (m) displacement, which is about 2.5 times larger.

### 3.2.2 Damping and restoring stiffness
Often base isolators or isolation bearings are designed to provide the required damping. Practically speaking, damping can be generated by several means.

The first kind of damping is material damping, such as high damping rubber bearings, lead-core rubber bearings as well as metallic dampers. The damping mechanism is generated through material deformations. Note that, high damping rubbers often do not provide sufficient damping.

The second is surface damping. The damping mechanism is generated through surface frictions of two moving parts, such as pendulum bearings, friction dampers. Note that, surface damping often is insufficient and can have a significant variation from time to time.

The third type of damping is viscous damping, which is often provided by hydraulic dampers. This kind of damping is more stable but with higher cost.

Closely related to damping generated by isolator is the restoring stiffness, since both of them are provided by bearings. In fact, the type of method of proving restoring stiffness and damping classifies the type of bearings.

The restoring stiffness can be generated by material deformations, including specific springs, which relies on the deformation to restore potential energy. Releasing of such potential energy can make the bearing return to its center position. Rubber bearings, including high damping rubber bearings, lead core rubbers bearings, steel-rubber-layered bearings fall into this category. Bearings with sliding surface and elastomeric springs also use restoring force generated by material deformation. Bearings with metal ribs which provide both stiffness and damping also fall into this category.

The restoring stiffness can also be generated by geometric shaped. Generally, when horizontal motion occurs, such bearings generate vertical movement which can also restore potential energy. Again, release of potential energy enables the bearing return to its center position. Pendulum sliding bearing falls into this category. Recently, a newly type of roller bearing is developed, which guarantees a low level constant horizontal acceleration, and will be described in Section 5.

## 4. Selected design considerations

### 4.1 Design windows

As mentioned earlier, the key issue in design of isolation system is an optimal compromise of the acceleration and the displacement, within an acceptable range of the period called "design window" period. The reason to consider the concept of design window is to avoid possible undesired displacement, which in general is independent to acceleration.

### 4.1.1 Lower bond of period

In order to reduce the acceleration, the period of the isolation system cannot be shorter than a given value. Although this value depends upon the site, it can be roughly estimated to be 1.5 to 2 seconds. This value actually establishes the left boundary of the design windows on the design spectrum. This lower limit is well understood. Because of the assumption of negligible damping, acceleration is taking as directly proportional to the displacement. Thus in typical design practice, it is assumed that displacement bound is given once the acceleration limit is established. This is not always true, because acceleration and displacement are in general independent.

### 4.1.2 Upper bond of period

In order to limit the displacement to a reasonable value, the period of the isolation system cannot be too long. This defines the right side boundary of the design window. In figure 4.1, the design spectrum and the design window for the isolation system is conceptually illustrated.

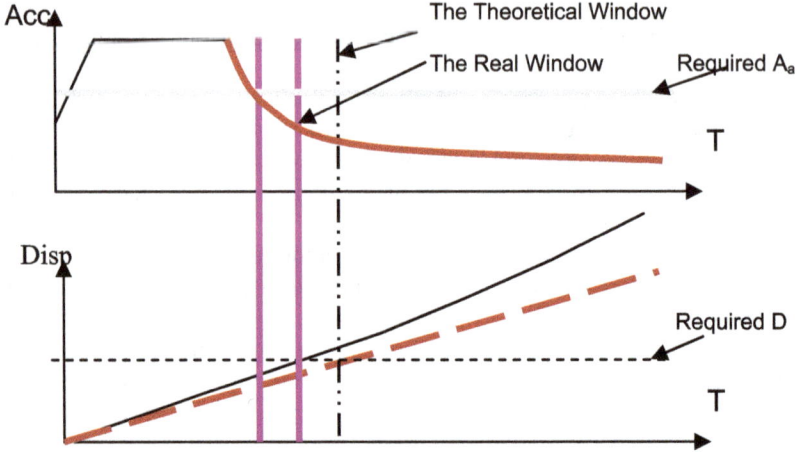

Fig. 4.1. Design Window for Isolation Systems

Figure 4.1 illustrates that if the design has a period that falls outside this "window," it is unacceptable. Within this window, reductions of the acceleration and tolerance displacement must be considered simultaneously. In this sense, a proper isolation system is an optimally compromised design.

It should be noted that, when damping is large, the reduction of acceleration will be reduced. When damping is too low, the displacement will be increased. Thus, the proper damping will have upper and lower period bonds from the viewpoint of isolation design.

### 4.2 Left boundary of the design window, acceleration related parameters
### 4.2.1 Design force

In a typical aseismic design, the base shear V typically needs to be designed by

$$s_A V \le [V] \tag{4.1}$$

where [V] is the allowable base shear and $s_A$ is a factor of safety. The base shear V can be determined by several ways. One is that given by (2.1), which is repeated here.

$$V = C_s W \tag{4.2}$$

To determine the seismic response factor $C_s$, the effective stiffness $K_{eff}$, or $K_{eff}$ in the case of MDOF system is required. The effective stiffness often should not be determined by using the secant stiffness, as previously explained.

At the same time, the base shear is equal to the product of the lateral stiffness $K_b$ and the displacement D of the bearing system plus the damping force $F_d$ measured at the maximum displacement.

$$V = K_b D + F_d \tag{4.3}$$

The specifications related to the damping force for isolators are provided by bearing vendors. For bi-linear damping, the damping force is roughly the dissipative force or the characteristic strength q (see fig. 3.2b). Using the maximum value of q, denoted by Q,

$$F_d = Q \tag{4.4}$$

For friction damping

$$F_d = \mu W \tag{4.5}$$

where $\mu$ is the friction coefficient.
For multi-story structures, base on the SRSS method,

$$V = \sqrt{\sum_{j=1}^{n} f_{Lj}^{2}} \tag{4.6}$$

where, $f_{Lj}$ is the lateral force of the $j^{th}$ story and

$$f_{Lj} = m_i \sqrt{\sum_{i=1}^{s} a_{ji}^{2}} \tag{4.7}$$

in which $m_j$ is the mass of the $j^{th}$ story and $a_{ji}$ is the absolute acceleration of the $i^{th}$ mode at the $j^{th}$ story. Note that, for an n-DOF system, there will be n- modal accelerations, and, usually the first few modes will contain sufficient vibration energy of the system. To be on the safe side, the desirable number of modes to be considered is

$$S = S_r + S_f \tag{4.8}$$

where $S_r$ is the number of the first few modes that contain 90~95% vibration energy and

$$S_f = 1\sim3, \tag{4.9}$$

as an extra safety margin particularly for irregular structures. This should not add too much computational burden.
To determine $S_r$, modal mass ratio $\gamma_i$ is often needed,

$$\gamma_i = \frac{\left(P_i^T M J\right)^2}{P_i^T M P_i} \tag{4.10}$$

and $S_r$ can be determined by

$$\sum_{i=1}^{S_r} \gamma_i > 90 \sim 95\% \tag{1.11}$$

In (4.7) the absolute acceleration $a_{ji}$ is the $j^{th}$ element of the $i^{th}$ acceleration vector $\mathbf{a}_i$, which is determined by

$$\mathbf{a}_i = \Gamma_i\, C_{si}\, \mathbf{P}_i \tag{4.12}$$

The term $C_{si}$ is the $i^{th}$ seismic coefficient, which can be determined through the building or bridge code. The modal participation factor $\Gamma_i$ is given by

$$\Gamma_i = \frac{\mathbf{P}_i^T \mathbf{M}\, \mathbf{J}}{\mathbf{P}_i^T \mathbf{M}\, \mathbf{P}_i} \tag{4.13}$$

Equations (4.10), (4.12) and (4.13) contain the mode shape function $\mathbf{P}_i$ which can be a triangular approximation, or more precisely, be given by the following eigen-equation with the normalization by letting the roof modal displacement equal to unity.

$$\omega_i{}^2\, \mathbf{P}_i = (\mathbf{M}^{-1}\mathbf{K})\, \mathbf{P}_i \tag{4.14}$$

Note that, the base shear calculated from (4.2) only considers the first mode, which may not be sufficiently accurate. Equation (4.3), is theoretically workable but practically speaking, the dissipative force $F_d$ is difficult to establish.

Note that, in (4.12), the mode shape $\mathbf{P}_i$ is obtained through (4.14), however, $\mathbf{P}_i$ is the displacement mode shape but not necessarily the acceleration mode shape. Therefore, cares must be taken by using (4.12) to calculate the modal acceleration; otherwise a design error can be introduced. For the limited space, the detailed explanation is not discussed in this chapter. Interested reader may see Liang et al 2012.

### 4.2.2 Overturning moment

Because of the potentially large P-Δ effect, base isolation design must carefully consider the overturning moment to ensure that the uplifting force is not magnified. Furthermore, most bearings cannot take tensions, whereas the overturning moment can generate uplift tension on bearings.

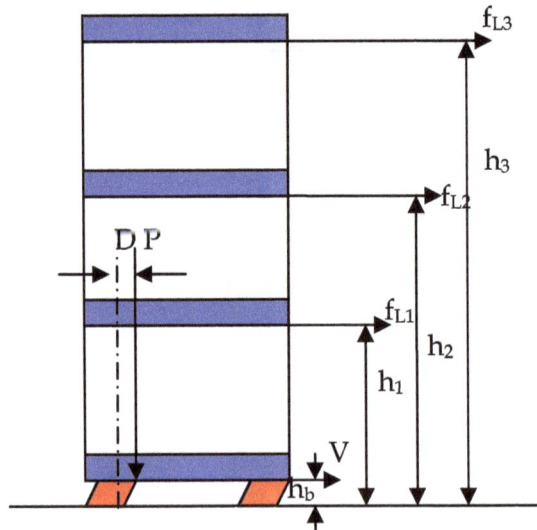

Fig. 4.2. Overturning moment

From figure 4.2, it is seen that the total overturning moment is given by

$$M_T = P\,D \, + Vh_b + \sum_{j=1}^{n} f_{Lj}h_j \qquad (4.15)$$

If the structure has notable plan irregularity, the additional moment due to the asymmetric distribution of mass must also be considered.
With the help of $M_T$ and knowing the geometric dimensions of the isolated structure, the uplift force can be calculated, and the corresponding criterion can be set up as

$$s_M\,M_T \leq [M_T] \qquad (4.16)$$

where $s_M$ is a safety factor and $[M_T]$ is allowable moment.
The above two criteria define the left side boundary of the design window (see fig 4.1).

## 4.3 Right boundary of design window: Independency of displacement
### 4.3.1 Independency of displacement
That the displacement D and the acceleration $A_a$ are independent parameters mentioned previously will be briefly explained here. For linear system, from (2.7) we have

$$d(t) = -\frac{C}{K_b}v(t) \, - \, \frac{M}{K_b}a_n(t) \qquad (4.17)$$

For large amount of damping, consider the peak values, we have

$$D \approx \sqrt{\left(\frac{C}{K_b}V\right)^2 + \left(\frac{M}{K_b}A_a\right)^2} \, > \, \frac{M}{K_b}A_a = \omega_b^2 A_a \qquad (4.18)$$

Additionally, for nonlinear system, softening behavior often occurs, so that the maximum force is not proportional to the maximum displacement. As a result, one cannot simply calculate D due to the nonlinearity of the most isolation systems, as

$$D = \frac{T_b^2}{4\pi^2}C_s\,g \qquad (4.19)$$

In design, the bearing displacement is as important as the acceleration related parameters. Due to the uncertainties illustrated above, a safety factor should be used before further research results are available.

$$s_D\,D \, \leq [D] \qquad (4.20)$$

where $s_D$ is a safety factor and $[D]$ is allowable displacement.

### 4.3.2 Right boundary due to displacement
Since the displacement needs to be regulated, the right boundary of the design window is defined. Because of the above-mentioned reasons, by using the safety factor $s_D$ , the right boundary will further shift leftwards.

It is seen that, the resulted window can be rather narrow.

## 4.4 Probability-base isolation design
In recent years, the probability-based design for civil engineering structures against natural and man-made load effects attracts more and more attentions. The basic idea is to treat both the loads and resistance of structures as random variables and to calculate the corresponding failure probability, based on which the load and resistant factors are specified, (see Nowak and Collins, 2000). Probability-based design for base isolation systems should be one of the frontier research areas for earthquake engineering researchers.

### 4.4.1 Failure probabilities of base-isolated structures
In the above discussions of seismic isolation, the design process is a deterministic approach because it is established on deterministic data.

A base-isolated building or bridge is a combination of civil engineering structures and mechanical devices, the bearings. In most cases of mechanical engineering devices, the safety factors are considerably larger than those used by civil engineers. The main reason, among many, is that the civil engineering structures have much larger redundancy. Mechanical devices, on the other hand, do not have such a safety margin. This raises an interesting issue on safety factors for seismically isolated structures.

Take the well known design spectrum, for example. Its spectral value is often generated by the sum of mean value plus one standard deviation. However, the maximum value can be much larger than the spectral value. For example, consider under 99 earthquake statistics of a structure with period = 2 second and damping ratio = 20%. The mean-plus-one standard deviation value of acceleration is 0.24 (g) whereas the maximum value is 0.47 (g). And the mean-plus-one standard deviation value of displacement is 0.22 (m), whereas the maximum displacement value is 0.45(m).

The question is: Given an isolation design, what is the chance that 0.47 (g) acceleration and/or 0.45 (m) displacement could occur? More specifically, if the allowed acceleration [A] and/or the allowed displacement [D] are preset, what is the chance that the acceleration and/or the bearing displacement can exceed the allowed design values? From the viewpoint of probability based design, the above-mentioned chance of exceeding is referred to as the failure probability. Therefore, a new concept of design criteria for seismic isolation may be stated by

$$p_{fA} = P(A \geq [A]) \leq [p_{fA}] \tag{4.21}$$

and

$$p_{fD} = P(D \geq [D]) \leq [p_{AD}] \tag{4.22}$$

Here, the subscripts $f_A$ and $f_D$ stand for failure of acceleration and failure of displacement respectively. Symbol [(.)] means the allowable value of (.).

### 4.4.2 Computation of failure probabilities
Figure 4.3 illustrates the probability distributions of a linear isolation system of $T_b = 2$ second and $\xi = 20\%$ under the excitation of 99 earthquake records. Figure 4.3(a) is the absolute acceleration and (b) is the relative displacement. The X-axes are the numerically

simulated values. The y axes are the probability density function. For example, in figure 4.3(a), this function is denoted as $f_A(a)$, namely the probability density curve is a function of the level of acceleration $a$.

Suppose the allowed acceleration

$$[A] = 0.25 \text{ (g)}$$

The failure probability can be calculated as

$$P_{fA} = \int_{0.25}^{\infty} f_A(a)\, da$$

The integral is the shadow area. From figure 4.3 (a), we can see that, it is about 16%. This failure probability is not small at all. Note that, in this case, the allowed value being 0.25 (g) is satisfied by using the mean-plus-one-standard-deviation value.

As another example, suppose the allowed displacement is

$$[D] = 0.29 \text{ (m)}$$

$$[D] = 0.29 \text{ (m)}$$

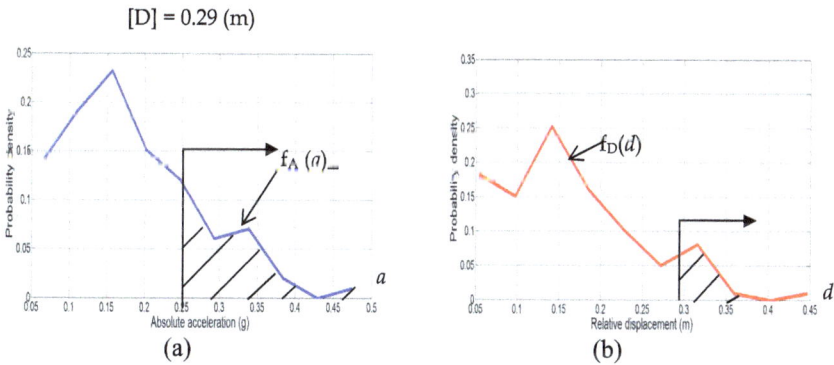

(a)                                                         (b)

Fig. 4.3. Probability distributions, $T_b = 2$ (sec), $\xi = 20\%$, 99 records, PGA = 0.4 (g).

The failure probability can be calculated as

$$P_{fD} = \int_{0.29}^{\infty} f_D(d)\, dd$$

The integral is also the shadow area. From figure 4.3 (b), we can also see that, this failure probability is not small either. Note that, in this case, the allowed value being 0.29 (m) is satisfied by using the mean-pulse-one-standard-deviation value.

Generally, the computations can be described as follows

$$P_{fA} = \int_{[A]}^{\infty} f_A(a)\, da \qquad (4.23)$$

for acceleration and

$$P_{fD} = \int_{[D]}^{\infty} f_D(d)\, dd \qquad (4.24)$$

for displacement.

The failure probability implies the chance of not meeting the design requirement. For acceleration, because of the structural redundancy of earthquake resilience, the allowed failure probability can be higher. Anyhow, 15% chance of failure may not acceptable by the owners of the structures (e.g. seismic isolation systems for nuclear reactors). For displacement, usually a bearing cannot tolerance virtually any exceedance. Thus, the allowed failure probability should also be lower.

This brief discussion above suggests that the design safety considerations for the combined civil engineering structures with mechanical devices may be examined by using a probability-based principle and approach.

### 4.5 Seismic isolation bearings

There are several types of isolation bearings commercially available, such as lead-rubber, high damping rubber as well as sliding bearings. Generally speaking, as long as the bearing system can satisfy the requirement of the above-mentioned design windows in the service life time of the isolation system, any bearing may be employed. Interested readers can consult Komodromos, P.I. (2000). In seismic isolation design, it is suggested that the allowed bearing displacement be used as the primary criterion; the effective period obtained via effective stiffness as the second criterion and providing the proper amount of damping as the third criterion for bearing selections. Because of the difficulty to choose correct type of bearings within the design window, a new type of bearing is introduced as described in Section 5. .

In the literature, certain non-traditional approaches of using relatively inexpensive bearings are reported. Many inexpensive bearings do not allow large displacements and have very limited service life. Therefore, they should not play an important role in seismic isolation design for bridges. Furthermore, the cost of the isolation bearing is often a smaller fraction of the total cost of bridge construction. In most cases, using inexpensive bearings may not be a good choice.

## 5. A new seismic isolation device

The above discussion on seismic isolation technology suggests that there are many aspects which require further research and development efforts, especially in the area to establish quantitative boundaries of design period in order to achieve optimal designs. In fact, among all commercially available base isolators, none seems to provide a good balance between acceleration reduction and reasonable cost. A recent development, the roller bearing, is briefly explained in this section. It has the promise to utilize damping to achieve desirable compromise in isolative design.

### 5.1 Concept of roller bearing

The core component of this device is the rolling assembly, which generates a desired low and constant acceleration of the superstructure regardless of the magnitude of the ground excitations. To explain the principle of constant acceleration, consider the simplified sketch of a roller traveling on an inclined surface shown in figure 5.1.

In figure 5.1, the inclined surface shown has a constant slope with an angle $\theta$ measured from the horizontal axis. The motion has an absolute acceleration in the horizontal direction, denoted by $a_h$. The roller has a relative acceleration $a_r$ along the sloping surface, which has a vertical component $a_v$ and a horizontal component $a_{rh}$. $A_g$ is the ground acceleration which is acting in the horizontal direction without a vertical component: In other words, the roller will have an absolute vertical acceleration, which is $a_v$.

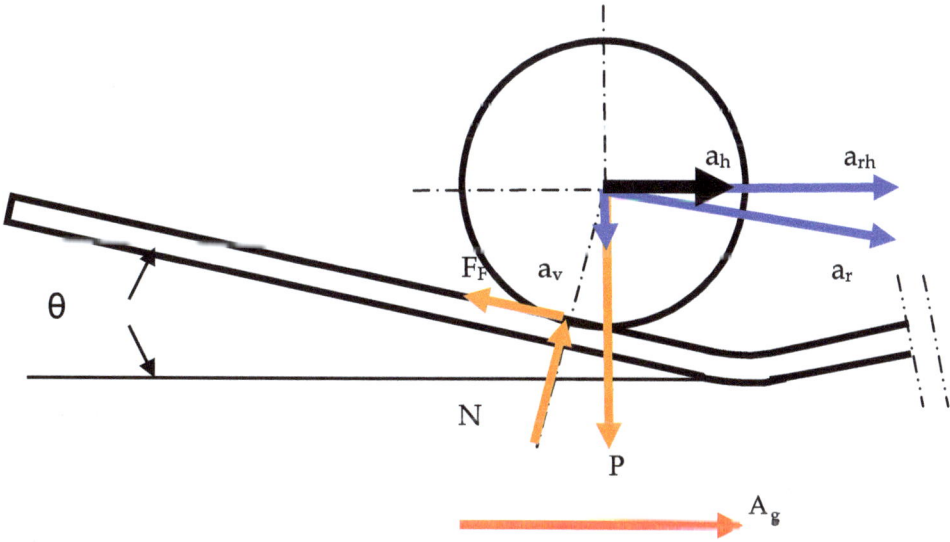

Fig. 5.1. Accelerations and forces of roller bearing

From figure 5.1, it is seen that, the amplitude of the friction force is

$$F_F = \mu\, N \tag{5.1}$$

where $\mu$ is the rolling friction coefficient.
The relative vertical and horizontal accelerations relationships are respectively given by

$$a_v = \tan\theta\; a_{rh} \tag{5.2}$$

and

$$a_h = a_{rh} + A_g \tag{5.3}$$

When the angle $\theta$ is sufficiently small, say,

$$\theta < 5^0 \tag{5.4}$$

and let the friction coefficient $\mu$ be less than 0.2:

$$\mu < 0.2 \tag{5.5}$$

in this case,

$$a_h \approx \frac{\sin\theta - \mu\cos\theta}{\cos\theta + \mu\sin\theta}\; g \tag{5.6}$$

Here, we assume

$$\frac{\sin\theta + \mu\cos\theta}{\cos\theta - \mu\sin\theta}\tan\theta \approx 0 \tag{5.7}$$

Because the quantities $\theta$, $\mu$ and g are all constant, the horizontal absolute acceleration is a constant. Note that the ground acceleration, $A_g$, does not appear in equation (5.6) [Lee et al 2007]. In other words, *the amplitude of the acceleration of the superstructure will be constant*. The first objective of having a constant and low amplitude acceleration of the superstructure thus may be achieved.

Next, if the friction is sufficiently small,

$$\mu << \theta$$

then,

$$a_h \approx \theta\, g \tag{5.8}$$

The rotation friction coefficient of the rolling motion will only be a fraction of one percent. However, with $\theta$ less than $5^0$, the angle will be several percent (unit in radians). That is, $\theta$ will be about ten times larger than $\mu$. The condition $\mu << \theta$ is then satisfied.

Next, the vertical acceleration can be given as

$$a_v \approx -A_g\,\tan\theta \tag{5.9}$$

Since $\tan\theta \approx \theta$ is for very small angles, the vertical acceleration due to the horizontal ground excitation $A_g$ is very small and can be shown quantitatively that it is negligible [Ou et al 2009].

## 5.2 Design parameters

The major design parameters for the roller isolation bearing are listed as follows:

Maximum vertical load per bearing: $P_r$ [kips];

Maximum permanent load: $L_{pm}$ [kips]

Lateral force (use the shear pins controlled the lateral force): $0.12L_{pm}$ [kips]

Roller diameter $D_r$ and length $l_r$:

$$l_r D_r \geq c.P_r \frac{E_s}{8\sigma_y^2}\left[in^2\right] \tag{5.10}$$

where

$\sigma_y$ = specified minimum yield strength of the weakest material at the contact surface

$E_s$ = Young's modulus for material

$P_r$ = load asymmetrical coefficient

c =1 for single direction rolling bearing

c = 0.7 for double direction rolling bearing

Fig. 5.2 shows the configuration of a prototype roller bearing assembly which was manufactured for laboratory studies. Details on the performance and design of roller isolation bearings may be found in publications by the authors and their colleagues that are given in the references.

### 5.3 Roller bearing implementation

With the support of a Federal Highway Administration grant, the new roller isolation bearing has been installed in an actual bridge to observe the constructability and other related issues in Rhode Island. As of August 2011, the installation is completed and the bridge is open to traffic. Certain monitoring systems are installed so that in time the performance of the roller isolation bearings under ambient live loads can be evaluated. Figure 5.3 contains a number of the New Street Bridge in RI which is the first bridge in the US implemented with the new roller bearings.

Fig. 5.2. Roller Bearing Assembly

Two Span Bridge Rehabilitation     Roller Isolation Bearing     Installation at Completion

Fig. 5.3. Bridge Rehabilitation in Rhode Island

## 6. Summary

Although seismic isolation is considered to be a relatively matured technology, there are several issues need to be further explored. The main concern is that in general, the targeted acceleration reduction and the bearing displacement are independent parameters. That is, giving one does not automatically define the other. While for most available isolation systems today, the design practice has been based on the assumption that they are dependent parameters. For this assumption works for many cases, it does not yield adequate design for special situations that can exist. Therefore, principles and approaches to deal with these two independent parameters are needed.

This chapter first discussed several design issues of seismic isolation followed by an explanation of the cases where current design approaches may introduce large errors. Certain key parameters for isolation design are discussed and some research needs are identified. The references provided are selected representative publications by many contributors. No intention was given in this chapter to provide a careful literature review on the subject of seismic isolation.

The main control parameter of the seismic induced vibration is the natural period or effective period of the isolation system. To reduce the acceleration, longer period is needed. However, the longer the period is the larger displacement will be resulted. Thus, isolation design is viewed from the perspective of compromising between acceleration reduction and displacement tolerance.

The second important parameter is the effective damping ratio, because displacement can only be regulated by damping. The issue becomes complicated because adding damping will affect the acceleration reduction.

A base-isolated structure is typically a combination of civil engineering structure with mechanical engineering devices. The former is often designed with sufficient safety redundancy. However, the isolation devices are "precision" mechanical parts that do not have large tolerance of displacement uncertainty. By using the concept of probability based design, the safety margin for base isolation systems may be established with more confidence.

This chapter emphasizes the importance of a design window to consider the demand of acceleration and the displacement. Since large displacement may induce many engineering problems, such as large artificial P-$\Delta$ effect, etc. the displacement has to be carefully regulated or controlled. This will narrow the available design window. In some cases, no window is available for a reasonable design of base isolation system, suggesting that for such a case it is not effective to use isolation bearings.

A new type of isolation bearing is briefly explained, which promises to deliver an optimally compromised acceleration reduction and displacement control through addition of damping.

## 7. Acknowledgement

Materials summarized in this chapter are based on studies carried out by the authors in recent years on earthquake protective systems under research grants from the National Science Foundation through the Multidisciplinary Center for Earthquake Engineering Research (ECE 86-07591 and ECE-97-01471) and the Federal Highway Administration (DTFH61-92-C-00106 and DTFH61-98-C-00094).

## 8. References

AASHTO (2010). *Guide Specifications for Seismic Isolation Design*. Interim 2000. Washington, D.C.

ASCE/SEI standard 41-06 (2007). *Seismic Rehabilitation of Existing Buildings*, American Society of Civil Engineers.

ASCE/SEI standard 7-10 (2010). *Minimum Design Loads of Buildings and Other Structures*, American Society of Civil Engineers.

ATC (1995). *Structural Response Modification Factors*, Report No. ATC-19, Applied Technology Council, Redwood City, California.

Austin, M.A., Lin, W.J. (2004). Energy Balance Assessment of Base-Isolated Structures. *J. Engineering Mechanics*, 130(3), pp. 347-358

Buckle, I., Constantinou, M., Dicleli, M. and Ghasemi, H. (2006). *Seismic Isolation of Highway Bridges*, Special Report MCEER-06-SP07, University at buffalo, The State University of New York.

Buckle, I.G. and Mayes, R.L. (1990). Seismic Isolation: History, Application, and Performance – A World View, *Earthquake Spectra*, 6(2), pp. 161-201.

Christopoulos, C. and Filiatrault, A. (2006). *Principles of Supplemental Damping and Seismic Isolation*. University Institute for Advanced Study (IUSS) Press, University of Pavia, Italy.

Clark, P. W., Whittaker, A. S., Aiken, L. D. and Egan, J. A. (1993). Performance Considerations for Isolation Systems in Regions of High Seismicity, *Proceedings of ATC-17-1 Seminar on Seismic Isolation, Passive Energy Dissipation, and Active Control*, San Francisco, CA, March 11-12, pp. 29-40.

Clough, R. and Penzien, J. (2003). *Dynamics of Structures*. 3rd Ed. Computers and Structures, Inc.

Constantinou, M.C., Caccese, J. and Harris, H.G. (1987). Frictional Characteristics of Teflon-Steel Interfaces under Dynamic Conditions, *Earthquake Engineering and Structural Dynamics*, 15, pp. 751-759.

Constantinou, M.C., Mokha, A S. and Reinhorn, A.M. (1990). Teflon Bearings in Base Isolation. Part 2: Modeling. *Journal of Structural Engineering*, ASCE, 116(2), pp. 455-474

Constantinou, M.C. (2001). New Developments in the Field of Seismic Isolation, *Proceedings of 2001 Structures Congress*, ASCE, Washington, DC.

European Committee for Standardization (2000). *Structural Bearings*, European Standard EN 1337-1, Brussels.

Fenz, D.M. and Constantinou, M.C. (2006). Behavior of the Double Concave Friction Pendulum Bearing, *Earthquake Engineering and Structural Dynamics*, Vol. 35, No. 11, pp. 1403-1424.

Fenz, D.M. and Constantinou, M.C. (2008). Spherical Sliding Isolation Bearings with Adaptive Behavior: Theory, *Earthquake Engineering and Structural Dynamics*, Vol. 37, No. 2, pp. 163-183.

Fenz, D.M. and Constantinou, M.C. (2008). Spherical Sliding Isolation Bearings with Adaptive Behavior: *Experimental Verification, Earthquake Engineering and Structural Vibration*, Vol. 37, No.2, pp. 185-205.

Hall, J.F., Heaton, T.H., Halling, M.W. and Wald, D.W.J. (1995). Near-Source Ground Motion and Its Effects on Flexible Buildings, *Earthquake Spectra*, 11(4), pp. 569-605

Hall, J.F. and Ryan, K.L. (2000). Isolated Buildings and the 1997 UBC Near-Source Factors, *Earthquake Spectra*, 16(2), pp. 393-411

Inaudi, J.A. and Kelly, J.M. (1993). Optimum Damping in Linear Isolation Systems, *Earthquake Engineering and Structural Dynamics*, 22, pp. 583-598

International Code Council (2000). *International Building Code*, Falls Church, Virginia.

Kelly, J.M. (1986). Aseismic Base Isolation: A Review and Bibliography, *Soil Dynamics and Earthquake Engineering*, 5, 202-216

Kelly, J.M. (1993). State-of-the-Art and State-of-the-Practice in Base Isolation, *Proceedings of ATC-17-1 Seminar on Seismic Isolation, Passive Energy Dissipation, and Active Control*, San Francisco, California, pp.9-28

Kelly, J.M. (1997). *Earthquake Design with Rubber*, Springer-Verlag, Inc.

Kelly, J.M. (1999). The Role of Damping in Seismic Isolation, *Earthquake Engineering and Structural Dynamics*, 28, pp. 3-20.

Kelly, J.M. (1999). The Current State of Base Isolation in United States, *Proceedings of the 2nd World Conference on Structural Control*, Kyoto, Japan, 1, 1043-1052

Kelly, J.M. (2004), Chapter 11, *Seismic Isolation, in Earthquake Engineering*, Borzognia & Bertero (eds), CRC Press LLC.

Komodromos, P.I. (2000). *Seismic Isolation for Earthquake-Resistant Structures*, Southampton, UK : WIT Press

Lee, G.C., Ou Y.-C, Liang Z., Niu T., Song J. (2007). *Principles and Performance of Roller Seismic Isolation Bearings for Highway Bridges*, MCEER Technical Report MCEER-07-0019, December 2007.

Lee, G.C., Ou, Y.-C., Niu, T., Song, J., and Liang, Z. (2010). Characterization of a Roller Seismic Isolation Bearing with Supplemental Energy Dissipation for Highway Bridges, *Journal of Structural Engineering*, ASCE, 136(5), 502-510, May 2010.

Liang, Z. and Lee, G. C. (1991). *Damping of structures, Part 1, Theory of complex Damping*. Technical Report NCEER-91-0004, State University of New York at Buffalo.

Liang, Z., Lee, G. C., Dargush, G. and Song, J. W. (2012). *Structural Damping: Applications in Seismic Response Modification*. CRC Press, 2012.

Lin, T.W., Hone, C.C. (1993). Base Isolation by Free Rolling Rods under Basement. *Earthquake Engineering and Structural Dynamics*, 22, pp. 261-273.

Mosqueda, G., Whittaker, A.S. and Fenves, G.L. (2004). Characterization and Modeling of Friction Pendulum Bearings Subjected to Multiple Components of Excitation. *Journal of Structural Engineering*, 130(3), pp. 433-442.

Mostaghel, N. and Khodaverdian, M. (1987). Dynamics of Resilient-Friction Base Isolator (R-FBI), *Earthquake Engineering and Structural Dynamics*, 11, pp. 729-748.

Naeim, F. and Kelly, J.M. (1999). *Design of Seismic Isolation Structures from Theory to Practice*, John Wiley & Sons Inc.

Nowak, A. S. and Collins, K. R. (2000) *Reliability of Structures* McGraw Hill

Ou, Y.-C., Song, J., and Lee, G.C. (2010). A parametric study of seismic behavior of roller seismic isolation bearings for highway bridges. *Earthquake Engineering and Structural Dynamics*, 39(5), 541-559,April 2010.

Robinson, W.H. (1982). Lead-Rubber Hysteretic Bearings Suitable for Protecting Structures During Earthquakes, *Earthquake Engineering and Structural Dynamics*, 10, pp. 593-604.

Roussis, P.C. and Constantinou, M.C. (2006). Uplift-Restraining Friction Pendulum Seismic Isolation System, *Earthquake Engineering and Structural Dynamics*, Vol. 35, No. 5, April 2006, pp. 577-593.

Skinner, R.I., Robinson, W.H. and McVerry, G.H. (1993). An Introduction to Seismic Isolation, John Wiley & Sons Ltd.

Soong, T.T. and Dargush, G. F. (1997) *Passive Dissipation Systems in Structural Engineering* John Wiley & Sons, New York,

Structural Engineers Association of Northern California (SEAONC) (1986). *Tentative Seismic Isolation Design Requirements*, Yellow Book.

Takewaki, I. (2009) *Building Control with passive Dampers, Optimal Performance-based Design for Earthquakes*, ohn Wiley &

Thompson, A.C.T., Whittaker, A.S., Fenves, G.L. and Mahin, S.A. (2000). Property Modification Factors for Elastomeric Seismic Isolation Bearings, *Proc. 12th World Conference on Earthquake Engineering*, New Zealand.

Tsopelas, P. and Constantinou, M.C. (1997). Study of Eiastoplastic Bridge Seismic Isolation System, *Journal of Structural Engineering*, Vol. 123, No.4, pp. 489-498.

Tsopelas, P., Constantinou, M.C., Kim, Y-S., and Okamoto, S. (1997). Experimental Study of FPS System in Bridge Seismic Isolation. *Structural Engineering and Structural Dynamics*, Vol. 25, No. 1, pp. 65-78.

Wang, J. (2005). *Seismic Isolation Analysis of a Roller Bearing Isolation System*, Ph.D. Dissertation, University at Buffalo, State University of New York, Buffalo, NY, July 2005.

Whittaker, A.S., Constantinou, M.C., and Tsopelas, P. (1998). Displacement Estimates For Performance-Based Seismic Design, *Journal of Structural Engineering*, Vol. 124, No.8, pp. 905-912.

Zayas, V.A., Low, S.S. and Mahin, S.A. (1990). A Simple Pendulum Technique for Achieving Seismic Isolation, *Earthquake Spectra*, 6, pp. 317-334

Zhou, Q., Lu, X.L., Wang, Q.M., Feng, D.G., Yao, Q.F. (1998). Dynamic Analysis on Structures Base-isolated by a Ball System with Restoring Property. *Earthquake Engineering and Structural Dynamics*, 27, pp. 773-791.

# 2

# Seismic Design Forces and Risks

Junichi Abe, Hiroyuki Sugimoto and Tadatomo Watanabe
*Hokubu Consultant Corporation, Hokkai Gakuen University*
*Japan*

## 1. Introduction

In recent years, seismic damages caused by giant earthquakes have occurred in many countries. For example, over 250,000 people were killed by the Haiti Earthquake in January 2010. In addition, over 15,000 people were killed by the Tohoku Japan Earthquake and the coasts of Tohoku Japan were devastated by the massive tidal wave in March 2011.

Meanwhile, The Japanese seismic design criteria for road and railway bridges provide that two levels of earthquake motions – Level 1, which is small in scale but is generated frequently, and Level 2, which is intensive but is not generated frequently – must be used for the verification of seismic performance. For Level 1 earthquake motions, the elastic limit value of a structure is usually adopted as the seismic performance. For Level 2 earthquake motions, on the other hand, the limit value with which a structure does not collapse or is repairable is adopted as the seismic performance depending on the importance of the intended structure.

Level 2 earthquake motions used for verification are based on the records of strong motion seismograms obtained from the Hyogoken-Nanbu and other earthquakes, and seismic waveforms are assigned according to ground type. The earthquake motions are assigned according to classification of the land area of Japan categorized into three types by degree of seismic risk and adjusting the seismic motions using regional correction factors of 1.0, 0.85 and 0.7 depending on the regional classification.

Meanwhile, studies to calculate seismic waveforms unique to the target region of seismic design have been conducted in recent years. Seismic waveforms calculated in these studies were determined by carefully examining past seismic records, ground data, source models and other data of the target region from the viewpoint of earthquake and geotechnical engineering.

In reality, however, earthquakes that generate ground motions stronger than Level 1 but do not exceed Level 2 may occur during the service life of a structure. In current seismic design, direct consideration was not given to changes in performance and risk with seismic motions through time or the importance of applying effective repair and reinforcement methods. These factors cannot be taken fully into account by simply verifying the elastic limit or the limit of reparability or collapse of a structure subject to Level 1 or 2 earthquake motions based on the current seismic design force.

Many seismic risk management studies, which evaluated the loss (seismic risk) caused by the damage or collapse of a structure, have also been conducted in recent years. In these papers, seismic risks were calculated using a hazard curve representing the probability of the generation of earthquake motions and a damage curve representing the probability of damage to the structure.

While this damage curve is calculated by statistical procedures using past damage records and analyses, it is necessary to define the damage to a structure with a single index, such as the top horizontal displacement or ductility factor. When damage is defined with an index, it is difficult to precisely associate the index with the repair for the damage. Therefore, these methods are considered difficult to apply them to the examination of seismic risks based on the definition of changes in the damage process and other details due to the difference in design.

To achieve these, it is first necessary to calculate design solutions reflecting the damage and collapse process of a structure under a uniform standard of value for various seismic forces. By calculating seismic risks for respective design solutions and comparing them for different seismic forces, it is possible to find the seismic force with which the total cost including the initial construction cost and seismic risk can be minimized. This is called the "target seismic design force" in this chapter. Although this method involves complex procedures, the necessity for target seismic design forces is expected to be higher for the design of long bridges and other structures that are highly important as lifelines from the viewpoint of the seismic risk management.

This chapter consists of the section presented below.

2. describes a design system with which design solutions are calculated using various seismic forces and the method for calculating the target seismic design forces. 3. explains the method for calculating seismic risks based on the definition of damage.

4. present the results of the analysis of an RC rigid-frame viaduct as Example of the calculation of the target seismic design forces.

While there is the possibility of loss caused by environmental and other effects besides those of earthquakes during the life cycle of a structure, such additional effects will be studied in the future and this paper limits its focus on the effects of earthquakes.

## 2. Target seismic design force

This section explains the method for calculating the target seismic design forces. In the case of design where the seismic risks of a variety of seismic forces are taken into account, it is assumed that the initial construction cost is low but the seismic risk is high for a structure designed for a low seismic force, while the seismic risk is low but the initial construction cost is high for a structure designed for a high seismic force. By quantifying this seismic risk based on the cost for the repair of damage and other factors to find the seismic risk cost, calculating the total cost by adding this to the initial construction cost and finding its relationship with the seismic force, the target seismic design force and the corresponding design solution can be obtained. Fig. 1 illustrates the flow of finding the target seismic design force and the corresponding design solution. Details of the flow are as described below.

### 2.1 Setting of the target structure and region
The type of the structure to be designed and the region where the structure will be constructed are set.

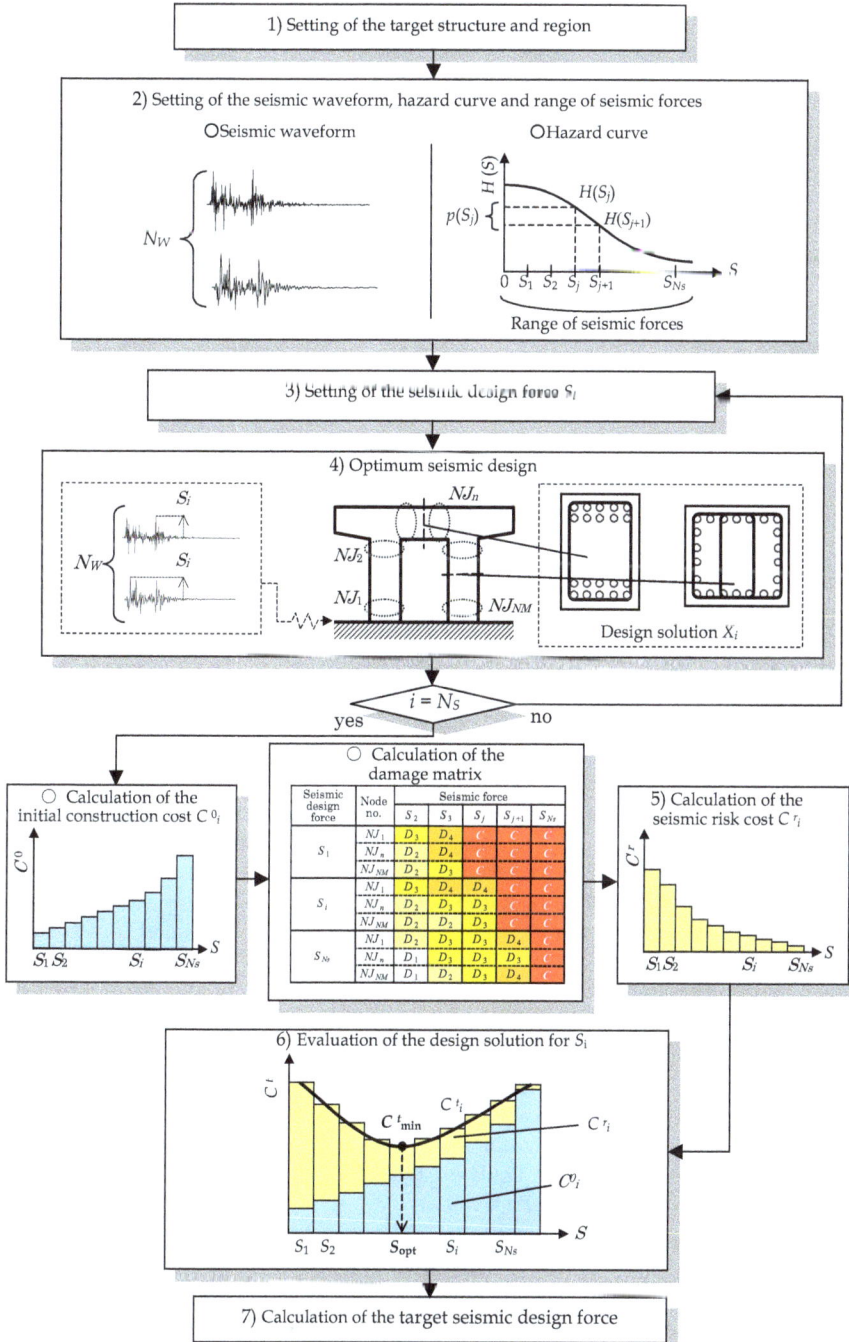

Fig. 1. Flow of calculation of the taqrget seismic design force.

## 2.2 Setting of the seismic waveform, hazard curve and range of seismic forces

Appropriate seismic waveform and hazard curve are set for the target region. The incremental value $\Delta S$ and division number $N_S$ of the seismic forces are also set as shown in Fig. 2.

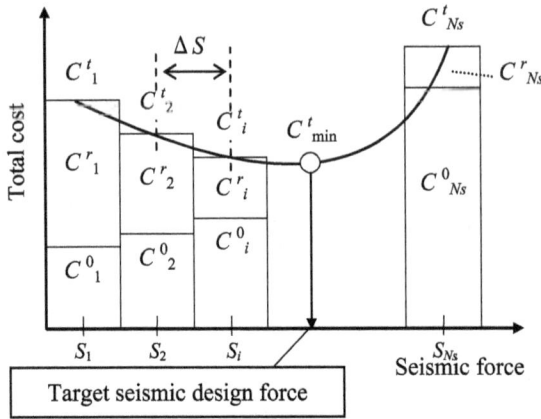

Fig. 2. Relationship between the totalcost and seismic force.

## 2.3 Setting of the seismic force

Based on the range of seismic forces set in 2), the seismic force for optimum seismic design $S_i (i=1\sim N_S)$ is set.

## 2.4 Optimum seismic design

Optimization of seismic design is performed for each seismic force $S_i$ ($i=1\sim N_S$). Details of the formulation of optimum seismic design will be presented later. Time history response analysis is performed by conducting amplitude adjustment to make the maximum amplitude for the seismic waveform set in 2) equal to the seismic force $S_i$. In this chapter, the optimum solution is calculated through the optimization of the response surface using the RBF network and Genetic Algorithm under the minimized initial construction cost. The initial construction cost of the optimum design solution obtained is presented as $C^0_i$ ($i=1\sim N_S$).

In this section, the optimum solution is calculated through the optimum seismic design system of the response surface using the RBF network and Genetic Algorithm by the authors.

## 2.5 Calculation of the seismic risk cost

The seismic risk cost $C^r_i$ ($i=1\sim N_s$) for each design solution found in 4) is calculated for the range of seismic forces set in 2). It means that analysis and verification are performed $N_s$ times for each design solution. The method for calculating the seismic risk cost is as mentioned below.

## 2.6 Evaluation of the design solution

The design solution for a seismic force $S_i$ is evaluated by the equation below, as the total cost $C^t_i$ ($i=1\sim N_s$) found by adding the initial construction cost $C^0_i$ of the design solution found in 4) to the seismic risk cost $C^r_i$ found in 5),

$$C^t_i = C^0_i + C^r_i \tag{1}$$

## 2.7 Calculation of the target seismic design force

The above calculation is performed to calculate the total cost $C^t_i$ for each $S_i$. Fig. 2 is a conceptual diagram of the relationship between the total cost $C^t_i$ and seismic force $S_i$. Of these $C^t_i$ values, the seismic force corresponding to the minimum total cost $C^t_{min}$ is the target seismic design force.

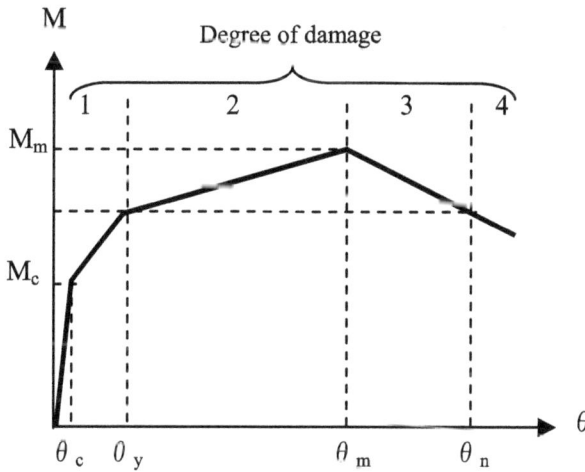

Fig. 3. Skeleton curve and degree of damage.

# 3. Seismic risk cost

As mentioned before, the total cost for each seismic force is calculated by totalling the initial construction and seismic risk costs. The seismic risk cost is usually calculated using damage and hazard curves. However, a damage matrix is constructed by evaluating damage to all elements where nonlinearity is taken into account instead of using a damage curve, and the seismic risk cost is found by calculating repair and other costs.

This section first defines the damage to an RC structure, and then describes the method for calculating seismic risk costs.

## 3.1 Definition of damage

In this chapter, damage is defined for all elements where nonlinearity is taken into account. The M-θ relationship of a tetra-linear model, which is represented by the thick black line in Fig. 3, is used as the relationship between the nonlinearity of RC elements and damage , in accordance with the method defined in the Design Code for Railway Structures and instruction manual (seismic design). In the figure, $M_c$ is the bending moment at the time of cracking, $M_y$ is the bending moment at the time of yield, $M_m$ is the maximum bending moment, $\theta_c$ is the angle of rotation at the time of cracking, $\theta_y$ is the angle of rotation at the time of yield, $\theta_m$ is the maximum angle of rotation to maintain $M_m$, and $\theta_n$ is the maximum angle of rotation to maintain $M_y$.

Classified degree of damage is defined as degree 1 if the maximum response angle of rotation found from time history response analysis is $\theta_y$ or smaller, degree 2 if it is $\theta_m$ or smaller, degree 3 if it is $\theta_n$ or smaller and degree 4 if it exceeds $\theta_n$.

The term "degree 1" represents a condition in which the cracks of concrete member have occurred. The term "degree 2" represents a condition in which the reinforcing bar in the axial direction has yielded. The term "degree 3" represents a condition in which the side of compression of concrete member has fractured. The term "degree 4 "represents a condition in which the flexure capacity has decreased by under the yield capacity.

## 3.2 Calculation of the damage matrix

To calculate the seismic risk cost, it is necessary to determine the damage of the structure for a certain seismic force and calculate repair and other costs. As mentioned before, this study uses a damage matrix instead of a damage curve, which is generally used to represent the relationship between the seismic force and damage of the structure.

Fig.4 presents the damage matrix using a single-layer portal rigid-frame structure. In the case of a rigid-frame structure, plastic hinges with the effect of nonlinearity are found at 6 sections in total – the upper and lower ends of each column member and the left and right ends of beam members. The table on fig.4 shows the node numbers displayed in the rows and seismic forces in the columns. It is a matrix notation of the damage at each node when various seismic forces are input for a certain design solution. In the table, "C" represents the collapse of the structure. This kind of damage matrix is developed for each of the design solution found for each seismic force.

## 3.3 Calculation of seismic risk costs

In this chapter, the seismic risk cost is calculated using a damage matrix representing the relationship between the seismic force and damage as shown in fig. 4 and a hazard curve representing the relationship between the seismic force and annual probability of excess as shown in Fig. 5. The seismic risk cost is calculated by the equation below,

$$C^r_i = \sum_{j=1}^{Ns} h(S_j) \cdot c_{ij} \cdot \Delta s (i = 1 \sim N_S) \tag{2}$$

where, $C^r_i$ is the seismic risk cost of the design solution designed for the i-th seismic force, h ($S_j$) is the annual probability of occurrence found from the hazard curve for the j–th seismic force $S_j$, $c_{ij}$ is the seismic loss cost for the damage of each element caused by the j-th seismic force when the design solution is designed for the i-th seismic force. While the seismic force $S_j$ is given as a discrete value in this study, the hazard curve shown in Fig. 5 is a continuous function. In this chapter, the annual probability of occurrence is converted into a discrete value by directly using the difference between the annual probabilities of excess corresponding to the seismic forces $S_j$ and $S_{j+1}$. It will be necessary in the future to study the influence on seismic risks in cases where the annual probability of excess is set with consideration to the range of incremental value $\Delta S$.

While there is the possibility of loss caused by repeated sequence earthquakes, such additional effects will be studied in the future and this chapter limits its focus on the effects of a single earthquake.

While there is the possibility of loss caused by repeated sequence earthquakes, such additional effects will be studied in the future and this chapter limits its focus on the effects of a single earthquake.

1) Determine the damage

(5) (6)

(2) (4)

(1) (3)

Desine solution $X_i$

$M$

$D_1$ $D_2$ $D_3$ $D_4$

$\theta$

$M$

$D_1$ $D_2$ $D_3$ $D_4$

$\theta$

2) Setting of the sesimic force $S_j$ ( $j = 1 \sim N_S$)

3) Amplitude adjustment $S_j$

4) Time history response analysis

$NJ_5$ $NJ_6$

$D_3$ $D_3$

$NJ_2$ $D_2$ $D_3$ $NJ_4$

$S_j$

$NJ_1$ $D_4$ $D_3$ $NJ_3$

$N_W$

5) Calculation of damage matrix for each $S_i$

| Node no. | $S_j$ |
|---|---|
| $NJ_1$ | $D_4$ |
| $NJ_2$ | $D_3$ |
| $NJ_3$ | $D_3$ |
| $NJ_4$ | $D_2$ |
| $NJ_5$ | $D_3$ |
| $NJ_6$ | $D_3$ |

6) $j = N_S$    no

yes

7) Calculation of damage matrix

| Node no. | Seismic force | | | | | |
|---|---|---|---|---|---|---|
| | $S_1$ | $S_2$ | $S_3$ | $S_j$ | $S_{j+1}$ | $S_{Ns}$ |
| $NJ_1$ | $D_1$ | $D_2$ | $D_3$ | $D_4$ | $D_4$ | $C$ |
| $NJ_2$ | $D_1$ | $D_2$ | $D_3$ | $D_3$ | $D_4$ | $C$ |
| $NJ_3$ | $D_1$ | $D_2$ | $D_2$ | $D_3$ | $D_4$ | $C$ |
| $NJ_4$ | $D_1$ | $D_2$ | $D_1$ | $D_2$ | $D_3$ | $C$ |
| $NJ_5$ | $D_1$ | $D_2$ | $D_2$ | $D_3$ | $D_3$ | $C$ |
| $NJ_6$ | $D_1$ | $D_2$ | $D_2$ | $D_3$ | $D_3$ | $C$ |

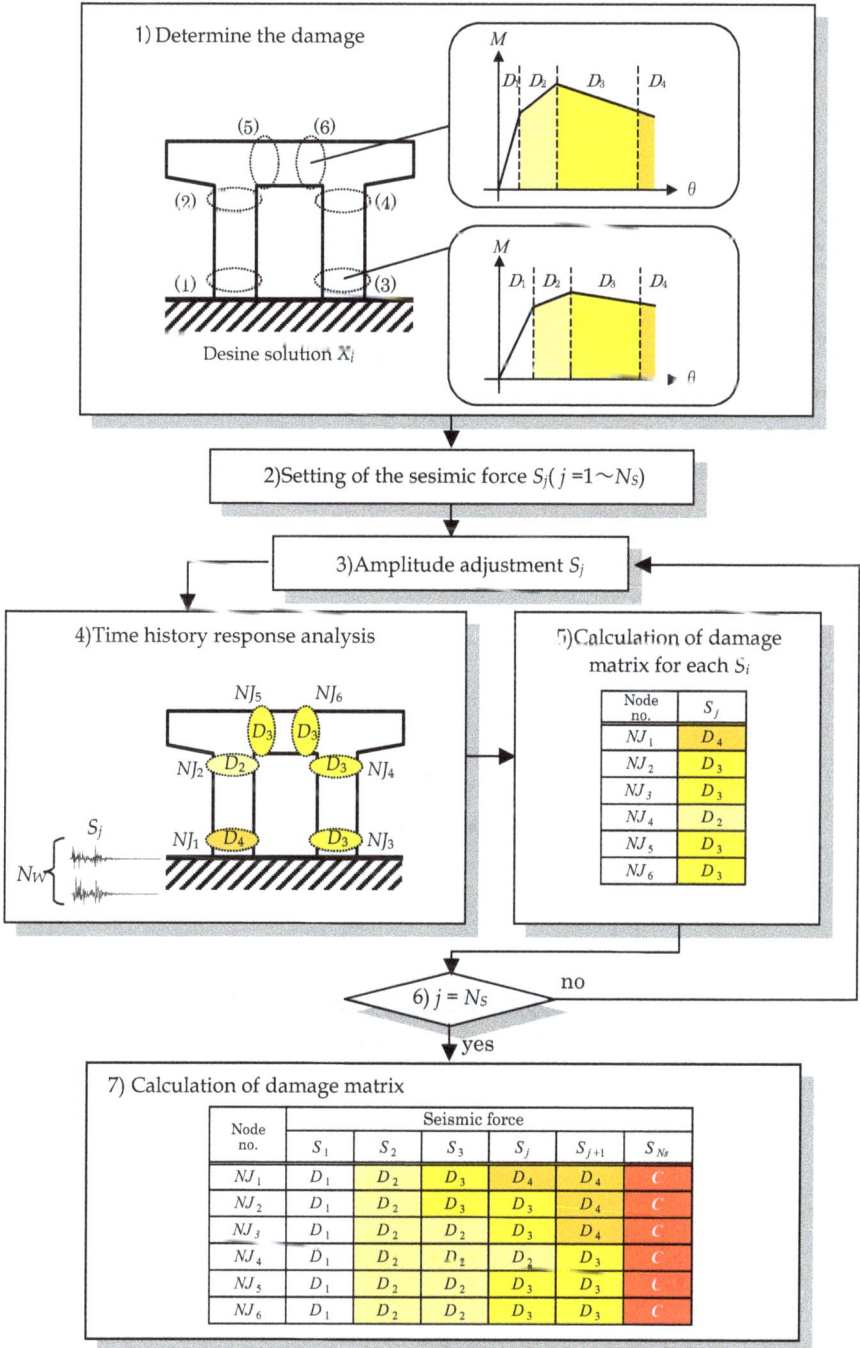

Fig. 4. Flow of the calculation of damage matrix.

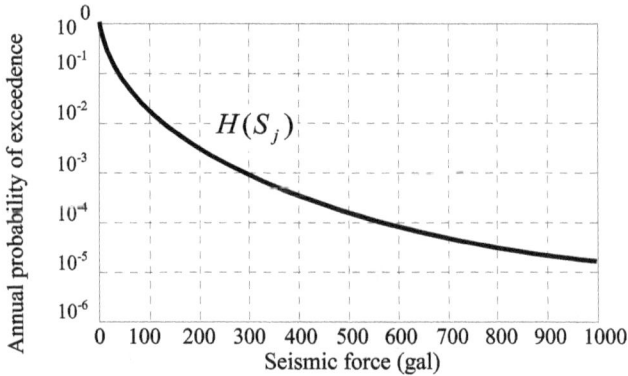

Fig. 5. An example of a hazard curve.

## 4. Example of the calculation of the target seismic design force

In this chapter, the target seismic design force of an RC rigid-frame railway viaduct is calculated. The optimum design problem and examples of numerical calculation will be presented below.

### 4.1 Optimum design problem

A standard single-layer RC rigid-frame railway viaduct with a spread foundation shown in Fig. 6 is used for calculation example. Non-linearity is taken into account for the columns and beam members.

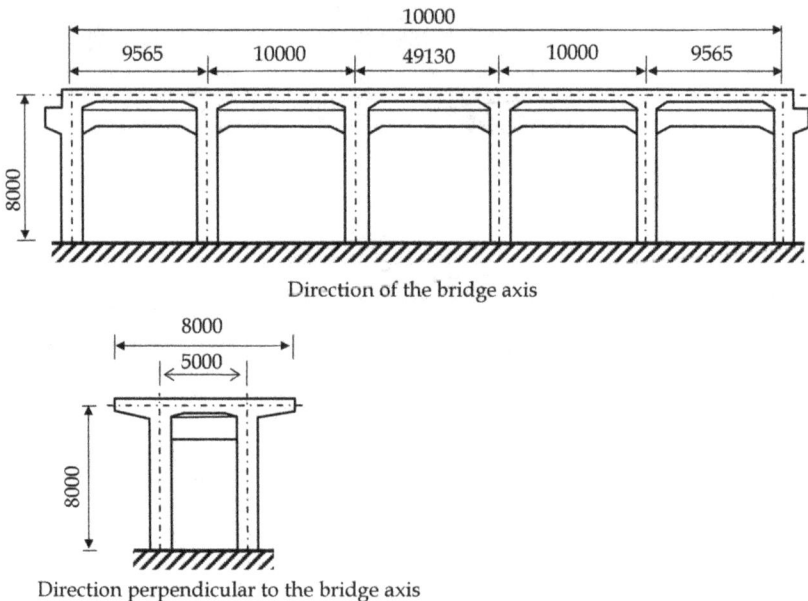

Direction of the bridge axis

Direction perpendicular to the bridge axis

Fig. 6. Structural model.

In the optimum design for a certain seismic force $S_i$, the initial construction cost, which is the total of the costs related to concrete and reinforcement, is used as the objective function. The objective function is calculated by the equation below

$$OBJ = C_i{}^o = C^c + C^s \to min \qquad (3)$$

where, $C^c$ is the concrete-related cost (unit) and $C^s$ is the reinforcement-related cost (unit). They are calculated by the equations (4) and (5), respectively,

$$C^c = a^c \cdot V^c \cdot K^c \qquad (4)$$

$$C^s = a^s \cdot V^s \cdot K^s \cdot G^s \qquad (5)$$

where, $a_c$ is the unit correction factor of concrete, $V_c$ is the amount of concrete (m³), $K_c$ is the cost per unit volume of concrete (=65.1unit/m³), $a_s$ is the unit correction factor of reinforcement, $V_s$ is the amount of reinforcement (m³), $K_s$ is the cost per unit weight of reinforcement (= 9.1unit/kN) and $G_s$ is the unit weight of reinforcement (=77kN/m³). In this study, $a_c$ and $a_s$ are both set as 1.0. The cost per unit volume of concrete and the cost per unit weight of reinforcement are found through conversion from the construction cost, including material cost, cost for scaffolding and personnel cost.

Constraints are found for the verifiability of the angle of rotation and shear force against the seismic force $S_i$, and are calculated by the equation below,

$$g^r{}_{Jk} = \frac{\theta^d{}_{Jk}}{\theta^m{}_{Jk}} - 1 \le 0 \, (J = 1 \sim N_m, k = 1 \sim 2) \qquad (6)$$

$$g^{SD}{}_J = \frac{V^d{}_J}{V^{rd}{}_J} - 1 \le 0 \qquad (=1 \sim N_m) \qquad (7)$$

where, $g^r{}_{Jk}$ is the angle of rotation, $g^{SD}{}_J$ is the constraint related to shear force, $\theta^d{}_{Jk}$ is the maximum response angle of rotation at the end k of the member J, $\theta^m{}_{Jk}$ is the maximum angle of rotation with which $M_m$ on the skeleton curve of the end k of the member J can be maintained, $V^d{}_J$ is the maximum response shear force of the element J, $V^{rd}{}_J$ is the permissible shear force of the member J and $N_m$ is the number of members.

The subjects of design are column and beam members. The cross sections of column members are square and those of beam members are rectangular. There are 7 design variables in total -- the section width B, section height H, number of reinforcing bars in the axial direction N, number of rows of reinforcing bars in the axial direction $J_N$, diameter of reinforcing bars in the axial direction D, placing of shear reinforcement $N_W$ and spacing of shear reinforcement $S_V$ .

Figs. 7 and 8 display the section specifications and arrangement of shear reinforcement, respectively. The spacing of shear reinforcement in section 2H of Fig. 8 is 100 mm.

Table 1. lists the potential values of design variables. By setting the minimum spacing of reinforcement as the diameter of reinforcement D × 2.5 (mm) and the maximum spacing of reinforcement as 250 mm, the maximum and minimum numbers of reinforcing bars, which are obtained based on the section width and diameter of reinforcing bars, are divided by 8 to find the design variable of the number of reinforcing bars in the axial direction N.

As materials, concrete with a design standard strength of 24N/mm² and SD345 reinforcement are used.

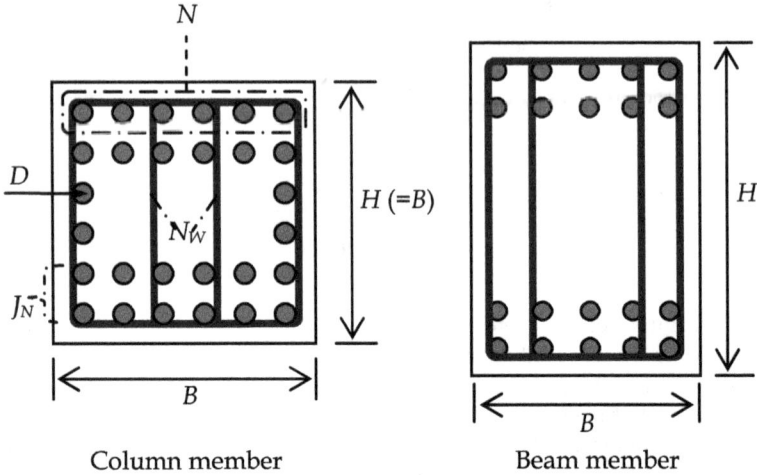

Column member          Beam member

Fig. 7. Details of dimensions and reinforcement for x-sections of columns and beams.

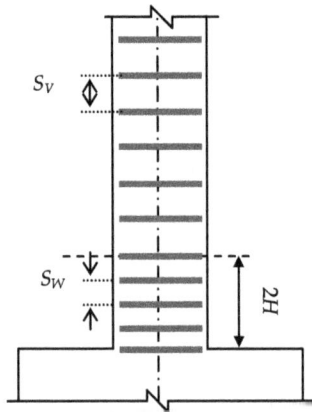

Fig. 8. Details of arrangement of shear reinforcement.

| | |
|---|---|
| $B$ (mm) | 500~1200 (100mm intervals) |
| $H$ (mm) | $B + 200$~800 (100mm intervals) |
| $N$ | 8 types depending on B and H |
| $J_N$ | 1 or 2 |
| $D$ (mm) | 19 or 25 or 29 or 32 |
| $N_W$ | 1~4 |
| $S_V$ (mm) | 100 or 200 |

Table 1. Potential values of design variables.

## 4.2 Seismic loss cost

The repair cost for damage is used as the seismic loss cost. The seismic loss cost is calculated by the equation below,

$$c_{ij} = \sum_{J=1}^{N_m} c^{rep}_{ijJ} \; (i = 1 \sim N_S, j = 1 \sim N_S) \tag{8}$$

where, $c_{ij}$ is the seismic loss cost for the damage of members caused by the j-th seismic force in a design solution designed for the i-th seismic force, and $c^{rep}_{ijJ}$ is the repair cost for the member J damaged by the j-th seismic force in a design solution designed for the i-th seismic force. The repair cost is determined depending on the repair method applicable to the considered section. In this chapter, different repair methods are adopted for the lower and upper ends of column members and upper beam sections.

Table 2. presents the damage conditions and repair methods corresponding to the damage of different members. Table 3. presents the calculation formulas of repair cost corresponding to the repair methods. Fig. 9 illustrates the calculation model of repair cost.

| Degree of damage | Damage condition | Repair method | | |
|---|---|---|---|---|
| | | Culumn(upper end) | Culumn(lower end) | Upper beam |
| 1 | Slight bending cracking | None | None | None |
| 2 | Yield of reinforcement in the axial direction<br>Bending and shear cracking | Scaffolding<br>Grouting of cracks | Excavation<br>Grouting of cracks | Scaffolding<br>Grouting of cracks |
| 3 | Flaking of concrete cover<br><br>Buckling of reinforcement in the axial direction | Scaffolding<br>Grouting of cracks<br>Adjustment of reinforcement<br>Repair of concrete cover | Excavation<br>Grouting of cracks<br>Adjustment of reinforcement<br>Repair of concrete cover | Track removal<br>Scaffolding<br>Grouting of cracks<br>Adjustment of reinforcement<br>Repair of concrete cover<br>Bridge-deck waterproofing<br>Track restoration |
| 4 | Damage of internal concrete<br>Break of reinforcement in the axial direction<br>Break of lateral ties | Temporary support of slab<br>Scaffolding<br>Concrete removal<br>Replacement of reinforcement<br>Concrete placement | Temporary support of slab<br>Excavation<br>Concrete removal<br>Replacement of reinforcement<br>Backfilling | Temporary support of slab<br>Track removal<br>Scaffolding<br>Concrete removal<br>Replacement of reinforcement<br>Concrete placement<br>Bridge-deck waterproofing<br>Track restoration |

Table 2. Damage conditions and repair methods.

(1) Culumn(upper end)

| Degree of damage | Repair method | | Unit | Unit price | Calculation formula |
|---|---|---|---|---|---|
| 1 | None | | - | - | - - - - - |
| 2 | Scaffolding | | m² | 2,380 | $\{(H+0.914\times2+0.4\times2)\times2+(H+0.4\times2)\times2\}\times2\times H1$ |
| | Grouting of cracks | | $\ell$ | 5,500 | $(H\times B\times H)\times2\times10$ |
| 3 | Scaffolding | | m² | 2,380 | $\{(H+0.914\times2+0.4\times2)\times2+(H+0.4\times2)\times2\}\times2\times H1$ |
| | Grouting of cracks | | $\ell$ | 5,500 | $(H\times B\times H)\times2\times10$ |
| | Repair of concrete cover | 1 | m³ | 22,410 | $(H\times B\times H)\times2\times0.35$ |
| | | 2 | m² | 7,090 | $(H\times H)\times4\times2$ |
| 4 | Temporary support of slab | | m³ | 4,680 | $10\times(B2+H)\times H$ |
| | Scaffolding | | m² | 2,380 | $\{(H+0.914\times2+0.4\times2)\times2+(H+0.4\times2)\times2\}\times2\times H1$ |
| | Concrete removal | | m³ | 32,000 | $(H\times B\times H)\times2$ |
| | Replacement of reinforcement | 1 | kg | 120 | $7850\times A_{SD}\times(H\times1.5)\times((N-1)\times4+2\times(N-2)\times(J_N-1))+7850\times A_{SD}\times N_W\times2\times$ $((1+0.4\times(NW-1))\times(B-2\times0.04-D)+N_W\times(H-2\times0.04-D))$ |
| | | 2 | | 2,700 | $4\times(N-1)+2\times(N-2)\times(J_N-1)+N_W$ |
| | Concrete placement | 1 | m³ | 22,410 | $(H\times B\times H)\times2$ |
| | | 2 | m² | 7,090 | $(H\times H)\times4\times2$ |

(2) Culumn(lower end)

| Degree of damage | Repair method | | Unit | Unit price | Calculation formula |
|---|---|---|---|---|---|
| 1 | None | | - | - | - - - - - |
| 2 | Excavation | | m³ | 6,720 | $\{(H+2)2-H2\}\times0.5\times2$ |
| | Grouting of cracks | | $\ell$ | 5,500 | $(H\times B\times H)\times2\times10$ |
| | Backfilling | | m³ | 1,112 | $\{(H+2)2-H2)\times0.5\times2$ |
| 3 | Excavation | | m³ | 6,720 | $\{(H+2)2-H2\}\times0.5\times2$ |
| | Grouting of cracks | | $\ell$ | 5,500 | $(H\times B\times H)\times2\times25$ |
| | Repair of concrete cover | 1 | m³ | 22,410 | $(H\times B\times H)\times2\times0.35$ |
| | | 2 | m² | 7,090 | $(H\times H)\times4\times2$ |
| | Backfilling | | m³ | 1,112 | $\{(H+2)2-H2)\times0.5\times2$ |
| 4 | Temporary support of slab | | m³ | 4,680 | $L1\times(B2+H)\times H$ |
| | Excavation | | m³ | 6,720 | $\{(H+2)2-H2\}\times0.5\times2$ |
| | Concrete removal | | m³ | 32,000 | $(H\times B\times H)\times2$ |
| | Replacement of reinforcement | 1 | kg | 120 | $7850\times A_{SD}\times(H\times1.5)\times((N-1)\times4+2\times(N-2)\times(J_N-1))+7850\times A_{SD}\times N_W\times2\times$ $((1+0.4\times(NW-1))\times(B-2\times0.04-D)+N_W\times(H-2\times0.04-D))$ |
| | | 2 | | 2,700 | $4\times(N-1)+2\times(N-2)\times(J_N-1)+N_W$ |
| | Repair of concrete cover | 1 | m³ | 22,410 | $(H\times B\times H)\times2$ |
| | | 2 | m² | 7,090 | $(H\times B\times H)\times4\times2$ |
| | Backfilling | | m³ | 1,112 | $\{(H+2)2-H2)\times0.5\times2$ |

(3) Upper beam

| Degree of damage | Repair method | | Unit | Unit price | Calculation formula |
|---|---|---|---|---|---|
| 1 | None | | - | - | - - - - - |
| 2 | Scaffolding | | m² | 2,380 | $(B\times2+B\times2)\times H1$ |
| | Grouting of cracks | | $\ell$ | 5,500 | $(H\times B\times H)\times2\times10$ |
| 3 | Track removal | | m | 50,000 | $L1\times2$ |
| | Scaffolding | | m² | 2,380 | $(B\times2+B\times2)\times H1$ |
| | Grouting of cracks | | $\ell$ | 5,500 | $(H\times B\times H)\times2\times25$ |
| | Repair of concrete cover | 1 | m³ | 22,410 | $(H\times B\times H)\times2\times0.35$ |
| | | 2 | m² | 7,090 | $\{(H-0.3)\times H\times2+(H\times B)\}\times2$ |
| | Bridge-deck waterproofing | | m² | 20,000 | $B1\times L1$ |
| | Track restoration | | m | 150,000 | $L1\times2$ |
| 4 | Temporary support of slab | | m³ | 4,680 | $L1\times(B2+H)\times H$ |
| | Track removal | | m | 50,000 | $L1\times2$ |
| | Scaffolding | | m² | 2,380 | $(B\times2+B\times2)\times H1$ |
| | Concrete removal | | m³ | 32,000 | $(h1\times B\times h1)\times2$ |
| | Replacement of reinforcement | 1 | kg | 120 | $7850\times A_{SD}^{*}\times(H\times1.5)\times((N-1)\times4+2\times(N-2)\times(J_N-1))+7850\times A_{SD}\times N_W\times2\times$ $((1+0.4\times(NW-1))\times(B-2\times0.04-D)+N_W\times(H-2\times0.04-D))$ |
| | | 2 | | 2,700 | $N\times J_N\times2+N_W$ |
| | Concrete placement | 1 | m³ | 22,410 | $(H\times B\times H)\times2$ |
| | | 2 | m² | 7,090 | $\{(H-0.3)\times H\times2+(H\times B)\}\times2$ |
| | Bridge-deck waterproofing | | m² | 20,000 | $B1\times L1$ |
| | Track restoration | | m | 150,000 | $L1\times2$ |

Table 3. Calculation formulas of repair cost.

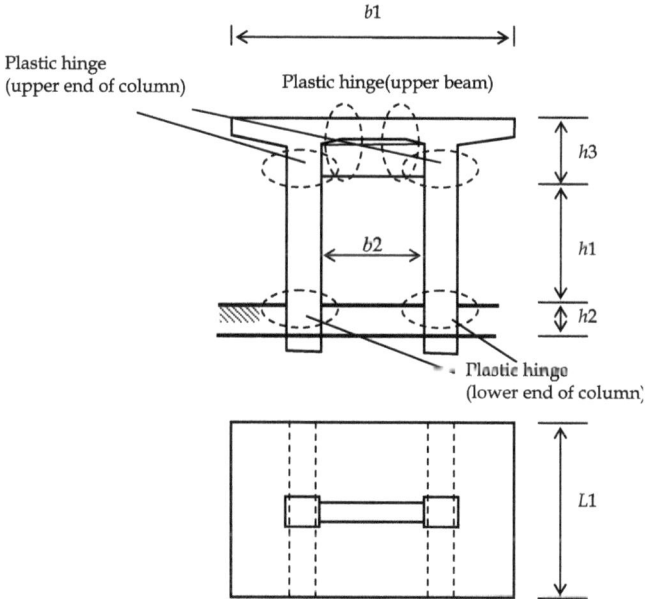

Fig. 9. Calculation model of repair cost.

If the lower ends of all the column members exceed the ultimate angle of rotation, it means that the structure has collapsed and the reconstruction cost replaces the repair cost, which is supposed to be 1.5 times the initial construction cost.

While this definition is based on bending fracture-type collapse, it is also necessary to take the shear fracture-type collapse of structures into account. However, since the seismic force causing bending fracture could be calculated using the damage matrix in this method, it is considered possible to perform analysis based on bending fracture-type collapse by the placement of shear reinforcement, which is not subject to shear fracture caused by the seismic force.

The acceleration waveform of an inland-type earthquake with Level 2 earthquake motion displayed in Fig. 10 is used as the input earthquake motion for time history response analysis and the calculation of the seismic risk cost, and 3 hazard curves (0.16, 0.50 and 0.84 in fractile) displayed in Fig. 11 are adopted.

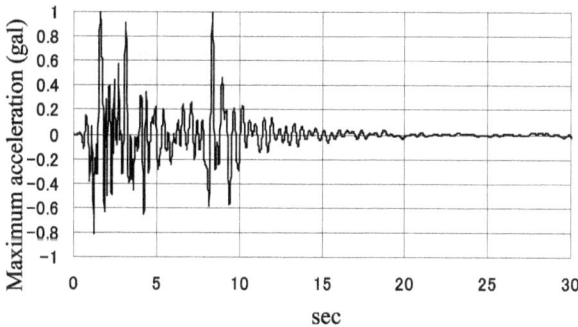

Fig. 10. Acceleration waveform.

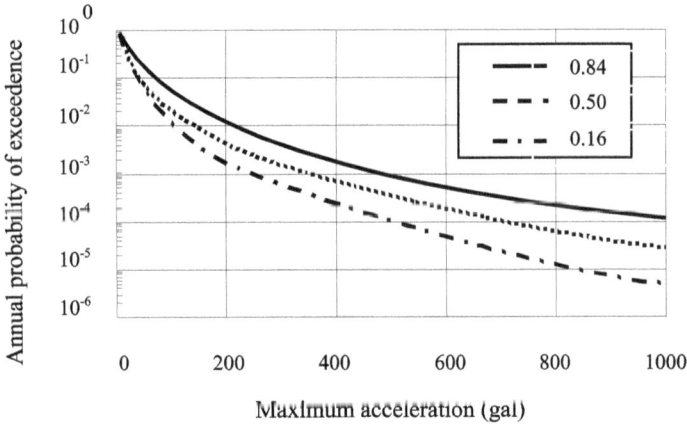

Fig. 11. Hazard curves.

## 4.3 Numerical results

The calculation results for the RC rigid-frame viaduct are presented. The calculation is performed for seismic forces of 50 to 1,000 gal on the assumption that the dividing width ΔS is 50 gal and the dividing number Ns is 20 for the seismic forces. Since the incremental value of design acceleration must be set taking the influence on design solutions into account, the value in this study is set as 50 gal, which is small enough not to have a significant influence on design solutions. The incremental value of design acceleration can be even smaller if necessary.

Table 4. lists the design solutions found for various seismic forces. In the table, $N_P$ and $N_B$ represent the numbers of reinforcing bars in the column and beam sections, respectively. Fig. 12 displays the relationship between the seismic force and initial construction cost. In the figure, the symbol ■ represents the initial construction cost.

When the seismic force is within the range of 50 to 400 gal, the initial construction cost is uniform. These are the design solutions with which the objective function becomes minimum by a combination of preset design variables. The initial construction cost tends to increase with increasing seismic force in design solutions of 400 gal or greater. The initial construction cost sharply increases between 750 and 800 gal. As shown in Table 4, this is because the design variables of the two design solutions, B = 900 mm and H = 1,200 mm of H, are necessary when the seismic force is 800 gal, while the seismic performance is satisfied with B = 600 mm and H = 800 mm at 50 gal.

Next, Table 5. presents the damage matrix of design solutions found for various seismic forces (Table 4.). The table shows the seismic forces in rows and input seismic forces for calculation of the damage matrix in columns. The structural model used has nonlinear performance at a total of 28 sections -- 22 in the direction of the bridge axis and 6 in the direction perpendicular to the bridge axis. Although damage is calculated for all members, the maximum values for columns and beams in two directions are presented for each design solution since it is difficult to display all the calculation results. In the table, $P_I$ is the column member in the direction of the bridge axis, $P_O$ is the column member in the direction perpendicular to the bridge axis, $B_I$ is the beam member in the direction of the bridge axis, and $B_O$ is the beam member in the direction perpendicular to the bridge axis. The right side of the thick line represents the cases where the input seismic force exceeds the value used for design.

| $S_i$ (gal) | $B$ (mm) | $H$ (mm) | $N_P$ | $N_B$ | $J_N$ | $D$ (mm) | $N_W$ | $S_V$ (mm) | $OBJ\ (C^0_i)$ (unit $\times 10^3$) |
|---|---|---|---|---|---|---|---|---|---|
| 50 | 500 | 700 | 3 | 5 | 1 | 22 | 1 | 200 | 6940 |
| 100 | 500 | 700 | 3 | 5 | 1 | 22 | 1 | 200 | 6940 |
| 150 | 500 | 700 | 3 | 5 | 1 | 22 | 1 | 200 | 6940 |
| 200 | 500 | 700 | 3 | 5 | 1 | 22 | 1 | 200 | 6940 |
| 250 | 500 | 700 | 3 | 5 | 1 | 22 | 1 | 200 | 6940 |
| 300 | 500 | 700 | 3 | 5 | 1 | 22 | 1 | 200 | 6940 |
| 350 | 500 | 700 | 3 | 5 | 1 | 25 | 1 | 200 | 6940 |
| 400 | 500 | 700 | 3 | 5 | 1 | 22 | 1 | 200 | 6940 |
| 450 | 500 | 700 | 4 | 6 | 1 | 22 | 1 | 200 | 7219 |
| 500 | 500 | 700 | 3 | 5 | 1 | 22 | 2 | 200 | 7389 |
| 550 | 500 | 700 | 3 | 5 | 1 | 25 | 2 | 200 | 7610 |
| 600 | 500 | 700 | 6 | 10 | 2 | 22 | 2 | 100 | 9646 |
| 650 | 600 | 800 | 4 | 6 | 1 | 22 | 2 | 200 | 10193 |
| 700 | 600 | 800 | 4 | 6 | 1 | 22 | 2 | 200 | 10193 |
| 750 | 600 | 800 | 5 | 7 | 2 | 25 | 2 | 100 | 12000 |
| 800 | 900 | 1200 | 16 | 23 | 1 | 22 | 2 | 200 | 24635 |
| 850 | 1000 | 1200 | 15 | 19 | 1 | 22 | 2 | 200 | 26833 |
| 900 | 1000 | 1200 | 16 | 21 | 1 | 22 | 2 | 200 | 27189 |
| 950 | 1000 | 1200 | 10 | 12 | 1 | 32 | 2 | 200 | 28087 |
| 1000 | 1100 | 1300 | 19 | 23 | 1 | 22 | 2 | 200 | 31852 |

Table 4. Design solution by seismic force ($S_i$).

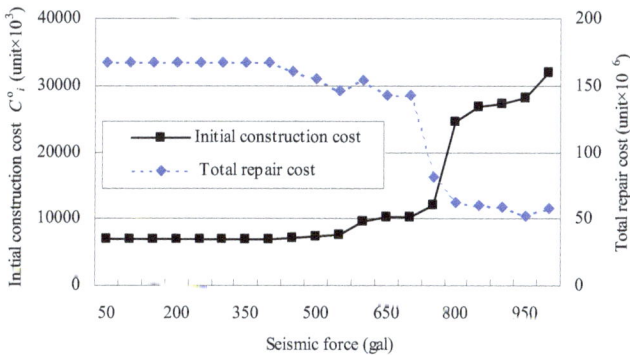

Fig. 12. Initial construction cost and total repair cost by seismic force.

| Seismic design force (gal) | member | Input seismic force (gal) | | | | | | | | | | | | | | | | | | | |
|---|---|---|---|---|---|---|---|---|---|---|---|---|---|---|---|---|---|---|---|---|---|
| | | 50 | 100 | 150 | 200 | 250 | 300 | 350 | 400 | 450 | 500 | 550 | 600 | 650 | 700 | 750 | 800 | 850 | 900 | 950 | 1000 |
| 50 | $P_I$ | 1 | 1 | 2 | 2 | 2 | 2 | 2 | 2 | 2 | 4 | c | c | c | c | c | c | c | c | c | c |
| | $B_I$ | 1 | 1 | 1 | 2 | 2 | 2 | 2 | 2 | 2 | 2 | c | c | c | c | c | c | c | c | c | c |
| | $P_O$ | 1 | 1 | 1 | 2 | 2 | 2 | 2 | 2 | 3 | c | c | c | c | c | c | c | c | c | c | c |
| | $B_O$ | 1 | 1 | 2 | 2 | 2 | 2 | 2 | 2 | 2 | c | c | c | c | c | c | c | c | c | c | c |
| 100 | $P_I$ | 1 | 1 | 2 | 2 | 2 | 2 | 2 | 2 | 2 | 4 | c | c | c | c | c | c | c | c | c | c |
| | $B_I$ | 1 | 1 | 1 | 2 | 2 | 2 | 2 | 2 | 2 | 2 | c | c | c | c | c | c | c | c | c | c |
| | $P_O$ | 1 | 1 | 1 | 2 | 2 | 2 | 2 | 2 | 3 | c | c | c | c | c | c | c | c | c | c | c |
| | $B_O$ | 1 | 1 | 2 | 2 | 2 | 2 | 2 | 2 | 2 | c | c | c | c | c | c | c | c | c | c | c |
| 150 | $P_I$ | 1 | 1 | 2 | 2 | 2 | 2 | 2 | 2 | 2 | 4 | c | c | c | c | c | c | c | c | c | c |
| | $B_I$ | 1 | 1 | 1 | 2 | 2 | 2 | 2 | 2 | 2 | 2 | c | c | c | c | c | c | c | c | c | c |
| | $P_O$ | 1 | 1 | 1 | 2 | 2 | 2 | 2 | 2 | 3 | c | c | c | c | c | c | c | c | c | c | c |
| | $B_O$ | 1 | 1 | 2 | 2 | 2 | 2 | 2 | 2 | 2 | c | c | c | c | c | c | c | c | c | c | c |
| 200 | $P_I$ | 1 | 1 | 2 | 2 | 2 | 2 | 2 | 2 | 2 | 4 | c | c | c | c | c | c | c | c | c | c |
| | $B_I$ | 1 | 1 | 1 | 2 | 2 | 2 | 2 | 2 | 2 | 2 | c | c | c | c | c | c | c | c | c | c |
| | $P_O$ | 1 | 1 | 1 | 2 | 2 | 2 | 2 | 2 | 3 | c | c | c | c | c | c | c | c | c | c | c |
| | $B_O$ | 1 | 1 | 2 | 2 | 2 | 2 | 2 | 2 | 2 | c | c | c | c | c | c | c | c | c | c | c |
| 250 | $P_I$ | 1 | 1 | 2 | 2 | 2 | 2 | 2 | 2 | 2 | 4 | c | c | c | c | c | c | c | c | c | c |
| | $B_I$ | 1 | 1 | 1 | 2 | 2 | 2 | 2 | 2 | 2 | 2 | c | c | c | c | c | c | c | c | c | c |
| | $P_O$ | 1 | 1 | 1 | 2 | 2 | 2 | 2 | 2 | 3 | c | c | c | c | c | c | c | c | c | c | c |
| | $B_O$ | 1 | 1 | 2 | 2 | 2 | 2 | 2 | 2 | 2 | c | c | c | c | c | c | c | c | c | c | c |
| 300 | $P_I$ | 1 | 1 | 2 | 2 | 2 | 2 | 2 | 2 | 2 | 4 | c | c | c | c | c | c | c | c | c | c |
| | $B_I$ | 1 | 1 | 1 | 2 | 2 | 2 | 2 | 2 | 2 | 2 | c | c | c | c | c | c | c | c | c | c |
| | $P_O$ | 1 | 1 | 1 | 2 | 2 | 2 | 2 | 2 | 3 | c | c | c | c | c | c | c | c | c | c | c |
| | $B_O$ | 1 | 1 | 2 | 2 | 2 | 2 | 2 | 2 | 2 | c | c | c | c | c | c | c | c | c | c | c |
| 350 | $P_I$ | 1 | 1 | 2 | 2 | 2 | 2 | 2 | 2 | 2 | 4 | c | c | c | c | c | c | c | c | c | c |
| | $B_I$ | 1 | 1 | 1 | 2 | 2 | 2 | 2 | 2 | 2 | 2 | c | c | c | c | c | c | c | c | c | c |
| | $P_O$ | 1 | 1 | 1 | 2 | 2 | 2 | 2 | 2 | 3 | c | c | c | c | c | c | c | c | c | c | c |
| | $B_O$ | 1 | 1 | 2 | 2 | 2 | 2 | 2 | 2 | 2 | c | c | c | c | c | c | c | c | c | c | c |
| 400 | $P_I$ | 1 | 1 | 2 | 2 | 2 | 2 | 2 | 2 | 2 | 4 | c | c | c | c | c | c | c | c | c | c |
| | $B_I$ | 1 | 1 | 1 | 2 | 2 | 2 | 2 | 2 | 2 | 2 | c | c | c | c | c | c | c | c | c | c |
| | $P_O$ | 1 | 1 | 1 | 2 | 2 | 2 | 2 | 2 | 3 | c | c | c | c | c | c | c | c | c | c | c |
| | $B_O$ | 1 | 1 | 2 | 2 | 2 | 2 | 2 | 2 | 2 | c | c | c | c | c | c | c | c | c | c | c |
| 450 | $P_I$ | 1 | 1 | 2 | 2 | 2 | 2 | 2 | 2 | 3 | 4 | c | c | c | c | c | c | c | c | c | c |
| | $B_I$ | 1 | 1 | 1 | 2 | 2 | 2 | 2 | 2 | 2 | 2 | 2 | c | c | c | c | c | c | c | c | c |
| | $P_O$ | 1 | 1 | 1 | 2 | 2 | 2 | 2 | 2 | 2 | 3 | c | c | c | c | c | c | c | c | c | c |
| | $B_O$ | 1 | 1 | 2 | 2 | 2 | 2 | 2 | 2 | 2 | 2 | c | c | c | c | c | c | c | c | c | c |
| 500 | $P_I$ | 1 | 1 | 2 | 2 | 2 | 2 | 2 | 2 | 2 | 2 | 2 | c | c | c | c | c | c | c | c | c |
| | $B_I$ | 1 | 1 | 1 | 2 | 2 | 2 | 2 | 2 | 2 | 2 | 2 | c | c | c | c | c | c | c | c | c |
| | $P_O$ | 1 | 1 | 1 | 2 | 2 | 2 | 2 | 2 | 2 | 3 | c | c | c | c | c | c | c | c | c | c |
| | $B_O$ | 1 | 1 | 2 | 2 | 2 | 2 | 2 | 2 | 2 | 2 | c | c | c | c | c | c | c | c | c | c |
| 550 | $P_I$ | 1 | 1 | 1 | 2 | 2 | 2 | 2 | 2 | 2 | 2 | 3 | c | c | c | c | c | c | c | c | c |
| | $B_I$ | 1 | 1 | 1 | 1 | 2 | 2 | 2 | 2 | 2 | 2 | 2 | 2 | c | c | c | c | c | c | c | c |
| | $P_O$ | 1 | 1 | 1 | 2 | 2 | 2 | 2 | 2 | 2 | 2 | 2 | c | c | c | c | c | c | c | c | c |
| | $B_O$ | 1 | 1 | 2 | 2 | 2 | 2 | 2 | 2 | 2 | 2 | 2 | c | c | c | c | c | c | c | c | c |
| 600 | $P_I$ | 1 | 1 | 1 | 2 | 2 | 2 | 2 | 2 | 2 | 2 | 2 | 3 | c | c | c | c | c | c | c | c |
| | $B_I$ | 1 | 1 | 1 | 1 | 1 | 1 | 1 | 1 | 1 | 1 | 1 | 1 | 1 | c | c | c | c | c | c | c |
| | $P_O$ | 1 | 1 | 1 | 1 | 2 | 2 | 2 | 2 | 2 | 2 | 2 | 2 | c | c | c | c | c | c | c | c |
| | $B_O$ | 1 | 1 | 1 | 1 | 1 | 1 | 1 | 1 | 1 | 1 | 1 | 1 | c | c | c | c | c | c | c | c |
| 650 | $P_I$ | 1 | 1 | 2 | 2 | 2 | 2 | 2 | 2 | 2 | 2 | 2 | 2 | 2 | 3 | c | c | c | c | c | c |
| | $B_I$ | 1 | 1 | 2 | 2 | 2 | 2 | 2 | 2 | 2 | 2 | 2 | 2 | 2 | 2 | c | c | c | c | c | c |
| | $P_O$ | 1 | 1 | 2 | 2 | 2 | 2 | 2 | 2 | 2 | 2 | 2 | 2 | 2 | c | c | c | c | c | c | c |
| | $B_O$ | 1 | 1 | 2 | 2 | 2 | 2 | 2 | 2 | 2 | 2 | 2 | 2 | 2 | c | c | c | c | c | c | c |
| 700 | $P_I$ | 1 | 1 | 2 | 2 | 2 | 2 | 2 | 2 | 2 | 2 | 2 | 2 | 2 | 2 | 3 | c | c | c | c | c |
| | $B_I$ | 1 | 1 | 2 | 2 | 2 | 2 | 2 | 2 | 2 | 2 | 2 | 2 | 2 | 2 | c | c | c | c | c | c |
| | $P_O$ | 1 | 1 | 2 | 2 | 2 | 2 | 2 | 2 | 2 | 2 | 2 | 2 | 2 | 2 | c | c | c | c | c | c |
| | $B_O$ | 1 | 1 | 2 | 2 | 2 | 2 | 2 | 2 | 2 | 2 | 2 | 2 | 2 | 2 | c | c | c | c | c | c |
| 750 | $P_I$ | 1 | 1 | 2 | 2 | 2 | 2 | 2 | 2 | 2 | 2 | 2 | 2 | 2 | 2 | 2 | 3 | 3 | c | c | c |
| | $B_I$ | 1 | 1 | 1 | 1 | 1 | 1 | 1 | 1 | 1 | 1 | 1 | 1 | 1 | 1 | 1 | 2 | 2 | c | c | c |
| | $P_O$ | 1 | 1 | 2 | 2 | 2 | 2 | 2 | 2 | 2 | 2 | 2 | 2 | 2 | 2 | 3 | 3 | c | c | c | c |
| | $B_O$ | 1 | 1 | 1 | 1 | 1 | 1 | 1 | 1 | 1 | 1 | 1 | 1 | 1 | 1 | 2 | c | c | c | c | c |
| 800 | $P_I$ | 1 | 1 | 1 | 1 | 2 | 2 | 2 | 2 | 2 | 2 | 2 | 2 | 2 | 2 | 2 | 2 | 3 | 3 | 3 | 3 |
| | $B_I$ | 1 | 1 | 1 | 1 | 1 | 1 | 2 | 2 | 2 | 2 | 2 | 2 | 2 | 2 | 2 | 2 | 2 | 2 | 2 | 2 |
| | $P_O$ | 1 | 1 | 1 | 1 | 2 | 2 | 2 | 2 | 2 | 2 | 2 | 2 | 2 | 2 | 3 | 3 | 3 | 3 | 3 | 3 |
| | $B_O$ | 1 | 1 | 1 | 1 | 1 | 1 | 1 | 2 | 2 | 2 | 2 | 2 | 2 | 2 | 2 | 2 | 2 | 2 | 2 | 2 |
| 850 | $P_I$ | 1 | 1 | 1 | 1 | 1 | 2 | 2 | 2 | 2 | 2 | 2 | 2 | 2 | 2 | 2 | 2 | 2 | 3 | 3 | 3 |
| | $B_I$ | 1 | 1 | 1 | 1 | 1 | 1 | 2 | 2 | 2 | 2 | 2 | 2 | 2 | 2 | 2 | 2 | 2 | 2 | 2 | 2 |
| | $P_O$ | 1 | 1 | 1 | 1 | 1 | 2 | 2 | 2 | 2 | 2 | 2 | 2 | 2 | 2 | 2 | 2 | 2 | 3 | 3 | 3 |
| | $B_O$ | 1 | 1 | 1 | 1 | 1 | 1 | 2 | 2 | 2 | 2 | 2 | 2 | 2 | 2 | 2 | 2 | 2 | 2 | 2 | 2 |
| 900 | $P_I$ | 1 | 1 | 1 | 1 | 1 | 2 | 2 | 2 | 2 | 2 | 2 | 2 | 2 | 2 | 2 | 2 | 2 | 3 | 3 | 3 |
| | $B_I$ | 1 | 1 | 1 | 1 | 1 | 1 | 1 | 2 | 2 | 2 | 2 | 2 | 2 | 2 | 2 | 2 | 2 | 2 | 2 | 2 |
| | $P_O$ | 1 | 1 | 1 | 1 | 1 | 1 | 2 | 2 | 2 | 2 | 2 | 2 | 2 | 2 | 2 | 2 | 2 | 3 | 3 | 2 |
| | $B_O$ | 1 | 1 | 1 | 1 | 1 | 1 | 2 | 2 | 2 | 2 | 2 | 2 | 2 | 2 | 2 | 2 | 2 | 2 | 2 | 2 |
| 950 | $P_I$ | 1 | 1 | 1 | 1 | 1 | 2 | 2 | 2 | 2 | 2 | 2 | 2 | 2 | 2 | 2 | 2 | 2 | 2 | 2 | 3 |
| | $B_I$ | 1 | 1 | 1 | 1 | 1 | 1 | 1 | 1 | 1 | 2 | 2 | 2 | 2 | 2 | 2 | 2 | 2 | 2 | 2 | 2 |
| | $P_O$ | 1 | 1 | 1 | 1 | 1 | 1 | 2 | 2 | 2 | 2 | 2 | 2 | 2 | 2 | 2 | 2 | 2 | 2 | 2 | 3 |
| | $B_O$ | 1 | 1 | 1 | 1 | 1 | 1 | 1 | 1 | 2 | 2 | 2 | 2 | 2 | 2 | 2 | 2 | 2 | 2 | 2 | 2 |
| 1000 | $P_I$ | 1 | 1 | 1 | 1 | 1 | 1 | 2 | 2 | 2 | 2 | 2 | 2 | 2 | 2 | 2 | 2 | 2 | 2 | 2 | 2 |
| | $B_I$ | 1 | 1 | 1 | 1 | 1 | 1 | 1 | 1 | 2 | 2 | 2 | 2 | 2 | 2 | 2 | 2 | 2 | 2 | 2 | 2 |
| | $P_O$ | 1 | 1 | 1 | 1 | 1 | 1 | 2 | 2 | 2 | 2 | 2 | 2 | 2 | 2 | 2 | 2 | 2 | 2 | 2 | 2 |
| | $B_O$ | 1 | 1 | 1 | 1 | 1 | 1 | 1 | 2 | 2 | 2 | 2 | 2 | 2 | 2 | 2 | 2 | 2 | 2 | 2 | 2 |

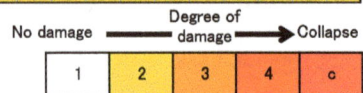

No damage ——— Degree of damage ——➤ Collapse

| 1 | 2 | 3 | 4 | c |

Table 5. Damage matrix.

Damage is examined for design solutions at 400 gal or more, with which the initial construction cost became the minimum. In design solutions between 400 gal and 800 gal, where the objective function increases sharply, collapse in the direction perpendicular to the bridge axis occurred with a seismic force 50 to     150 gal stronger than the seismic force used for design, while collapse in the direction of the bridge axis occurred with a seismic force 100 to 200 gal stronger. It can thus be seen that the seismic performance in the direction perpendicular to the bridge axis is lower than that in the direction of the bridge axis when the seismic force is stronger than that used for design. In design solutions at 800 gal or more, on the other hand, collapse does not occur even with a seismic force of 1,000 gal.

Next, the symbol ◆ in Fig. 12 represent the total repair cost for each design solution calculated from the damage matrix in Table 5. The total repair cost is found by totalling the repair costs for all the seismic forces (columns in Table 5.) between 50 and 1,000 gal for each design solution. The total repair cost of each design solution tends to be in inverse proportion to the initial construction cost. The difference in total repair cost is small although the initial construction cost of the design solution at 750 gal is almost double that of the design solution at 800 gal. This is because the damage level of the beam member using the design solution at 800 gal is 2 at 350 gal, while the beam member using the design solution at 750 gal is undamaged until the seismic force reached 800 gal. Since the repair of beam members requires scaffolding and other works even if damage is minor, the repair cost is higher compared with that for column members. Also, since collapse would not occur even with a seismic force of 1,000 gal in the case of a design solution for a seismic force of 800 gal or more, the total repair cost is approximately half of that for other design solutions with collapse, except for that at 750 gal.

Figs. 13 to 15 display the relationship between the total cost and seismic force in the case where the repair cost for each design solution, which is calculated using the hazard curve in Fig. 11 and based on the damage matrix in Table 5. , is used as the seismic risk cost. In the figures, the horizontal and vertical axes represent the seismic force and total cost and the white and blue parts indicate the initial construction cost and seismic risk cost, respectively. Each figure presents the results for a 0.16, 0.50 or 0.84 fractile hazard curve. The arrow in each figure indicates the section where the total cost is the lowest, or the target seismic design force.

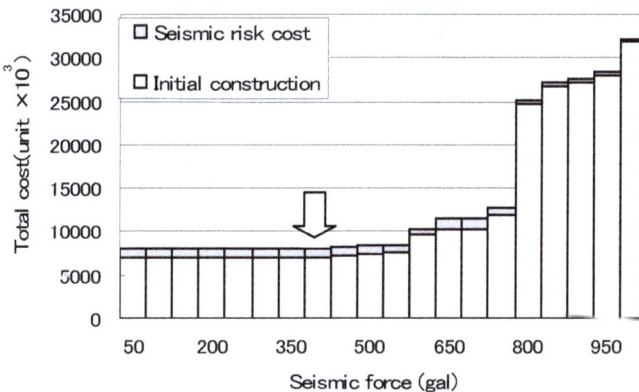

Fig. 13. Relationship between the total cost and seismic force (0.16 fractile hazard curve) .

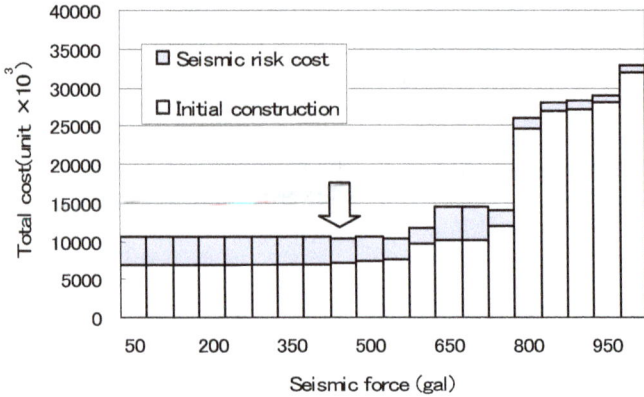

Fig. 14. Relationship between the total cost and seismic force (0.50 fractile hazard curve).

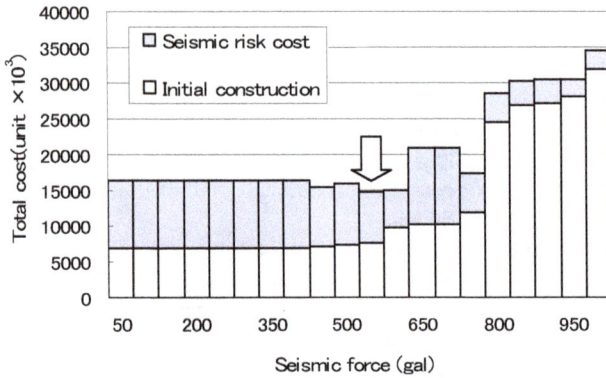

Fig. 15. Relationship between the total cost and seismic force (0.84 fractile hazard curve) .

The target seismic design force is 400, 450 and 550 gal for 0.16, 0.50 and 0.84 fractile hazard curves, respectively. It is confirmed that, even with the same structural model, the target seismic design forces would vary with differences in the occurrence probability of earthquakes. In the relationship between the total cost and seismic force in 0.50 and 0.85 fractile hazard curves, the total cost at 750 gal is locally low. This is because the seismic risks are extremely high at 650 and 700 gal. It can be seen from the damage matrix that damage to beam members started at 150 gal in design solutions designed for 650 and 700 gal. Because the seismic force causing damage is lower compared with other design solutions and the repair cost for beam members is higher, the estimated seismic risk became higher. As a result, the total cost at 750 gal is locally low.

## 5. Conclusion

The current seismic design criteria are based on the verification of seismic performance using Level-1 and -2 seismic forces. However, since earthquake motions that are stronger than Level 1 but do not exceed Level 2 may be generated through time during the service life of a structure in reality. Against such a background, this chapter examined target seismic design forces taking seismic risks into account as an attempt to apply seismic risk management to seismic design methods.

The results obtained in this chapter are as listed below.

1. A method for calculating seismic forces with which the total cost can be minimized is presented. The proposed method has the following characteristics:

- The total cost is the total of the initial construction and seismic risk costs. The seismic risk cost includes the costs associated with the damage and collapse of structures.
- The damage of members is calculated by using the nonlinear characteristics related to the damage of members.
- To find the damage and collapse processes of structures, a damage matrix based on the damage conditions of all members with nonlinearity is used to reflect the influence of the repair cost depending on differences in structural type and damage conditions as precisely as possible.

2. The proposed method for calculating target seismic design forces is applied to RC rigid-frame railway viaduct. As a result of calculation using three hazard curves with different fractile values, the following knowledge is obtained:

- In calculation example, the target seismic design forces vary with difference in the occurrence probability of earthquakes. When the probability is higher, the target forces also become higher.

A method is presented for the calculation of target seismic design forces, for which the seismic risks of damage and collapse caused by various seismic forces are taken into account. By applying hazard curves unique to this region and seismic waveforms taking regional ground and other properties closely into account to the method presented in this study, the target seismic design force with minimum total cost including seismic risk can be found from the occurrence probability of earthquakes in the target region and damage unique to the target structure. While social consensus based on the accumulation of this kind of study is necessary for the setting of seismic forces to use in seismic design, the authors will be pleased if these studies serve as references for future studies of seismic forces in seismic design.

## 6. References

USGS/EERI Advance Reconnaissance Team (2010). *THE M<sub>W</sub>7.0 HAITI EARTHQUAKE of JANUARY 12,2010 : TEAM REPORT V.1.1*

2011 Great East Japan Earthquake-Japan Society of Civil Engineers, Available from http://committees.jsce.or.jp/s_iad/node/9

Railway Technical Research Institute (1999). *Design Criteria for Railway Structures and instruction manual – seismic design*, ISBN 4-621-04650-0,Maruzen Corporation, Japan.

Japan Road Association(2002). *Specifications for Highway Bridges and instruction manual – V seismic design*, ISBN 4-88950-248-3, Maruzen Corporation, Japan.

Japan Society of Civil Engineers (2002). *Standard Specifications for Concrete Structures - seismic performance verification*, ISBN 978 4810601954, Japan Society of Civil Engineers, Japan.

Japan Society of Civil Engineers (2001). *Technology for the Seismic Performance Verification of Concrete Structures – Present Status and Future Prospects*, *Concrete Technology Series 68*, ISBN 4-8106-0506-x, Japan Society of Civil Engineers, Japan.

Special Committee on Seismic Design Methods for Civil Engineering Structures(2000). *Third Proposal on Seismic Design Methods for Civil Engineering Structures and instruction manual*, Japan.

Earthquake Resistance Standard Subcommittee, Earthquake Engineering Committee Japan Society of Civil Engineers (2003). *New Ideas of Level 1 in the Seismic Design of Civil Engineering Structures*, Japan.

Sugimoto, H. ; Abe, J. ; Furukawa, K. & Arakawa, M.( 2004). *On Optimum Seismic Design of Steel Piers by RBF Method , The Third China-Japan-Korea Joint Symposium on Optimization of Structural and Mechanical Systems (CJK-OSM 3) ,* pp.463-468 , Japan.

Abe, J.; Watanabe, T. & Sugimoto, H (2006). *On Optimum Seismic Design of RC Piers by Constraints Approximation Method Applying RBF Network*, JSCE Journal of Structural Mechanics and Earthquake Engineering, Vol. 62, No.2, pp.405-418, Japan.

Abe, J.; Sugimoto, H & Watanabe, T. (2007). *A Study On Calculation of Seismic Design Forces for Structures Considering their Seismic Risks*, JSCE Journal of Structural Mechanics and Earthquake Engineering, Vol. 63, No.4, pp.780-794, Japan.

Sugimoto, H.; Lu, B. and Yamamoto, H (1993). *A Study on an Improvement of Reliability of GA for the Discrete Structural Optimization*, JSCE Journal of Structural Mechanics and Earthquake Engineering, No.471/I-24, pp.67-76,Japan.

Japan Society of Civil Engineers (2007). *Standard Specifications for Concrete Structures - designe verification*, ISBN 978-4-8106-0414-6, Japan Society of Civil Engineers, Japan.

Railway Technical Research Institute (2004). *Design Criteria for Railway Structures and instruction manual – concrete structures*, ISBN 978-4621074121,Maruzen Corporation, Japan.

Sugimoto, H.; Watanabe, T. and Saito, H. (2000). *A Study of the Optimization of the Earthquake Resistance of an RC Rigid-Frame Viaduct*, JSCE Journal of Structural Engineering I, Vol. 46A, pp.385-394,Japan.

Watanabe, T.; Sugimoto; H. & Asahi, K.(2002). *Minimum Cost Design of RC Structures Considering Initial Construction Cost and Repair Cost for Damages Sustained by L2 Earthquake Motions*, JSCE Journal of Materials, Concrete Structures and Pavements, No.718/V-57, pp. 81-93, Japan.

Abbas Moustafa.(2011). *Damage-based design earthquake loads for SDOF inelastic structures*, Journal of Structural Engineering (ASCE).137(3):456-467.

Abbas Moustafa & Iziri TAKEWAKI (2010). *Modeling Critical Ground-Motion Sequencesfor Inelastic Structures,Advances in Structural EngineeringVolume 13 No. 4 2010* 137(3):456-467.

Subcommittee for Studies of Level 2 Earthquake Motions, Earthquake Engineering Committee, Japan Society of Civil Engineers (2000) *Report on the Results of Activities of the Subcommittee for Studies of Level 2 Earthquake Motions*, Available from http://www.jsce.or.jp/committee/eec2/index.html

Kameda, H.; Ishikawa, H;, Okumura, T. & Nakajima, M.(1997). *Probabilistic Scenario Earthquakes -Definition and Engineering Applications-*, JSCE Journal of Structural Mechanics and Earthquake Engineering, No.577/I-41, pp.75-87, Japan.

Nagao, T.; Yamada, M; & Nozu, A.(2005.) *Probabilistic Seismic Hazard Analysis With Focus on Fourier Amplitude and Group Delay Time*, JSCE Journal of Structural Mechanics and Earthquake Engineering, No.801/I-73, pp. 141-158, Japan.

National Research Institute for Earth Science and Disaster Prevention(2004.) *Report of the Review Committee on Engineering Use of Earthquake Hazard Maps*, Japan.

Yoshikawa, H.(2008). Seismic design and seismic risk analysis of RC structures. ISBN 978-4-621-07955-3, Maruzen Corporation, Japan.

# 3

# Seismic Bearing Capacity of Shallow Foundations

Francesco Castelli[1] and Ernesto Motta[2]
[1]*Kore University of Enna, Faculty of Engineering and Architecture*
[2]*University of Catania, Department of Civil and Environmental Engineering*
*Italy*

## 1. Introduction

The seismic risk mitigation is one of the greatest challenges of the Civil Engineering and an important contribution toward this challenge can be given by the Geotechnical Earthquake Engineering. Lesson learned by recent destructive earthquakes (January 2010 Port-au-Prince region of Haiti and March 2011 Tohoku Japan), confirms that local soil conditions can play a significant role on earthquake ground motions.

Earthquake-induced damage in Port au-Prince was devastating and widespread. Yet, there were clearly areas of the city where little to no damage occurred, and areas of the city where an overwhelming majority of the buildings were severely damaged or destroyed.

These types of damage patterns are common in earthquakes, and a wide number of factors need to be considered in order to conclusively piece together the causes.

For a given earthquake, these factors include, but are not limited to: (*a*) relative distance from the fault rupture plane, (*b*) construction type and quality, (*c*) local soil conditions (i.e. strength/stiffness of the soil foundation, depth to bedrock, impedance contrasts, geology), (*d*) topography (topographic and basin effects), and (*e*) near fault effects (rupture directivity, fling step, hanging wall effects, polarity effects, etc.). Often several of these factors work together and it can be difficult to identify the primary cause of damage.

Design of foundations in seismic areas needs special considerations compared to the static case. The inadequate performance of structures during recent earthquakes has motivated researchers to revise existing methods and to develop new methods for seismic-resistant design. This includes new design concepts, such as, performance-based design (PBD) (Priestley et al., 2005) and new measures of the structure performance based on energy concepts and damage indexes (Park et al., 1987; Moustafa, 2011).

Similarly, the widespread damage and inadequate performance of code-designed structures during the 1994 Northridge (California) and the 1995 Kobe (Japan) earthquakes have prompted seismologists and engineers of the essential importance of characterizing and modelling near-field ground motions with impulsive nature (Moustafa & Takewaki, 2010).

For foundations of structures built in seismic areas, the demands to sustain load and deformation during an earthquake will probably be the most severe in their design life.

As stressed by Hudson (1981) the soil-structure interaction is a crucial point for the evaluation of the seismic response of structures.

Due to seismic loading, foundations may experience a reduction in bearing capacity and increase in settlement. Two sources of loading must be taken into consideration: "inertial" loading caused by the lateral forces imposed on the superstructure, and "kinematic" loading caused by the ground movements developed during the earthquake.

Part 5 of Eurocode 8 (2003) states that foundations shall be designed for the following two loading conditions :

a.  inertia forces on the superstructure transmitted on the foundations in the form of axial, horizontal forces and moment ;

b.  soil deformations arising from the passage of seismic waves.

In the last years the seismic action has increased in many National Codes according to recent records which show values up to 0.8 g for very destructive earthquakes. The upgrading of the seismic action requires accurate analyses taking into account all the boundary conditions including the presence of surcharges, sloping ground, depth factors and so on.

With the aim to investigate the influence of these factors on the seismic stability of a shallow foundation, a model based on the limit equilibrium method has been developed.

Many analytical and numerical solutions are today available to evaluate seismic bearing capacity of shallow foundations, and cover area such as the limit equilibrium method, limit analysis, methods of characteristics, finite element analysis and other areas for the computation of the seismic bearing capacity factors required for the design of a foundation. Nevertheless, pseudo-static approaches are more attractive because they are simple, when compared to difficult and more complex dynamic analyses.

Thus, a pseudo-static model to account for reduction in bearing capacity due to earthquake loading is presented. In this model the loading condition consists in normal and tangential forces on the foundation and inertial forces into the soil. An upper bound solution of the limit load of the shallow foundation is found.

Results of the proposed analysis are given in terms of the ratios between seismic and static bearing capacity factors $N_c^*/N_c$ , $N_q^*/N_q$ and $N_\gamma^*/N_\gamma$. Results are also compared with those deduced by other authors using different methods of analysis.

## 2. Method of analysis

The prediction of the bearing capacity of a shallow foundation is a very important problem in Geotechnical Engineering, and in the last decades solutions using limit analysis, slip-line, limit equilibrium and, recently, numerical methods (i.e. finite element and difference finite methods) have been developed.

The problem of static bearing capacity of shallow foundations has been extensively studied in the past by Terzaghi (1943), Meyerhoff (1963), Vesic (1973) and many others. The ultimate load that the foundation soil can sustain is expressed by the linear combination of the three bearing capacity factors $N_c$ , $N_q$ and $N_\gamma$ which depend uniquely on the friction angle of the soil. Further solutions for the bearing capacity were given successively in a more general form, taking into account, by means of corrective factors, of the shape of the foundation, of the load and ground inclination and of the depth and inclination of the bearing surface.

In all these studies, the bearing capacity evaluation is based on the assumption that a failure surface can develop beneath the foundation, according to the well known failure surfaces given by the limit equilibrium method or by the limit analysis.

Most foundation failures during earthquakes occur due to liquefaction phenomena, even if failures due to reduction in bearing capacity have been observed during Naigata earthquake (1964) Japan and Izmit earthquake (1999) in Turkey (Day, 2002).

Liquefiable soils are categorized by all seismic codes as extreme ground conditions, where, following a positive identification of this hazard, the construction of shallow footings is essentially allowed only after proper soil treatment. More specifically, liquefaction-induced shear strength degradation of the foundation subsoil may result in post-shaking static bearing capacity failure, while excessive seismic settlements may also accumulate. However, the accurate estimation of the degraded bearing capacity and the associated dynamic settlements could potentially ensure a viable performance-based design of shallow footings. Richards et al. (1993) observed seismic settlements of foundations on partially saturated dense or compacted soils. These settlements were not associated with liquefaction or densification and could be easily explained in terms of seismic bearing capacity reduction.

In fact, the inertial forces applied on the foundation and in the soil mass reduce the static bearing capacity. Thus, many authors have investigated the seismic bearing capacity giving results in terms of the ratio of the seismic to the static bearing capacity factors $N_c^*/N_c$, $N_q^*/N_q$ and $N_\gamma^*/N_\gamma$.

The pseudo-static approach is being used to determine bearing capacity of the foundations subjected to seismic loads in non-liquefying soils, considering also the depth effects for an embedded footing and the effect of a sloping ground located at some distance from the footing. Dynamic nature of the load and other factors which affect the dynamic response are not being accounted for.

Ground factors and bearing capacity ratios $N_c^*/N_c$, $N_q^*/N_q$ and $N_\gamma^*/N_\gamma$ are presented as a function of the friction angle of soil $\phi$, of the ratio $H/B$ between the embedment depth $H$ and the width of the footing $B$, of a slope angle $\beta$ and of the ratio $d/B$ being $d$ the distance from the edge of the slope. The inertial and kinematic effects due to seismic loading have been analyzed in the evaluation of the seismic bearing capacity.

## 2.1 Limit equilibrium analysis

The method of analysis is based on the limit equilibrium technique. The failure mechanism, as shown in Figure 1, is a circular surface which from the foundation propagates until the ground surface is reached (Castelli & Motta, 2010; 2011).

A similar model was proposed by Castelli & Motta (2003) for a bearing capacity analysis of a strip footing resting on a soil of limited depth.

The seismic forces are considered as pseudo-static forces acting both on the footing and on the soil below the footing. The ultimate load can be found by a moment equilibrium respect to the centre of the circular surface.

Referring to Figure 1 a moment equilibrium can be written and the mobilizing moment is :

$$M_{mob} = \sum_{i=1}^{n_{tot}} W_i (1 - k_v) h_{wi} + q_{lim} \sum_{i=1}^{n_1} \Delta x_i b_{qli} + q_{lim} \sum_{i=1}^{n_1} k_{h1} \Delta x_i b_{qlhi} + \sum_{i=1}^{n_{tot}} k_{h2} W_i b_{whi} \qquad (1)$$

The resisting moment given by the shear strength $S_i$ acting along the base of the slices is :

$$M_{res} = R \sum_{i=1}^{n_{tot}} S_i = R \sum_{i=1}^{n_{tot}} c \Delta x_i / \cos \alpha_i + R \sum_{i=1}^{n_{tot}} N_i \tan \phi_i \qquad (2)$$

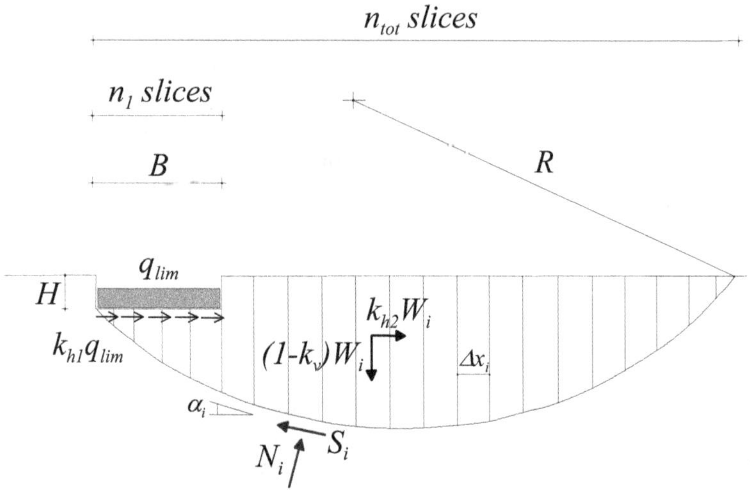

Fig. 1. Failure mechanism and applied forces adopted in the analysis

being :

$$S_i = c\Delta x_i \, / \cos\alpha_i + N_i \tan\phi_i \tag{3}$$

The force $N_i$ resultant of the normal stress distribution acting at the base of the slice can be derived by the Bishop's method of slices (1955) with an equilibrium equation in the vertical direction, so one obtains (see Figure 1), for the slices under the footing where $i$ = 1 to $n_1$ :

$$N_i = \frac{q_{\lim}\Delta x_i + W_i(1-k_v) - c\Delta x_i \tan\alpha_i}{\cos\alpha_i + \sin\alpha_i \tan\phi} \tag{4}$$

and for the remaining slices ($n_1 + 1 \le i \le n_{tot}$) :

$$N_i = \frac{W_i(1-k_v) - c\Delta x_i \tan\alpha_i}{\cos\alpha_i + \sin\alpha_i \tan\phi} \tag{5}$$

Thus :

$$M_{res} = R\sum_{i=1}^{ntot} c\Delta x_i \, / \cos\alpha_i + \begin{aligned}&R\sum_{i=1}^{n_1}\frac{q_{\lim}\Delta x_i + W_i(1-k_v) - c\Delta x_i \tan\alpha_i}{\cos\alpha_i + \sin\alpha_i \tan\phi}\cdot\tan\phi +\\&+ \sum_{i=n1+1}^{ntot}\frac{W_i(1-k_v) - c\Delta x_i \tan\alpha_i}{\cos\alpha_i + \sin\alpha_i \tan\phi}R\tan\phi\end{aligned} \tag{6}$$

where :
- $q_{lim}$ = vertical limit load acting on the footing;
- $c$ = soil cohesion;
- $\Delta x_i$ = width of the i[th] slice;
- $W_i$ = weight of the i[th] slice;
- $R$ = radius of the circular failure surface;

- $\alpha_i$ = angle of the base of the $i^{th}$ slice;
- $n_1$ = number of slices under the footing;
- $n_{tot}$ = total number of slices;
- $k_{h1}$ = horizontal seismic coefficient for the limit load;
- $k_{h2}$ = horizontal seismic coefficient for the soil mass;
- $k_v$ = vertical seismic coefficient for the soil mass;
- $b_{wi}$ − distance of the weight $W_i$ of the $i^{th}$ slice to the centre of the circular failure surface;
- $b_{whi}$ = distance of the inertia force $k_{h2}W_i$ of the $i^{th}$ slice to the centre of the circular failure surface;
- $b_{qli}$ = distance of the limit load $q_{lim}$ acting on the $i^{th}$ slice to the centre of the circular failure surface;
- $b_{qhli}$ − distance of the shear limit force $k_{h1}q_{lim}$ acting on the $i^{th}$ slice to the centre of the circular failure surface.

Substituting the following terms :

$$a_1 = R \sum_{i=1}^{n_{tot}} c\Delta x_i / \cos\alpha_i$$

$$a_2 = R\tan\phi \sum_{i=1}^{n_1} \frac{\Delta x_i}{\cos\alpha_i + \sin\alpha_i \tan\phi}$$

$$a_3 = R\tan\phi \sum_{i=1}^{n_1} \frac{W_i(1-k_v) - c\Delta x_i \tan\alpha_i}{\cos\alpha_i + \sin\alpha_i \tan\phi}$$

$$a_4 = R\tan\phi \sum_{i=n_1+1}^{n_{tot}} \frac{W_i(1-k_v) - c\Delta x_i \tan\alpha_i}{\cos\alpha_i + \sin\alpha_i \tan\phi}$$

$$a_5 = \sum_{i=1}^{n_{tot}} W_i(1-k_v)b_{wi}$$

$$a_6 = \sum_{i=1}^{n_{tot}} k_{h2}W_i b_{whi}$$

$$a_7 = \sum_{i=1}^{n_1} \Delta x_i b_{qli}$$

$$a_8 = \sum_{i=1}^{n_1} k_{h1}\Delta x_i b_{qlhi}$$

and equating $M_{mob} = M_{res}$ the limit load is given by :

$$q_{lim} = \frac{a_5 + a_6 - a_1 - a_3 - a_4}{a_2 - a_7 - a_8} \tag{7}$$

Even if the failure mechanism adopted is quite simple, it allows to investigate a variety of loading and geometric conditions that could have been troublesome using other failure mechanisms and results are in a very good agreement with those obtained by other authors. In fact, referring to the kinematic effect due to the inertia of the soil mass on the seismic bearing capacity, Figure 2 shows a comparison between the results of the present study (for $k_v = 0$), those produced by the method proposed by Paolucci & Pecker (1997) and those found by Cascone et al. (2004) with the method of characteristics.

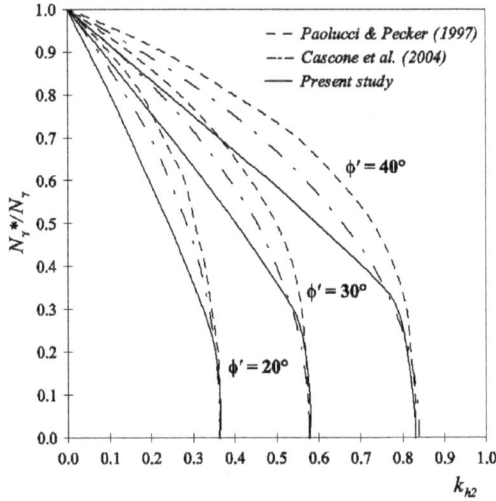

Fig. 2. Seismic ratios as a function of the soil mass inertia

The reduction of the bearing capacity is presented in terms of the ratio $N_\gamma^*/N_\gamma$ as a function of the seismic coefficient $k_{h2}$ in the soil mass.

Despite of the different methods, the results obtained are in good agreement. However, for low values of $k_{h2}$, the limit equilibrium approach seems to give the greatest reduction thus it is on the safe side.

## 3. Parametric analysis

To investigate the influence of the depth factor on the seismic stability of a shallow foundation, the model proposed has been applied and an upper bound solution of the limit load is found. Results of the analysis are given in terms of the ratios between seismic and static bearing capacity factors $N_c^*/N_c$, $N_q^*/N_q$ and $N_\gamma^*/N_\gamma$.

Ground factors and bearing capacity ratios are presented as a function of the friction angle of soil $\phi$ and of the ratio $H/B$ between the embedment depth $H$ and the width of the footing $B$. The inertial and kinematic effects due to seismic loading have been analyzed in the evaluation of the seismic bearing capacity.

For a shallow foundation resting on a cohesionless soil, with horizontal ground surface and in absence of surcharge, the limit load can be expressed by :

$$q_{\lim} = 1/2\, B\gamma N_\gamma i_{\gamma_i} i_{\gamma_k} d_\gamma \tag{8}$$

where:

- $B$ = width of the footing;
- $\gamma$ = unit weight of soil;
- $N_\gamma$ = bearing capacity factor;
- $i_{\gamma i}$ = load inclination factor due to the inertia of the structure;
- $i_{\gamma k}$ = reduction factor due to the inertia of the soil mass (*kinematic interaction factor*);
- $d_\gamma$ = depth factor.

The load inclination factor related to the inertia of the structure ($i_{\gamma i}$) has been discussed by some authors (Pecker & Salencon, 1991; Budhu & Al-Karni, 1993; Dormieux & Pecker, 1995; Paolucci & Pecker, 1997; Fishmann et al., 2003), while less information are available on the depth factor ($d_\gamma$) and on the reduction factor due to the inertia of the soil mass ($i_{\gamma k}$ = kinematic interaction factor)

Conventionally, the depth factor ($d_\gamma$) is assumed equal to unit (Brinch Hansen, 1970) Nevertheless, in an analysis in which the effects due to the inertia of the soil mass are taken into consideration, it is also necessary to take into account the inertia of the soil mass corresponding to the embedment depth $H$ of the footing.

## 3.1 Depth factor evaluation

In static conditions the depth factor $d_\gamma$ has been evaluated by a parametric analysis, for both drained ($\phi$ = 20°, 30°, 40°) and undrained conditions ($\phi$ = 0), varying the ratio $H/B$ between the embedment depth $H$ and the width of the footing $B$.

In the present analysis, the depth factor $d_\gamma$ for drained conditions is defined as the ratio between the bearing capacity factors $N_\gamma'$ of a shallow foundation with embedment $H$ and the conventional bearing capacity $N_\gamma$ of a shallow foundation with an embedment equal to 0 :

$$d_\gamma = N_\gamma' / N_\gamma \qquad (9)$$

Similarly, for undrained conditions the depth factor $d_\gamma^\circ$ is defined as the difference between the bearing capacity factors $N_\gamma^{\circ\prime}$ of a shallow foundation with an embedment $H$ and the conventional bearing capacity $N_\gamma^\circ$ of a shallow foundation with an embedment equal to 0 :

$$d_\gamma^\circ = (N_\gamma^{\circ\prime} - N_\gamma^\circ) = N_\gamma^{\circ\prime} \qquad (10)$$

being $N_\gamma^\circ$ in undrained conditions, as known, equal to 0.

With reference to equation (10) in Table 1 are reported, as an example, the values of the depth factor $d_\gamma^\circ$ for the undrained conditions.

| $H/B$ | $d_\gamma^\circ$ (eq.10) |
|-------|--------------------------|
| 0     | 0                        |
| 0.25  | 0.562                    |
| 0.5   | 1.25                     |
| 0.75  | 2.062                    |
| 1.0   | 3.0                      |

Table 1. Values of the depth factor $d_\gamma^\circ$ for the undrained conditions

In Figure 3 are reported the values of the depth factor $d_\gamma$ for drained conditions versus $H/B$. For the values of the friction angle of soil $\phi$ taken into consideration, curves shown approximately a linear trend, thus it is possible to express the depth factor $d_\gamma$ as a linear function of the ratio $H/B$ according to the equation :

$$d_\gamma = 1 + [(0.85H / B)\cot g\varphi'] \qquad (11)$$

For undrained conditions the results obtained (Figure 4) can be conveniently expressed by the following linear equation :

$$d_\gamma^\circ = (2H / B) - 0.25 \qquad (12)$$

that, obviously, is valid for $H/B > 0.125$.

## 3.2 Kinematic interaction factor evaluation

The kinematic interaction factor $i_{\gamma k}$ has been evaluated by a parametric analysis only for drained conditions, varying the friction angle of soil in the range $\phi$ = 20°, 30° and 40° and the horizontal seismic coefficient for the soil mass $k_{h2}$ between 0.1 up to 0.3.

The kinematic interaction factor $i_{\gamma k}$ is defined in this study as the ratio between the bearing capacity factor $N_\gamma^*$ derived for a given value of the horizontal seismic coefficient $k_{h2}$, and the conventional bearing capacity factor $N_\gamma$ :

$$i_{\gamma k} = N_\gamma^* / N_\gamma \qquad (13)$$

The numerical analyses have been carried out assuming a vertical seismic coefficient for the soil mass $k_v$ equal to ½ $k_{h2}$. In Figure 5 are reported the values of the kinematic interaction factor $i_{\gamma k}$ obtained for the soil friction angles taken into consideration.

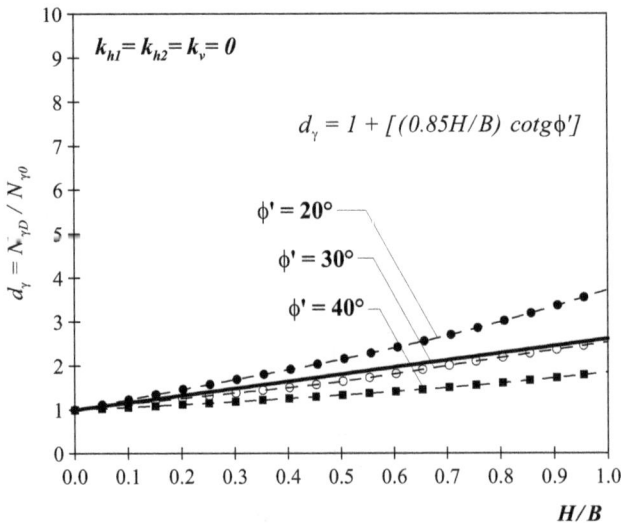

Fig. 3. Depth factor $d_\gamma$ for drained conditions

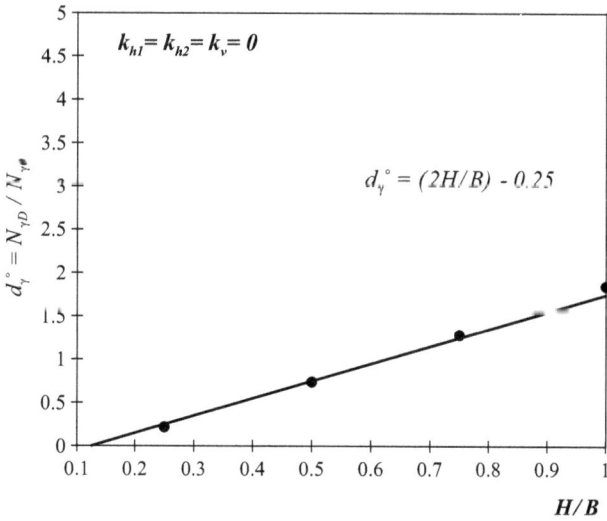

Fig. 4. Depth factor $d_\gamma^\circ$ for undrained conditions

Curves shown approximately a linear trend, thus it is possible to express the kinematic interaction factor $i_{\gamma k}$ as a linear function of $k_{h2}$ by the following equation :

$$i_{\gamma k} = 1 - k_{h2} \cot g \phi'$$ (14)

It is simple to verify that for $k_{h2} = \tan \phi$ the kinematic interaction factor $i_{\gamma k}$ is equal to 0.

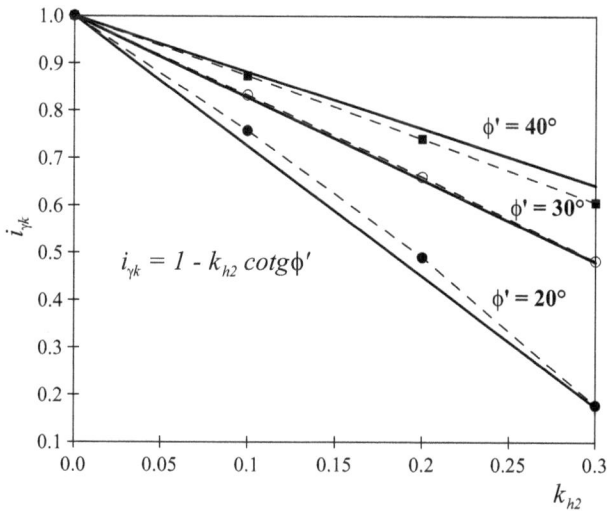

Fig. 5. Values of the kinematic interaction factor $i_{\gamma k}$

## 4. Bearing capacity of strip footings near slopes

When a shallow foundation is placed near the edge of a sloping ground the bearing capacity may be reduced both in static (De Buhan & Gaernier, 1988; Saran et al., 1989; Narita & Yamaguchi, 1990; Shields et al., 1990; Jao et al., 2001) and seismic conditions (Sawada et al., 1994; Pecker, 1996; Sarma & Iossifelis, 1990; Sarma & Chen, 1996; Kumar & Rao, 2003).

In this case, the failure mechanism is influenced by the distance $d$ of the foundation from the edge of the sloping ground (Figure 6). If the shallow foundation is far enough from the edge, the failure mechanism will be not affected by the slope.

In the last decades extensive studies have been made for two dimensional problems of a strip footing resting on inclined slope surface so that different methods of analysis are available.

In these studies, the bearing capacity evaluation is based on the assumption that a failure surface can develop beneath the foundation, according to a general shear failure surfaces given by the limit equilibrium method or by kinematic mechanisms.

General shear failure is characterized by the existence of a well-defined failure pattern (Terzaghi, 1943), which consists of a continuous slip surface from one edge of the footing to the horizontal or inclined ground surface. Referring to Figure 6, the mobilizing moment is :

$$M_{mob} = \sum_{i=1}^{n_{tot}} W_i(1-k_v)b_{wi} + q_{lim}\sum_{i=1}^{n_1}\Delta x_i b_{qli} + \sum_{i=n_1+1}^{n_{tot}} q_v(1-k_v)b_{qvi}\Delta x_i +$$

$$+q_{lim}\sum_{i=1}^{n_1}k_{h1}\Delta x_i b_{qlhi} + \sum_{i=1}^{n_{tot}}k_{h2}W_i b_{whi} + \sum_{i=n_1+1}^{n_{tot}} k_{h3}q_v b_{qhi}$$

(15)

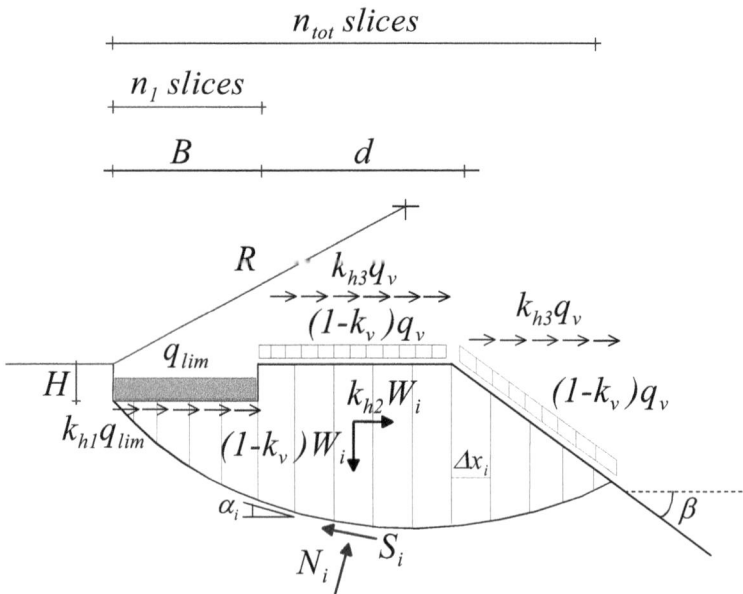

Fig. 6. Strip footings near slopes: failure mechanism and applied forces

The resisting moment $M_{res}$ and the shear strength $S_i$ acting along the base of the slices are expressed by equation (2) and (3) respectively.

For the slices $i = 1$ to $n_1$ the force $N_i$ resultant of the normal stress distribution acting at the base of the slice, derived by the Bishop's method of slices (1955) with an equilibrium equation in the vertical direction, is given by equation (4), while for the remaining slices ($n_1 + 1 \leq i \leq n_{tot}$) is :

$$N_i = \frac{q_v(1 - k_v)\Delta x_i + W_i(1 - k_v) - c\Delta x_i \tan \alpha_i}{\cos \alpha_i + \sin \alpha_i \tan \phi} \tag{16}$$

Thus :

$$M_{res} = R\sum_{i=1}^{n_{tot}} c\Delta x_i / \cos \alpha_i + R\sum_{i=1}^{n_1} \frac{q_{\lim}\frac{\Delta x_i}{} + W_i(1 - k_v) - c\Delta x_i \tan \alpha_i}{\cos \alpha_i + \sin \alpha_i \tan \phi} \tan \varphi +$$
$$+ \sum_{i=n_1+1}^{n_{tot}} \frac{q_v(1 - k_v)\Delta x_i + W_i(1 - k_v) - c\Delta x_i \tan \alpha_i}{\cos \alpha_i + \sin \alpha_i \tan \phi} R \tan \phi \tag{17}$$

where :
- $q_{lim}$ = vertical limit load acting on the footing;
- $c$ = soil cohesion;
- $\phi$ = friction angle of soil;
- $\Delta x_i$ = width of the $i^{th}$ slice;
- $W_i$ = weight of the $i^{th}$ slice;
- $q_v$ = vertical surcharge;
- $R$ = radius of the circular failure surface;
- $\alpha_i$ = angle of the base of the $i^{th}$ slice;
- $n_1$ = number of slices under the footing;
- $n_{tot}$ = total number of slices;
- $k_{h1}$ = horizontal seismic coefficient for the limit load;
- $k_{h2}$ = horizontal seismic coefficient for the soil mass;
- $k_{h3}$ = horizontal seismic coefficient for the surcharge;
- $k_v$ = vertical seismic coefficient for the soil mass and the surcharge;
- $b_{wi}$ = distance of the weight $W_i$ of the $i^{th}$ slice to the centre of the circular failure surface;
- $b_{whi}$ = distance of the inertia force $k_{h2}W_i$ of the $i^{th}$ slice to the centre of the circular failure surface;
- $b_{qvi}$ = distance of the surcharge force $q_v\Delta x_i$ of the $i^{th}$ slice to the centre of the circular failure surface;
- $b_{qhi}$ = distance of the horizontal surcharge force $k_{h3}q_v\Delta x_i$ of the $i^{th}$ slice to the centre of the circular failure surface;
- $b_{qli}$ = distance of the limit load $q_{lim}$ acting on the $i^{th}$ slice to the centre of the circular failure surface;
- $b_{qhli}$ = distance of the shear limit force $k_{h1}q_{lim}$ acting on the $i^{th}$ slice to the centre of the circular failure surface.

Substituting the following terms :

$$a_1 = R\sum_{i=1}^{n_{tot}} c\Delta x_i / \cos \alpha_i \tag{18}$$

$$a_2 = R \tan \phi \sum_{i=1}^{n_1} \frac{\Delta x_i}{\cos \alpha_i + \sin \alpha_i \tan \phi} \tag{19}$$

$$a_3 = R \tan \phi \sum_{i=1}^{n_1} \frac{W_i(1-k_v) - c\Delta x_i \tan \alpha_i}{\cos \alpha_i + \sin \alpha_i \tan \phi} \tag{20}$$

$$a_4 = R \tan \phi \sum_{i=n_1+1}^{n_{tot}} \frac{q_v(1-k_v)\Delta x_i + W_i(1-k_v) - c\Delta x_i \tan \alpha_i}{\cos \alpha_i + \sin \alpha_i \tan \phi} \tag{21}$$

$$a_5 = \sum_{i=1}^{n_{tot}} W_i(1-k_v)b_{wi} \tag{22}$$

$$a_6 = \sum_{i=1}^{n_{tot}} k_{h2}W_i b_{whi} \tag{23}$$

$$a_7 = \sum_{i=n_1+1}^{n_{tot}} q_v(1-k_v)\Delta x_i b_{qvi} \tag{24}$$

$$a_8 = \sum_{i=n_1+1}^{n_{tot}} k_{h3}q_v\Delta x_i b_{qhi} \tag{25}$$

$$a_9 = \sum_{i=1}^{n_1} \Delta x_i b_{qli} \tag{26}$$

$$a_{10} = \sum_{i=1}^{n_1} k_{h1}\Delta x_i b_{qlhi} \tag{27}$$

and equating $M_{mob} = M_{res}$ the limit load is given by :

$$q_{lim} = \frac{a_5 + a_6 + a_7 + a_8 - a_1 - a_3 - a_4}{a_2 - a_9 - a_{10}} \tag{28}$$

For example, referring to the ground slope factor $g_\gamma$ taking into account the effect of the sloping ground surface, in Figure 7 the values derived assuming the distance $d$ equal to zero and the slope angle $\beta > 0$ (angle that the ground surface makes with the horizontal), have been evaluated for three different friction angles ($\phi$ = 20°, 30° and 40°) and compared with those obtained by the well known Brinch Hansen's solution (1970).

The angle $\beta$ is positive when the ground slopes down and away from the footing. According to Brinch Hansen (1970) we have :

$$g_\gamma = (1 - 0.5 \tan \beta)^5 \tag{29}$$

while in the present study we obtain :

$$g_\gamma = (1 - 0.5 \tan \beta)^{4.5} \tag{30}$$

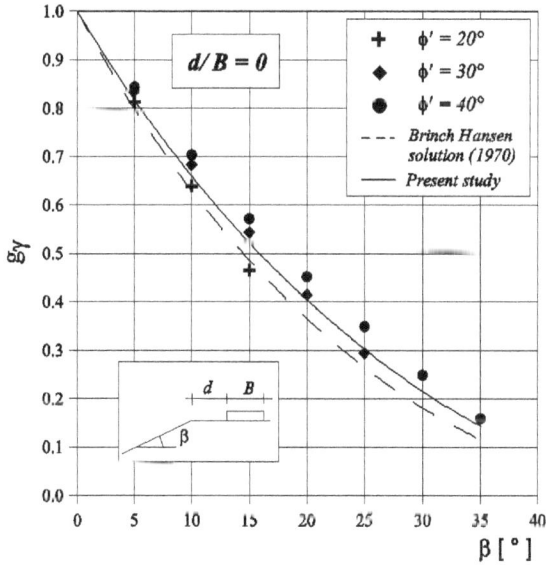

Fig. 7. Values of the ground factors $g_\gamma$ and comparison with Brinch Hansen's solution (1970)

## 4.1 Seismic analysis

The recommendations of Eurocode 8 - Part 5 (2003) state that in the calculation of the bearing capacity of a shallow foundation one should include the load inclination and eccentricity arising from the inertia forces of the structure, as well as the possible effects of the inertia in the soil.

Thus, the seismic analysis was carried out considering the following seismic coefficients: $k_{h1}$ = 0.1 and 0.2 for the inertia of the structure; $k_{h2}$ = 0.2 and 0.4 for the inertia of the soil mass.

The value of $k_{h1}$ was chosen lower than $k_{h2}$ because the Eurocode 8 (2003) allows to reduce the seismic action by a behaviour factor associated with the ductility classification of the structures.

This consideration takes to the conclusion that the kinematic effect, and the consequent reduction in bearing capacity due to the soil inertia, cannot be neglected and in some circumstances it's reduction could be more significant than the reduction due to the inertia of the structure (Cascone et al., 2006). In this study the seismic coefficient $k_{h3}$ of the surcharge was assumed equal to $k_{h1}$.

The friction angle of soil was chosen in the range 0° up to 40°, while the angle of the slope near the footing was varied in the range 5° to 35°. In the seismic analysis the angle of the sloping ground is affected by further limitations, because simple equilibrium considerations, for a cohesionless soil ($c' = 0$), take to the following :

$$\tan \beta < \frac{(1 - k_v) \tan \phi' - k_{h,i}}{1 - k_v + k_{h,i} \tan \phi'} \ (i = 2, 3) \tag{31}$$

or in a simpler form :

$$\beta < \phi' - \theta \tag{32}$$

where :

$$\theta = \tan^{-1}\left[\frac{k_{h,i}}{1-k_v}\right] \text{ (i = 2, 3)} \tag{33}$$

In the following, Table 2 shows some limit values of $\beta$, for $k_v = 0$ and $\phi = 20°$, $30°$ and $40°$.

| $\phi$ | $k_{h2}$ , $k_{h3}$ | | | |
|---|---|---|---|---|
| | 0.1 | 0.2 | 0.3 | 0.4 |
| 20° | 14.29° | 8.69° | 3.30° | - |
| 30° | 24.29° | 18.69° | 13.30° | 8.19° |
| 40° | 34.29° | 28.69° | 23.30° | 18.19° |

Table 2. Limit values of $\beta$ for a vertical seismic coefficient $k_v = 0$

In Figures 8 to 16 the results of the parametric analysis are shown in a synthetic form. The seismic bearing capacity ratios $N_c^*/N_c$, $N_q^*/N_q$, $N_\gamma^*/N_\gamma$ are represented as a function of $d/B$ for various slope angles and for different values of the friction angle of soil.
The threshold distance $(d_t)$ at which the sloping ground does not affect anymore the bearing capacity mainly increases with the increasing of the angle of friction and secondarily with the increasing of the seismic coefficient and with the increasing of the slope angle $\beta$.
The embedment depth of the footing does not play a significant role on the threshold distance, however it may produce a considerable increasing of the bearing capacity.
Referring to the $N_c^*/N_c$ ratios, we can observe values of the normalized threshold distances varying between about $d_t/B = 1$, for an undrained analysis ($\phi_u = 0°$), and $d_t/B = 5$ for $\phi = 40°$.
For the $N_q^*/N_q$ ratios, we determined values of the normalized threshold distances varying between about $d_t/B = 2$, for $\phi = 20°$ and about $d_t/B = 4$ for $\phi = 40°$.
Finally for the $N_\gamma^*/N_\gamma$ ratios, we determined values of the normalized threshold distances varying between about $d_t/B = 1.5$ for $\phi = 20°$ and about $d_t/B = 4$ for $\phi = 40°$.
No significant difference in the threshold distance was found when the inertia of the structure or the inertia of the soil mass is considered.
Furthermore the combined effects of soil and structure inertia can be taken into account by using the superposition of the effects principle.
In this case, at the same way as found by Paolucci & Pecker (1997) and Cascone et al. (2004), the bearing capacity of the soil self weight under both the seismic loading due to the coefficients $k_{h1}$ and $k_{h2}$ , can be evaluated through the following equation :

$$q_{\lim} = \frac{1}{2}B\gamma N_{\gamma_e} \approx \frac{1}{2}B\gamma N_\gamma e_{\gamma_i} e_{\gamma_k} \tag{34}$$

where :
  $N_{\gamma e}$ = bearing capacity factor reduced by both acting the coefficients $k_{h1}$ and $k_{h2}$ ;
  $N_\gamma$ = static bearing capacity factor;
  $e_{\gamma i} = N_{\gamma 1}^*/N_\gamma$ bearing capacity ratio for structure inertia only ($k_{h1} > 0$, $k_{h2} = k_{h3} = 0$);
  $e_{\gamma k} = N_{\gamma 2}^*/N_\gamma$ bearing capacity ratio for soil mass inertia only ($k_{h2} > 0$, $k_{h1} = k_{h3} = 0$).

Figure 17 shows a comparison between the $N_{\gamma e}/N_\gamma$ ratio and the product $e_{\gamma i} \cdot e_{\gamma k}$ for, as an example, $k_{h1}$ = 0.1, 0.2 and 0.3 and $k_{h2}$ = 0.1, 0.2 and 0.3.

In particular, Figure 17 shows that when the seismic coefficients $k_{h1}$ and $k_{h2}$ are small, there is not a significant difference between the $N_{\gamma e}/N_\gamma$ ratio and the product $e_{\gamma i} \cdot e_{\gamma k}$.

On the contrary, when the seismic coefficients are high enough to produce a great reduction of the limit load, one can find a great difference in using the $N_{\gamma e}/N_\gamma$ ratio instead of the product $e_{\gamma i} \cdot e_{\gamma k}$.

As example, for $\phi$ = 20°, $k_{h1}$ = 0.1 and $k_{h2}$ = 0.1, we have :
$N_{\gamma e}/N_\gamma$ = 0.565 and $e_{\gamma i} \cdot e_{\gamma k}$ = 0.587
while for $\phi$ = 20°, $k_{h1}$ = 0.3 and $k_{h2}$ = 0.3, we have :
$N_{\gamma e}/N_\gamma$ = 0.05 and $e_{\gamma i} \cdot e_{\gamma k}$ = 0.096.

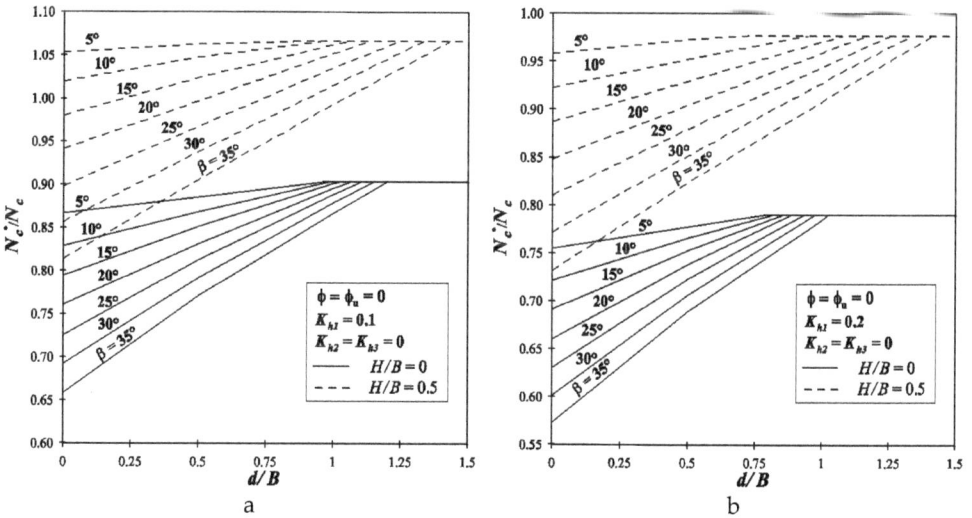

Fig. 8. $N_c^*/N_c$ ratios as a function of the normalized $d/B$ slope distance (undrained analysis $\phi = \phi_u = 0$) for $k_{h1}$ = 0.1 (a) and $k_{h1}$ = 0.2 (b)

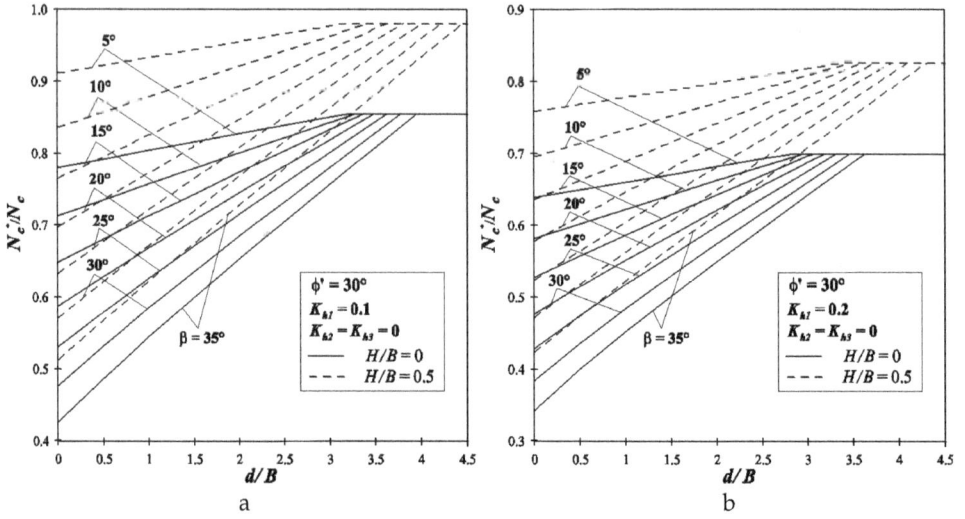

Fig. 9. $N_c^*/N_c$ ratios as a function of the normalized $d/B$ slope distance when $\phi' = 30°$ and $k_{h1} = 0.1$ (a) and $k_{h1} = 0.2$ (b)

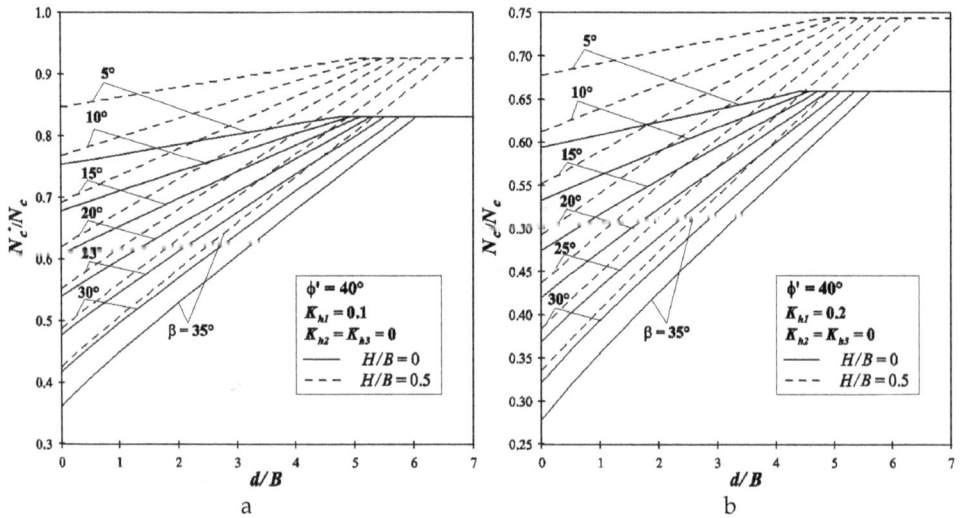

Fig. 10. $N_c^*/N_c$ ratios as a function of the normalized $d/B$ slope distance when $\phi' = 40°$ and $k_{h1} = 0.1$ (a) and $k_{h1} = 0.2$ (b)

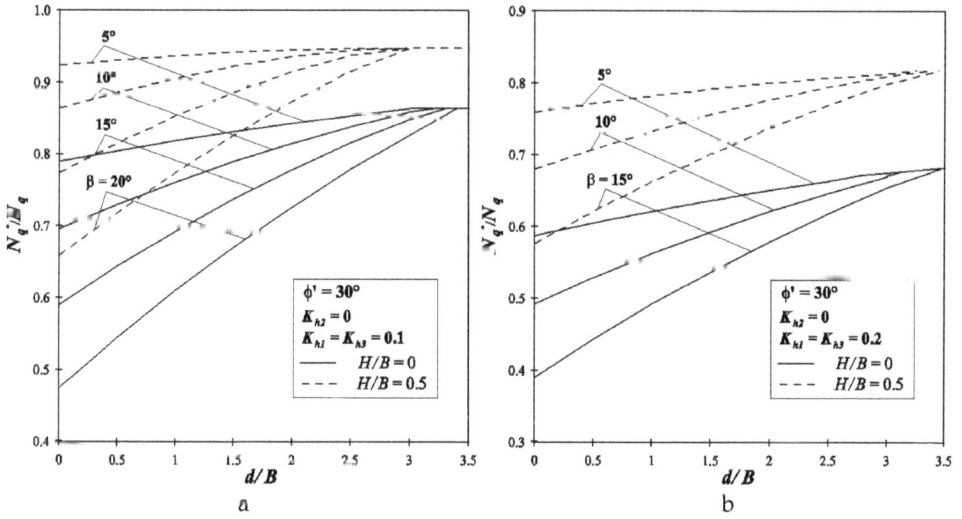

Fig. 11. $N_q^*/N_q$ ratios as a function of the normalized $d/B$ slope distance when $\phi' = 30°$ and $k_{h1} = k_{h3} = 0.1$ (a) and $k_{h1} = k_{h3} = 0.2$ (b)

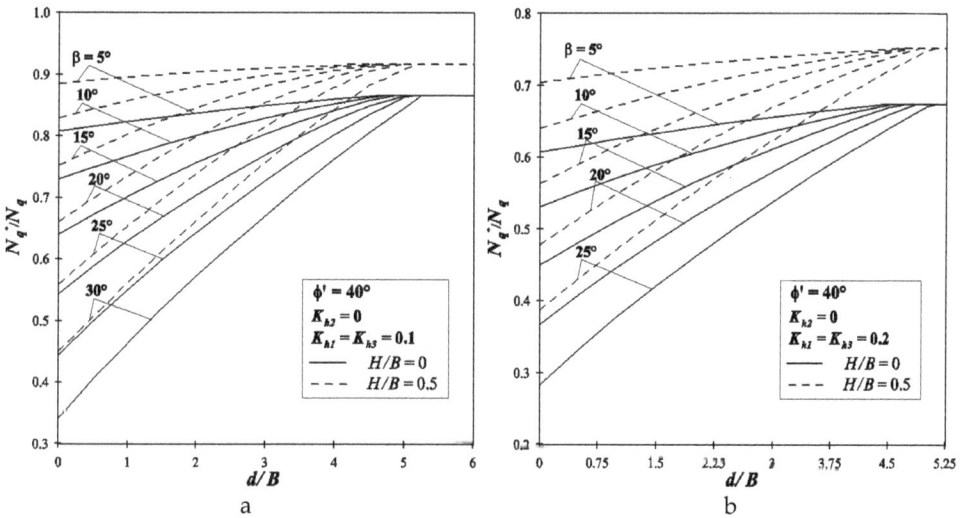

Fig. 12. $N_q^*/N_q$ ratios as a function of the normalized $d/B$ slope distance when $\phi' = 40°$ and $k_{h1} = k_{h3} = 0.1$ (a) and $k_{h1} = k_{h3} = 0.2$ (b)

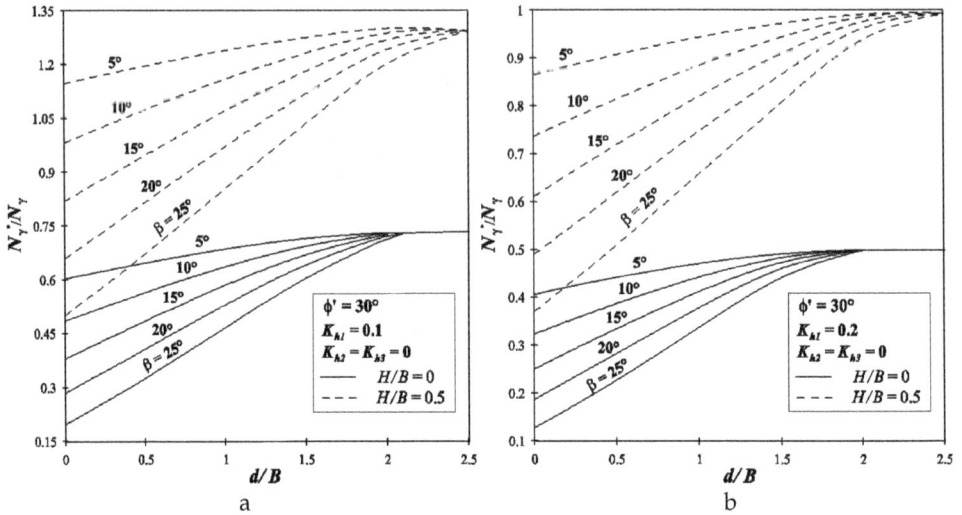

Fig. 13. $N_\gamma^*/N_\gamma$ ratios for structural inertia as a function of the normalized $d/B$ slope distance when $\phi' = 30°$ and $k_{h1} = 0.1$ (*a*) and $k_{h1} = 0.2$ (*b*)

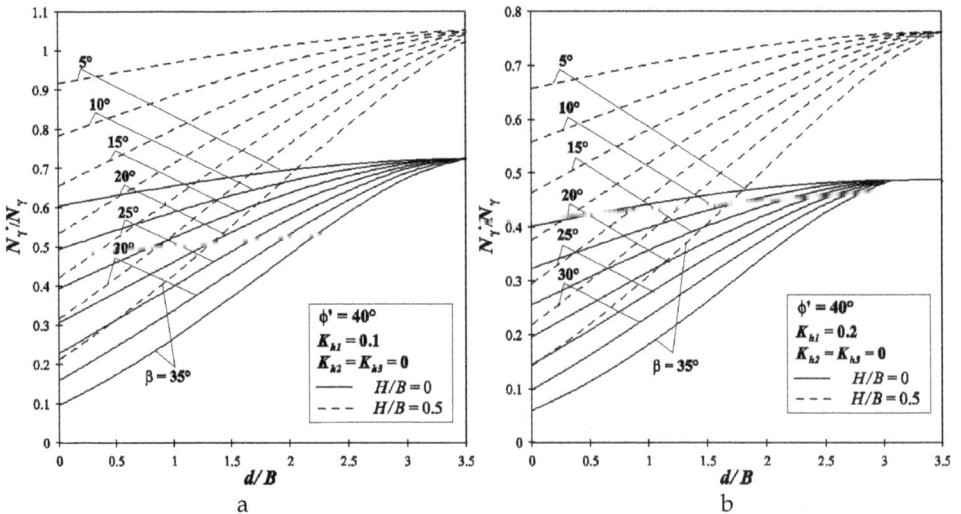

Fig. 14. $N_\gamma^*/N_\gamma$ ratios for structural inertia as a function of the normalized $d/B$ slope distance when $\phi' = 40°$ and $k_{h1} = 0.1$ (*a*) and $k_{h1} = 0.2$ (*b*)

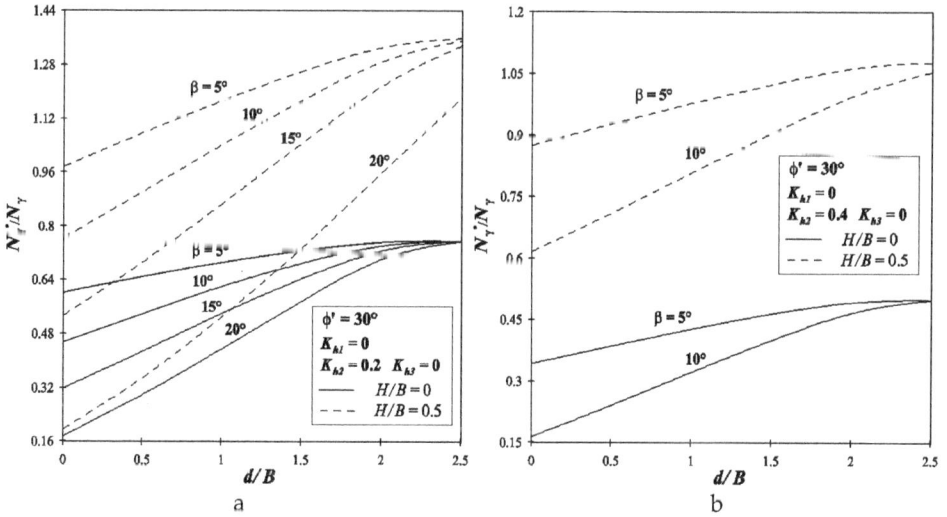

Fig. 15. $N_\gamma^*/N_\gamma$ ratios for soil inertia as a function of the normalized $d/B$ slope distance when $\phi' = 30°$ and $k_{h2} = 0.2$ ($a$) and $k_{h2} = 0.4$ ($b$)

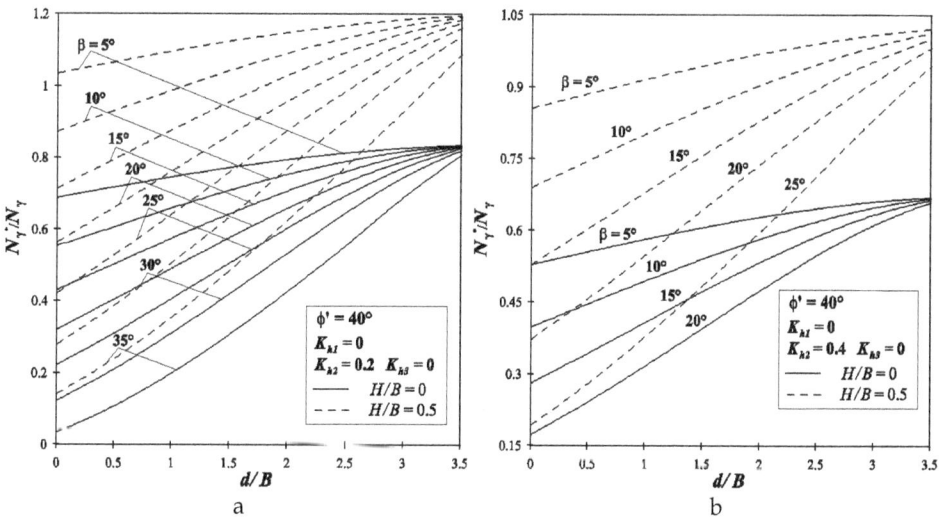

Fig. 16. $N_\gamma^*/N_\gamma$ ratios for soil inertia as a function of the normalized $d/B$ slope distance when $\phi' = 40°$ and $k_{h2} = 0.2$ ($a$) and $k_{h2} = 0.4$ ($b$)

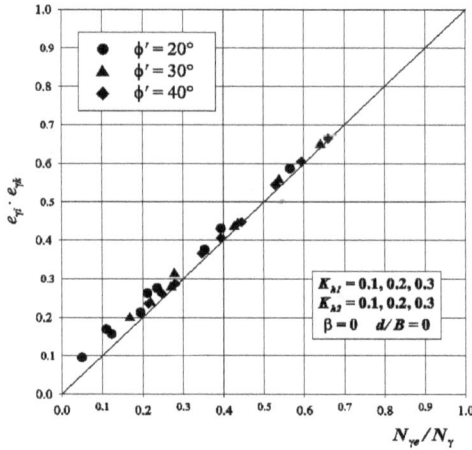

Fig. 17. Comparison between the ratio $N_{\gamma e}/N_\gamma$ and the product $e_{\gamma 1} \cdot e_{\gamma 2}$

This happens because, when the limit load approaches to zero, the critical surface associated to the simultaneous presence of both the inertial and the kinematic effects is significantly different from that deduced when one considers separately the inertial and the kinematic effects. In this case the superposition of the effects principle may lead to an unconservative design, being the product $e_{\gamma i} \cdot e_{\gamma k}$ significantly greater than the $N_{\gamma e}/N_\gamma$ ratio.

## 5. Conclusions

The design of shallow foundations subject to different static loadings has been an important area of research for geotechnical engineers. The devastating effects of recent earthquakes on shallow foundations has increased the complexity of the problem. Consequently, it is useful to obtain closed-form solutions for the earthquake resistant design of foundations.

Many analytical and numerical solutions are available for the computation of the seismic bearing capacity factors required for the design of shallow foundations.

In the present study the seismic bearing capacity of shallows foundation has been evaluated with the limit equilibrium method.

Numerical analysis shows that, by considering pseudo-static seismic forces, design solutions can be found for the computing of seismic bearing capacity factors for shallow foundations embedded in both horizontal and sloping ground.

Seismic bearing capacity factors with respect to cohesion, surcharge and unit weight components have been computed for a wide range of variation in parameters such as soil friction angle ($\phi$), horizontal and vertical seismic coefficients ($k_h$ and $k_v$).

An "upper bound" approach of the limit load was adopted to evaluate the seismic reduction factors to take into account the embedment depth of the footing ($i_{\gamma i}$) and the inertia of the soil mass ($i_{\gamma k}$), as well as, the bearing capacity ratio for structure inertia only ($e_{\gamma i}$) and the bearing capacity ratio for soil mass inertia only ($e_{\gamma k}$).

Some considerations can be formulated:

- In the evaluation of the bearing capacity due to the soil weight it has been observed that the depth of the embedment depth may play a significant role especially for low values of the friction angle.

- In some cases the seismic reduction in the bearing capacity for the soil inertia (*kinematic effect*) cannot be ignored, being of about the same amount of that produced by the inertia of the structure.
- The bearing capacity factors decrease appreciably with increases in both $k_h$ and $k_v$. Bearing capacity decreases as ground inclination $\beta$ increases and as the embedment depth $H$ increases.
- The superposition of the effects principle can be applied to determine the reduced bearing capacity caused by both the seismic actions. However, when the seismic reduction is great, due to high seismic coefficients $k_{h1}$ and $k_{h2}$, the superposition of the effects principle may lead to an unconservative design.
- By the simple limit equilibrium method modified bearing capacity factors and simple relations have been proposed which can be used for the practical design of shallow foundations embedded in both horizontal and sloping soil. In many cases the solutions obtained compare well with the previous static results and available results for the seismic conditions.

## 6. Acknowledgments

Funding for this research was provided by "DPC-ReLUIS Research Project 2010 - 2013", Task 2.1 - MT2 "Shallow and Deep Foundations".

## 7. References

Bishop, A.W. (1955). The use of the slip circle in the stability analysis of slopes. *Geotechnique*, Vol. 5, No.1, pp. 7-17

Brinch Hansen, J. (1970). A revised and extended formula for bearing capacity. *Danish Geotechnical Institute* - Bulletin no.98, Copenhagen, pp. 5-11

Budhu, M. & Al-Karni, A. (1993). Seismic bearing capacity of soils. *Geotechnique*, Vol.43, No.1, pp. 181-187

Cascone, E.; Maugeri, M. & Motta, E. (2006). Effetto dell'azione sismica sulla valutazione del fattore $N_\gamma$. *Proceedings V CNRIG*, Bari (Italy), 15 September, 2006, pp. 83-97

Cascone, E.; Carfi, G.; Maugeri, M. & Motta, E. (2004). Effetto dell'inerzia del terreno sul fattore di capacità portante $N_\gamma$. *Proceedings IARG 2004*, Trento (Italy), 7- 9 July, 2004

Castelli, F. & Motta, E. (2011). Effetto dell'affondamento sul fattore $N_\gamma$ per il calcolo del carico limite di una fondazione superficiale in condizioni sismiche, *Proceedings ANIDIS 2011*, Bari (Italy), 18-22 September 2011

Castelli, F. & Motta, E. (2010). Bearing capacity of strip footings near slopes. *Geotechnical and Geological Engineering Journal*, Vol. 28, No.2, pp. 187-198

Castelli, F. & Motta, E. (2003). Bearing capacity of shallow foundations resting on a soil layer of limited depth. *Proceedings International Symposium on Shallow Foundations, FONDSUP2003*, Paris, 5-7 November, 2003, 8 p

De Buhan, P. & Gaernier, D. (1988). Three dimensional bearing capacity analysis of a foundation near a slope. *Soils and Foundations*, Vol. 38, No.3, pp. 153-163

Dormieux, L. & Pecker, A. (1995). Seismic bearing capacity of foundation on cohesionless soil. *Journal Geotechnical Engineering*, ASCE, Vol.121, No.3, pp. 300-303

EN1998 - 1. Eurocode 8 (2003). Design of structures for earthquake resistance Part 1: General Rules, seismic actions and rules for buildings. *CENTC250*, Brussels, Belgium

EN1998 - 5. Eurocode 8 (2003). Design of structures for earthquake resistance Part 5: Foundations, retaining structures and geotechnical aspects. *CENTC250*, Brussels, Belgium

Fishman, K.L.; Richards, R. & Yao, D. (2003). Inclination factors for seismic bearing capacity. *Journal of Geotechnical and Geoenvironmental*, ASCE, Vol. 129, No.9, pp. 861-865

Kumar, J. & Rao, V.B.K. (2003). Seismic bearing capacity of foundations on slopes. *Geotechnique*, Vol.53, No.3, pp. 347-361

Hudson, D.E. (1981). The role of Geotechnical Engineering in earthquake problems. *Proceedings International Conference on Recent Advances in Geotechnical Earthquake Engineering and Soil Dynamics*, St. Louis, MO 1981

Jao, M.; Agrawal, V. & Wang, W.C. (2001). Performance of strip footings on slopes. *Proceedings 15th ICSMFE*, Vol.1, pp. 697-699

Meyerhof, G. G. (1963). Some recent research on the bearing capacity of foundations. *Canadian Geotechnical Journal*, Vol.1, No.1, pp. 16–26

Moustafa, A. (2011). Damage-based design earthquake loads for single-degree-of-freedom inelastic structures. *Journal of Structural Engineering*, ASCE, Vol.137, No.3, pp. 456-467

Moustafa, A. & Takewaki, I. (2010). Deterministic and probabilistic representation of near-field pulse-like ground motion. *Soil Dynamics and Earthquake Engineering*, Vol.30, pp. 412-422

Narita, K. & Yamaguchi, H. (1990). Bearing capacity analysis of foundations on slopes by use of log-spiral sliding surfaces. *Soils and Foundations*, Vol. 30, No.3, pp. 144-152

Paolucci, R. & Pecker, A. (1997). Seismic bearing capacity of shallow strip foundations on dry soils. *Soils and Foundations*, Vol.37, No.3, pp. 95-105

Park, Y. J.; Ang, A. H.-S., & Wen, Y. K. (1987). Damage-limiting aseismic design of buildings. *Earthquake Spectra*, Vol.3, No.1, pp. 1-26

Pecker, A. & Salencon, J. (1991). Seismic Bearing Capacity of Shallow Strip Foundations on Clay Soils. *Proceedings of the International Workshop on Seismology and Earthquake Engineering*, pp. 287-304

Pecker, A. (1996). Seismic bearing capacity of shallow foundations. *Proceedings XI World Conference on Earthquake Engineering*, Acapulco, Mexico, 23-28 June 1996, paper No. 2076

Priestley, M.J.N.; Calvi, G.M. & Kowalsky, M.J. (2005). Displacement-Based Seismic Design of Structures. *IUSS Press*, Pavia, Italy

Richards, R.; Elms, D.G. & Budhu, M. (1993). Seismic Bearing Capacity and Settlement of Foundations. *Journal of Geotechnical Engineering Division*, ASCE, Vol.119, No.4, pp. 662-674

Saran, S.; Sud, V.K. & Handa, S.C. (1989). Bearing capacity of footings adjacent to slopes. *Journal of Geotechnical Engineering*, ASCE, Vol.115, GT4, pp. 553-573

Sarma, S.K. & Chen, Y.C. (1996). Bearing capacity of strip footing near sloping ground during earthquakes. *Proceedings XI World Conference on Earthquake Engineering*, Acapulco, Mexico, 23-28 June 1996, paper No.2078

Sarma, S.K. & Iossifelis, I.S. (1990). Seismic bearing capacity factors of shallow strip footings. *Geotechnique*, Vol. 40, No.2, pp. 265-273

Shields, D.H.; Chandler, N. & Garnier, J. (1990). Bearing capacity of foundation in slopes. *Journal of Geotechnical Engineering*, ASCE, Vol.116, GT3, pp. 528-537

Sawada, T.; Nomachi, S. & Chen, W. (1994). Seismic bearing capacity of a mounted foundation near a down hill slope by pseudo-static analysis. *Soils and Foundations*, Vol.34, No.1, pp. 11-17

Terzaghi, K. (1943). Bearing capacity. *Theoretical soil mechanics*, Chapter 8, NY 1943, 118-143

Vesić, A.S. (1973). Analysis of ultimate loads of shallow foundations. *Journal of the Soil Mechanics and Foundations Division*, ASCE, Vol. 99, SM1, pp. 45-73

# 4

# Advanced Base Isolation Systems for Light Weight Equipments

Chong-Shien Tsai

*Department of Civil Engineering, Feng Chia University, Taichung, Taiwan*

## 1. Introduction

This chapter is intended to introduce the earthquake proof technology particularly in the area of base isolation systems that have been used to protect light weight structures, such as motion sensitive equipment, historic treasures, and medical instruments, etc., from earthquake damage. This chapter presents theoretical background, experimental studies, numerical analyses, and the applications of the advanced isolation systems consisting of rolling- and sliding-type isolation systems for light weight structures. The efficiency of these isolators in reducing the seismic responses of light weight equipment was also investigated in this study. In addition, the results from theoretical and experimental studies for these isolators are compared and discussed.

### 1.1 General background

One of the greatest casualties in recorded history is the Huaxian earthquake that occurred in China in 1556, causing over 830,000 deaths (Kanamori, 1978). The Tangshan great earthquake that struck the northeastern part of China in 1976 killed 242,769 people, according to official sources, although some estimates of the death toll are as high as 650,000 (Kanamori, 1978). The Mw 7.0 earthquake (Eberhard et al., 2010) that struck the Republic of Haiti on January 12, 2010, resulted in a death toll, as reported by the Government of Haiti, that exceeds 217,000, with an additional 300,000 people injured. The earthquake damaged nearly 190,000 houses, of which 105,000 were completely destroyed, and left long term suffering for the residents of the country. The moment magnitude 9.0 Tohoku earthquake (Takewakin et al., 2011) that struck eastern Japan on March 11, 2011, is one of the most five powerful earthquakes in the world since modern measurements began in 1900, killing more than 20,000 people and causing huge damage and economic loss that cannot be ignored.

Traditionally designed structures have used the strength and ductility of their structural members to resist the seismic forces or dissipate earthquake induced energy. However, many past earthquakes have proven that structures collapse or lose their functionality when the ductility capacity of the structures is consumed during the earthquake. Even if the structures survive earthquakes through excellent designs to provide more strength or ductility to the structures, the vibration sensitive equipment located in the structures may still lose its functionality due to floor accelerations.

Several techniques exist to minimize earthquake effects on structures, such as light-weight structure design, improving the ductility capacities of structures, and structural control (earthquake proof technology), etc. Structural control technology has been recognized as an effective tool in seismic mitigation, and can be classified as active, passive, hybrid and semi-active controls, which can be clarified by the following equation:

$$M\ddot{u} + C\dot{u} + Ku = -MB\ddot{u}_g + F(\ddot{u}, \dot{u}, u, t) \tag{1}$$

where M, C, and K are the mass, the damping, and stiffness matrices, respectively, of a structure, which are the natural characteristics of a structure; $\ddot{u}$, $\dot{u}$ and u denote the vectors of the relative acceleration, velocity, and displacement with respect to supports, respectively, which are structural responses during earthquakes; $\ddot{u}_g$ is the ground acceleration; B is the displacement transformation matrix; and $F(\ddot{u}, \dot{u}, u, t)$ depicts the control force that is an external force provided by various types of power and control systems.

Active control technology has a control force used to activate the control system. The control force is generated through the control signal which is based on the results calculated from the measured responses of the structure and the specified control algorithm. The structural responses can be lessened by changing its characteristic through the control force in the second term on the right hand side of Eq. (1) which could be proportional to the measured displacement, velocity and acceleration of the structure during earthquakes. From a mathematical point of view, we can then move the control force from the right hand side to the left hand side of Eq. (1) to combine with the corresponding terms depending on the values of the control fore proportional to. As a result, the mass, damping, and stiffness matrices of a structure are modified by the control force, but not by an actual device.

In the passive control, there is no external control force, $F(\ddot{u}, \dot{u}, u, t)$, in the system, which means that there is no second term on the right hand side of Eq. (1). The mass, damping, or stiffness which are the first three terms on the left hand side are modified by adding actual devices to the structure (Soong and Dragush, 1997; Takewakin, 2009). The device used to modify the mass matrix is named the tuned mass damper. Any actual devices used to modify the second and third terms on the left hand side of Eq. (1) are called energy absorbing systems (or dampers). A device such as a fluid damper producing an internal force that is strongly dependent on the relative velocity between the two ends of the device is called a velocity dependent device (damper). On the other hand, a device such as a friction and yielding dampers producing an internal force that is strongly dependent on the relative displacement between the two ends of the device is called a displacement dependent device (damper). Usually, the velocity dependent device produces minimum internal forces at the moments of maximum displacement due to zero velocity, which means that this type device provides no damping effect to the structure while the structure deforms at critical moments of earthquakes. On the other hand, the displacement dependent device produces maximum internal forces at maximum displacements. This means that this type device can provide maximum damping effect at the moments of maximum displacement and immediately reduce the structural responses at the most critical time of structural responses. The discussions of the advantages and disadvantages of these two types of devices are out of the scope of this chapter.

A system called the base (seismic) isolation system inserts a soft layer or device (base isolator) between the structure and its foundation to isolate earthquake-induced energy trying to penetrate into the structure, thereby protecting the structure from earthquake damage (Skinner et al., 1993; Naeim and Kelly, 1999). A base isolation system is used to minimize the seismic force which is the first term on the right hand side of Eq. (1) in two ways: (i) by reflecting the seismic energy by lengthening the natural period of the entire system including the structure and the base isolator and (ii) by absorbing the seismic energy through the hysteretic loop of the isolator displacement and the force induced in the isolator. The combination of the active and passive control is called hybrid control, which also needs a large control force for controlling structural responses. By contrast, semi-active control uses substantially smaller control force in the manner of an on and off switch to improve the efficiency of the passive control system through an active control algorithm, but not massive control force. In conclusion, structural control technology protects structures through mechanisms that are used to prohibit the seismic energy from transmitting into major members such as the beams, columns and walls of a structure, which are used for supporting structural weight.

In the past, there have been a lot of papers and reports concerning the use of vibration isolation technology to increase the precision of machines by isolating vibration sources resulting from the environment, such as moving vehicles, or for improving human comfort by isolating vibration sources that result from machines and moving vehicles (Rivin, 2003). Recently, the isolation technology has been acknowledged as an effective technique to promote the earthquake resistibility of the structures by controlling structural responses during earthquakes on the basis of theoretical and experimental results and earthquake events (Naeim and Kelly, 1999). Several theoretical studies have been made on the applications of the base isolation technology to critical equipment in seismic mitigation (Alhan and Gavin, 2005; Chung et al., 2008). However, little attention has been given to the experimental study of the efficiency of base isolation on the protection of vibration sensitive equipment in the events of earthquakes, especially for experimental investigations under tri-directional seismic loadings (Tsai et al., 2005b, 2007, 2008a; Fan et al., 2008).

This chapter is aimed at the seismic isolation, especially for the equipments in hospitals and facilities of emergency departments used for saving peoples' lives. These are of extreme importance and should be kept functional during and after earthquakes.

## 1.2 Background of rolling types of bearings

To the best of the author's knowledge, the isolation system with doubled spherical concave surfaces and a rolling ball located between these two concave surfaces was first patented by Touaillon in 1870, as shown in Fig. 1(a). Several similar isolation systems with a ball located between two spherical concave surfaces were also proposed (Schär, 1910; Cummings, 1930; Bakker, 1935; Wu, 1989), as shown in Figs. 1(b)-1(e). In 1997, Kemeny propounded a ball-in-cone seismic isolation bearing that includes two conical concave surfaces and a ball seated between the conical surfaces, as shown in Fig. 1(f). The dynamic behavior of the ball-in-cone isolation system has been investigated (Kasalanati et al., 1997). In addition, Cummings (1930) also proposed a seismic isolation system with a rolling rod of a cylinder sandwiched between two concave surfaces, as shown in Fig. 1(c). Lin and Hone (1993), Tsai et al. (2006b) and M. H. Tsai et al. (2007) conducted research on the effectiveness of this type of base isolation system in seismic mitigation, as shown in Figs. 2(a) and 2(b). Kim (2004) proposed a seismic isolation system that has rollers of a bowling shape to roll in the friction channel, as shown in Fig. 2(c). Tsai (2008a, b) revealed seismic isolation systems each consisting of shafts rolling in the concave slot channels, as shown in Figs. 2(a) and 2(d). These devices are capable of resisting the uplift while the vertical force in the isolator becomes negative under severe earthquakes.

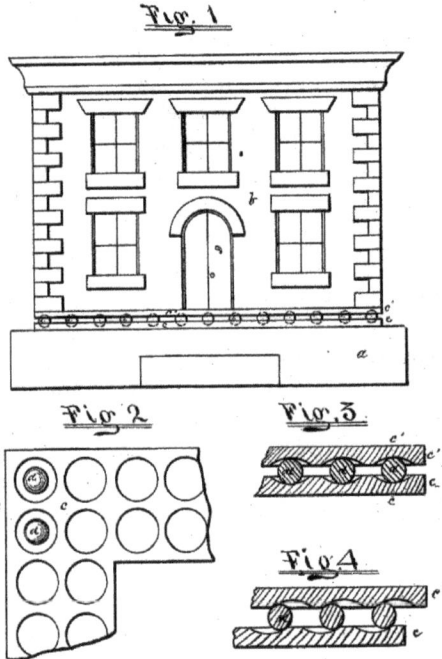

Fig. 1(a). Touaillon's original patent (1870)

F. SCHÄR.
FOUNDATION FOR BUILDINGS.
APPLICATION FILED JULY 27, 1909.

951,028.

*Fig. 1*

Patented Mar. 1, 1910.

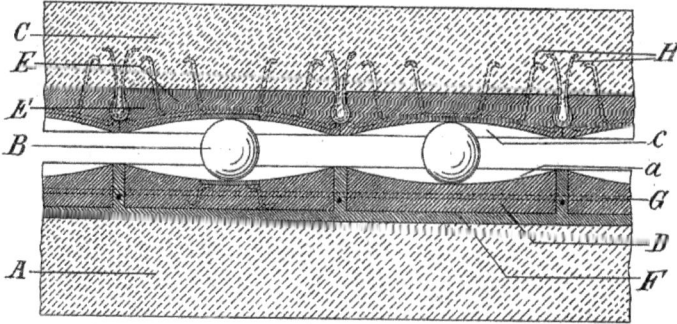

*Fig. 2*

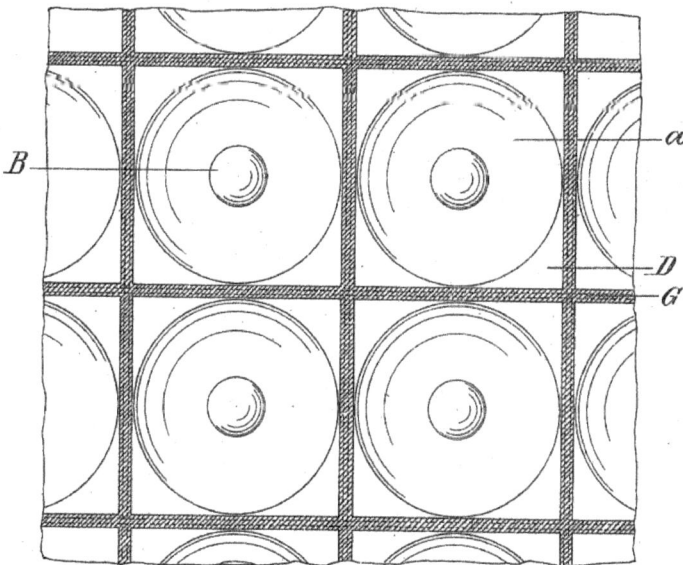

WITNESSES
w. P. Burk
A. F. Heuman

INVENTOR
Ferdinand Schär
by Mc Wallace White
ATT'Y

Fig. 1(b). Schär's original patent (1910)

June 3, 1930.                    F. D. CUMMINGS                    1,761,659

BUILDING CONSTRUCTION (QUAKEPROOF BUILDING)

Filed Jan. 18, 1928

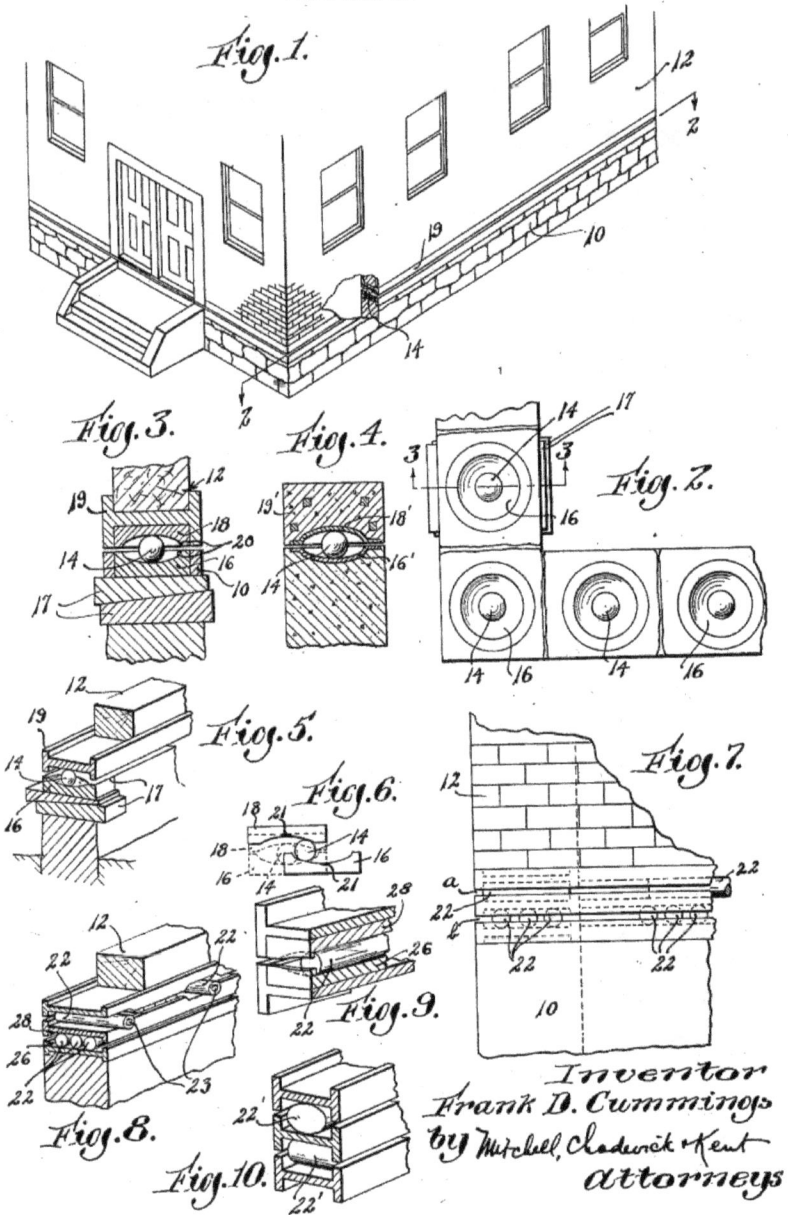

Fig. 1(c). Cummings' original patent (1930)

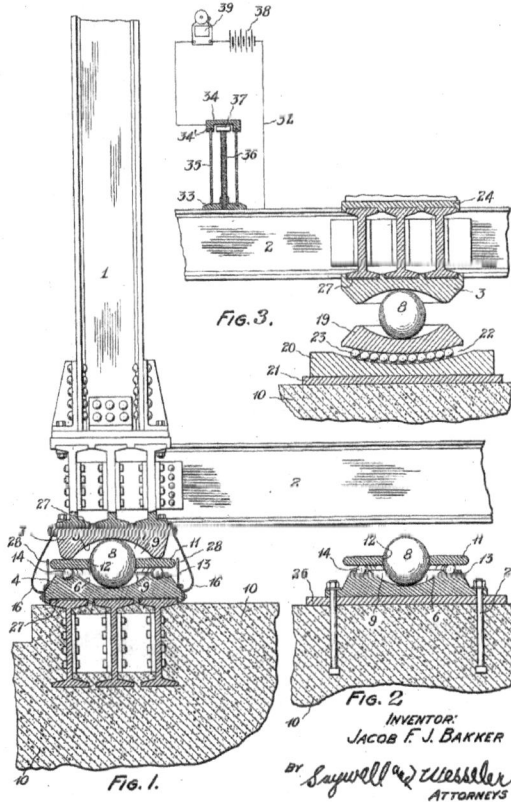

Fig. 1(d). Bakker's original patent (1935)

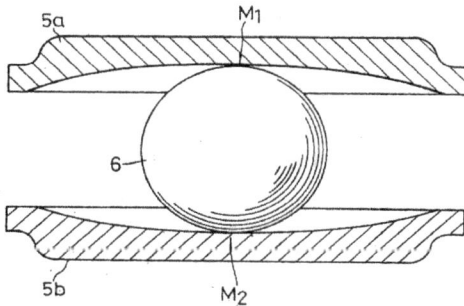

Fig. 1(e). Wu's original patent (1989)

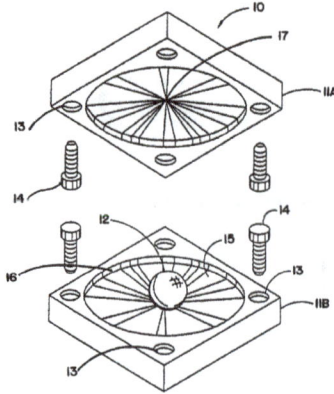

Fig. 1(f). Kemeny's original patent (1997)

Fig. 2(a). Rolling rod isolation system

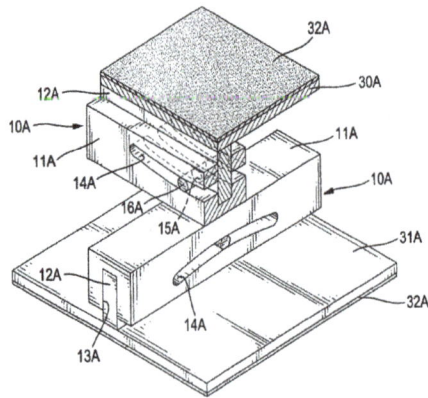

Fig. 2(b). Tsai's original patent (2008a)

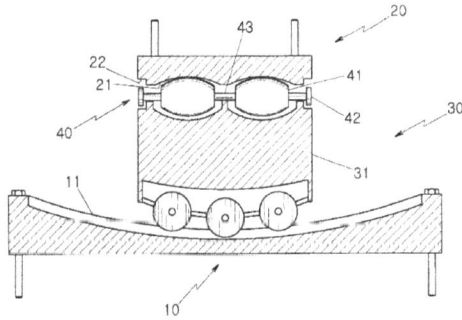

Fig. 2(c). Kim's original patent (2004)

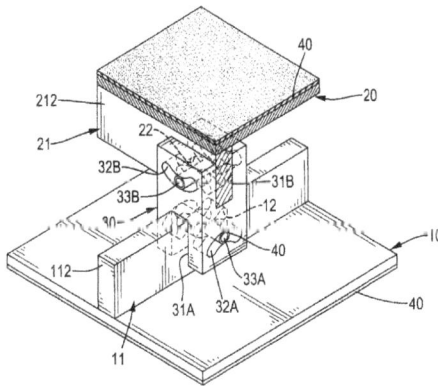

Fig. 2(d). Tsai's original patent (2008b)

The isolation system with two concave surfaces and a rolling ball (Touaillon, 1870; Wu, 1989; Kemeny, 1997) possesses some shortcomings even under small loadings like equipment and medical instruments, such as negligible damping provided by the system, a highly concentrated stress resulted from the weight of the equipment on the rolling ball and the concave surfaces due to the small contact area, and scratches and damage to the concave surfaces caused by the ball rolling motions during earthquakes. The rolling ball has a tendency to move even under environmental loadings such as human activities during regular services. In addition, the bearing size is large because of the large bearing displacements under seismic loadings due to insufficient damping provided by the rolling motion of the ball on the concave surfaces in the system.

To supply more damping to the isolation system and simultaneously reduce the bearing size as a consequence of smaller bearing displacements during earthquakes, Tsai et al. (2006a) proposed a ball pendulum system (BPS). As shown in Figs. 3(a) and 3(b), this system comprises two spherical concave surfaces and a steel rolling ball covered with a special damping material to provide horizontal and vertical damping to tackle the problems mentioned above. A series of shaking table tests conducted by Tsai et al. (2006a) have proven that the BPS isolator can enhance the seismic resistibility of vibration sensitive equipment under severe earthquakes with smaller displacements compared to an isolation system with negligible damping. However, the special material covering the steel ball that supports the weight of the vibration sensitive equipment for a long period of time in its service life span might result in permanent deformation due to plastic deformation in the damping material. It may damage or flat the contacting surface of the special damping material after sustaining a certain period of service loadings and affect the isolation efficiency.

Fig. 3(a). Open-up view of ball pendulum isolation system

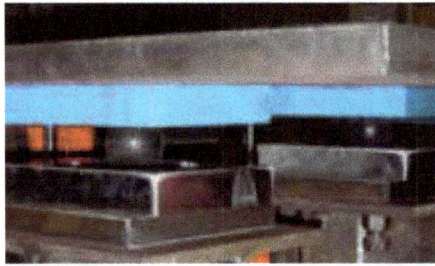

Fig. 3(b). Test set-up for ball pendulum isolation system

An alternative approach for increasing damping and lessening the isolator displacement is to add a damping device to the isolation system (Fan et al., 2008). Fathali and Filiatrault (2007) presented a spring isolation system with restraint which is a rubber snubber to play a role of displacement restrainer to limit the isolator displacement. In general, the displacement restrainer will involve impact mechanisms as a result of contact made with isolated equipment, which lead to amplified acceleration responses and large dynamic forces.

To increase damping for a rolling bearing and to prolong the service life of a bearing, an isolation system called the static dynamics interchangeable–ball pendulum system (SDI-BPS) shown as Figures 4(a) and 4(b) was proposed by Tsai et al. (2008a). The SDI-BPS system consists of not only two spherical concave surfaces and a steel rolling ball covered with a special damping material to provide supplemental damping and prevent any damage and scratches to the concave surfaces during the dynamic motions induced by earthquakes but also several small steel balls that are used to support the static weight to prevent any plastic deformation or damage to the damping material surrounding the steel rolling ball during the long term of service loadings. Because the concave surfaces are protected by the damping material covering the steel ball from damage and scratches, they may be designed as any desired shapes in geometry, which can be spherical, conical or concave surfaces with variable radii of curvature. The natural period of the SDI-BPS isolator depends only on the radii of curvature of the upper and lower concave surfaces, but not a function of the vertical loading (static weight). It can be designed as a function of the isolator displacement, and predictable and controllable for various purposes of engineering practice.

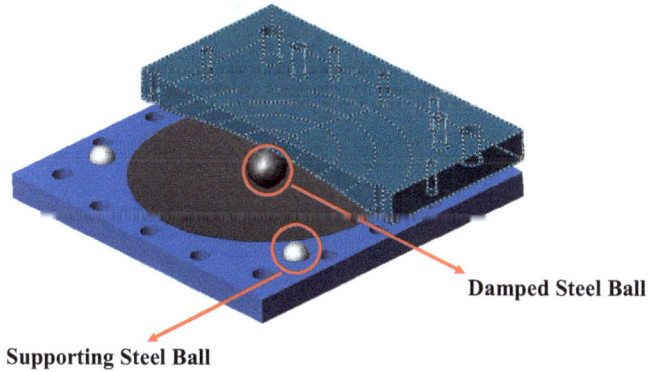

Fig. 4(a). Exploded perspective view of static dynamics interchangeable–ball pendulum system (SDI-BPS)

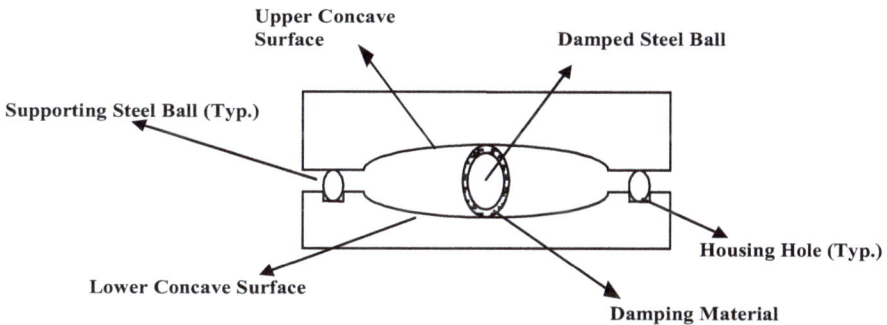

Fig. 4(b). Cross-sectional view of SDI-BPS

## 1.3 Background of sliding types of bearings

Most sliding types of isolation systems are suitable not only for light weight structures such as equipment and medical instruments but also for very heavy structures such as buildings and bridges. These types of bearings provide damping through frictional mechanism between sliding surfaces. As shown in Fig. 5(a), Penkuhn (1967) proposed a sliding isolation system including a concave sliding surface and a universal joint to accommodate the rotation resulted from the superstructure and the sliding motion of the universal joint, and suggested that the superstructure be supported by a rigid supporting base which was in turn supported by three proposed isolation bearings. Zayas (1987) and Zayas et al. (1987)

proposed a friction pendulum system (FPS) with a concave sliding surface and an articulated slider, as shown in Figs 5(b) and 5(c). Through extensive experimental and numerical studies, the FPS isolator has proven to be an efficient device for reducing the seismic responses of structures (Zayas et al., 1987; Al-Hussaini et al., 1994).

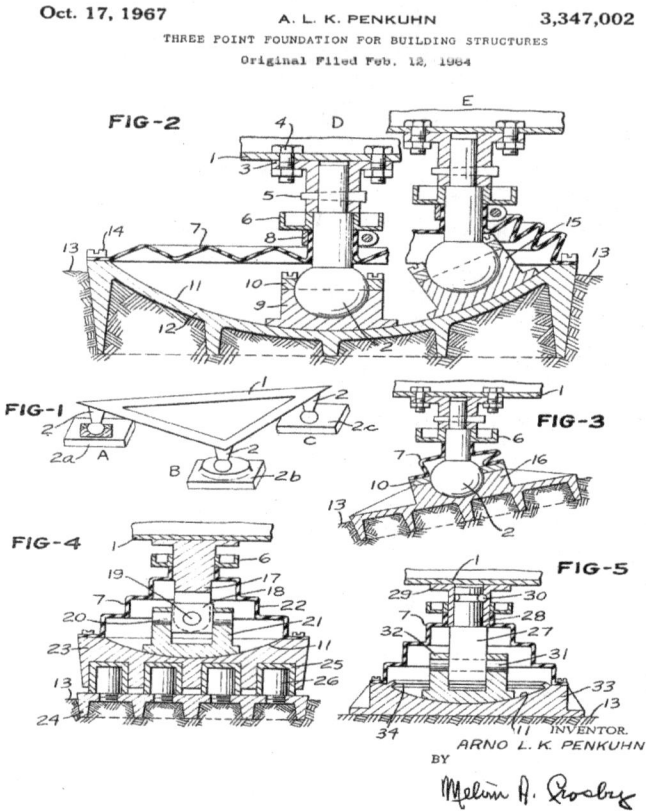

Fig. 5(a). Penkuhn's original patent (1967)

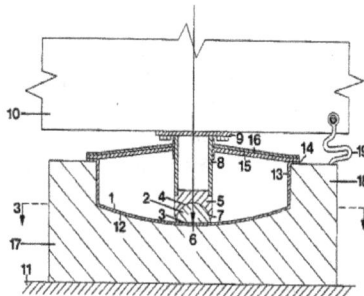

Fig. 5(b). Cross-sectional view of the friction pendulum system

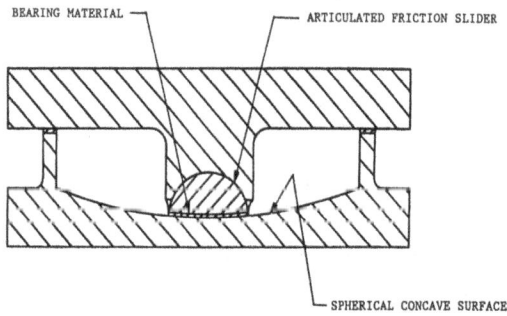

BEARING MATERIAL — 
— ARTICULATED FRICTION SLIDER
— SPHERICAL CONCAVE SURFACE

Fig. 5(c). Zayas's original patent (1987)

To avoid the possibility of resonance of the isolator with long predominant periods of ground motions, Tsai et al. (2003a) presented an analytical study for a variable curvature FPS (VCFPS). In order to enhance the quakeproof efficiency and reduce the size of the FPS isolator, Tsai (2004a,b) and Tsai et al. (2003b, 2005a,b, 2006c) proposed a sliding system called the multiple friction pendulum system (MFPS) with double concave sliding surfaces and an articulated slider located between the concave sliding surfaces, as shown in Figs. 6(a)-6(f). Based on this special design, the displacement capacity of the MFPS isolator is double of the FPS isolator that only has a single concave sliding surface, and the bending moment induced by the sliding displacement for the MFPS isolator is an half of that for the FPS isolator. Moreover, the fundamental frequency is lower than that of the FPS as a result of the series connection of the doubled sliding surfaces in the MFPS isolation system, and the bearing is a completely passive apparatus, yet exhibits adaptive stiffness and adaptive damping by using different coefficients of friction and radii of curvature on the concave sliding surfaces to change the stiffness and damping to predictable values at specifiable and controllable displacement amplitudes. Hence, the MFPS device can be given as a more effective tool to reduce the seismic responses of structures even subjected to earthquakes with long predominant periods, and be more flexible in design for engineering practice. In addition, Fenz and Constantinou (2006) conducted research and published their results on this type of base isolation system with double sliding surfaces. Kim and Yun (2007) reported the seismic response characteristics of bridges using an MFPS with double concave sliding interfaces.

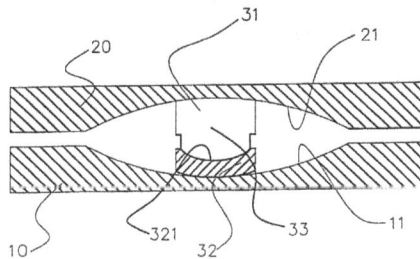

Fig. 6(a). Tsai's original patent (2004a)

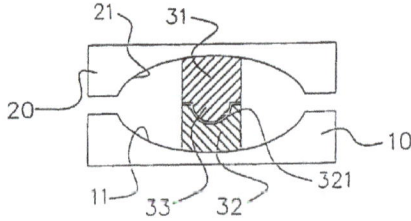

Fig. 6(b). Tsai's original patent (2004a)

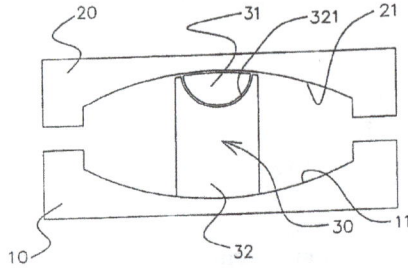

Fig. 6(c). Tsai's original patent (2004a)

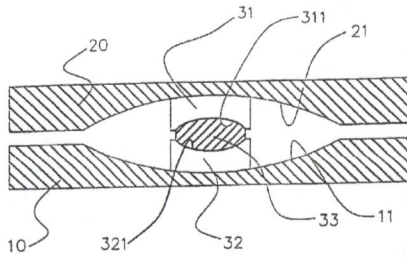

Fig. 6(d). Tsai's original patent (2004b)

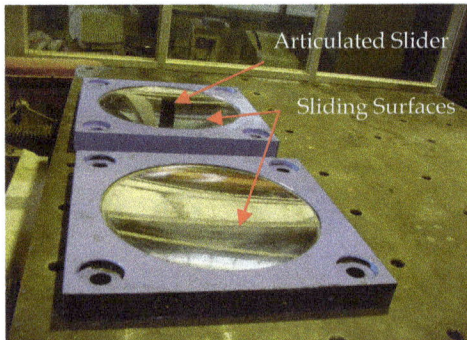

Fig. 6(e). Open-up view of MFPS isolator

Fig. 6(f). Assembled MFPS isolator

Furthermore, Tsai (2003) proposed several other types of MFPS isolators, as shown in Figs. 7(a)-7(f), each with multiple sliding interfaces, which essentially represent that each FP isolation system above and below the slider has multiple sliding interfaces connected in series (Tsai et al., 2008b, 2010a). Fenz and Constantinou (2008a, b) published their research on the characteristics of an MFPS isolator with four sliding interfaces under unidirectional loadings. Morgan and Mahin (2008, 2010) investigated the efficiency of an MFPS isolator with four concave sliding interfaces on seismic mitigation of buildings. As shown in Figs. 7(g) and 7(h), Tsai et al. (2010a, b) proposed an MFPS isolator with numerous sliding interfaces (any number of sliding interfaces). As explained earlier, these types of bearings, each having N number of sliding interfaces, possess adaptive features of stiffness and damping by adopting different coefficients of friction and radii of curvature on the concave sliding surfaces to result in changeable stiffness and damping at specified displacement amplitudes. Tsai et al. (2011a) published experimental investigations on the earthquake performance of these types of friction pendulum systems. The efficiency of the MFPS isolator with multiple sliding interfaces in mitigating structural responses during earthquakes has been proven through a series of shaking table tests on a full scale steel structure.

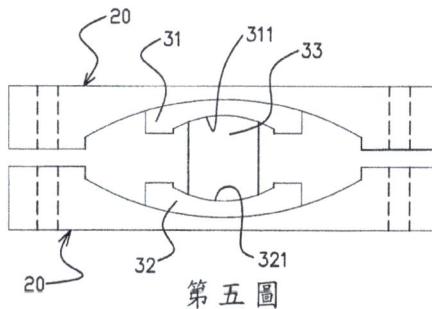

Fig. 7(a). Tsai's original patent (2003)

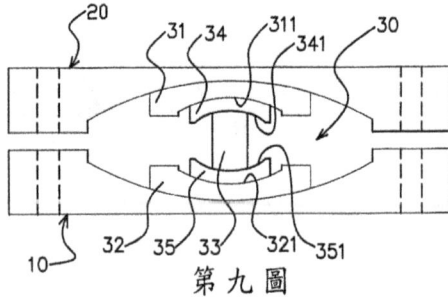

第 九 圖

Fig. 7(b). Tsai's original patent (2003)

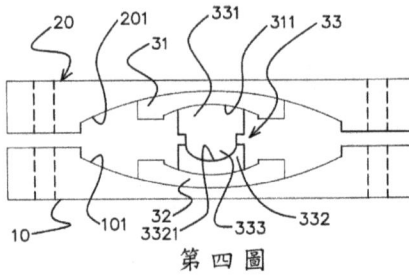

第 四 圖

Fig. 7(c). Tsai's original patent (2003)

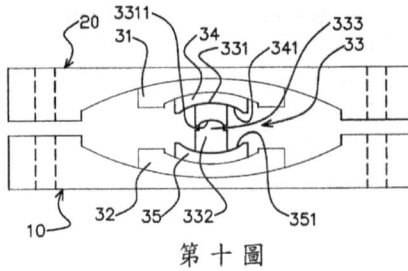

第 十 圖

Fig. 7(d). Tsai's original patent (2003)

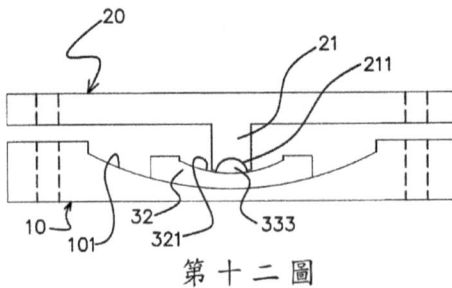

第 十 二 圖

Fig. 7(e). Tsai's original patent (2003)

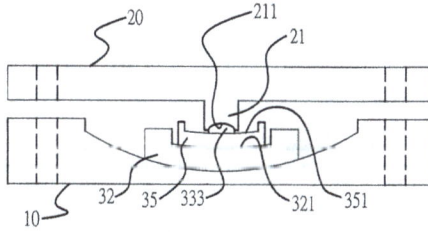

第二十圖

Fig. 7(f). Tsai's original patent (2003)

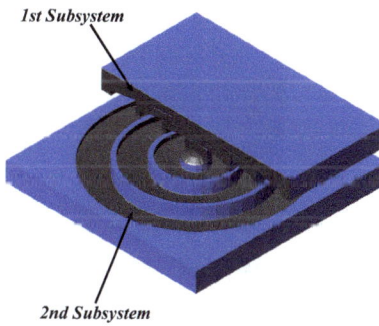

Fig. 7(g). Exploded view of MFPS isolator with six sliding surfaces

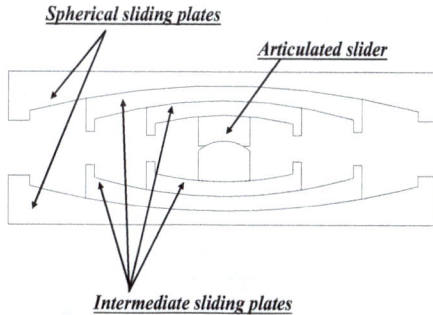

Fig. 7(h). Cross-sectional view of MFPS isolator with six sliding surfaces

An isolation system, as shown in Fig. 8(a) and called the XY-FP isolator, consisting of two orthogonal concave beams interconnected through a sliding mechanism has been published by Roussis and Constantinou (2005). The FP isolator (XY-FP) possesses the uplift-restraint property by allowing continuous transition of the bearing axial force between compression and tension, and has different frictional interface properties under compressive and tensile normal force in the isolator. A device, as shown Fig. 8(b), similar to the design concept but without the uplift-restraint property was also proposed by Tsai (2007). This device has an articulated slider seated between the FP bearings in the X and Y directions to accommodate the rotation as a result of sliding motion of the articulated slider and to maintain the isolated structure standing vertically during earthquakes.

Fig. 8(a). X-Y (Adatped from Roussis and Constantinou 2005)

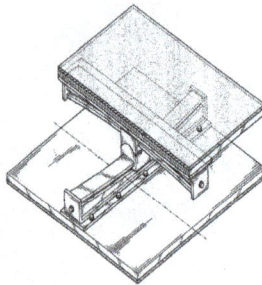

Fig. 8(b). Tsai's original patent (2007)

Tsai et al. (2010c) proposed a trench friction pendulum system (TFPS), as shown in Figs. 9(a) and 9(b), that consists of one trench concave surface in each of two orthogonal directions, and an articulated slider situated between the trench concave surfaces to accommodate the rotation induced by the sliding motion of the slider. The TFPS possesses independent characteristics such as the natural period and damping effect in two orthogonal directions, which can be applied to a bridge, equipment or a structure with considerably different natural periods in two orthogonal directions.

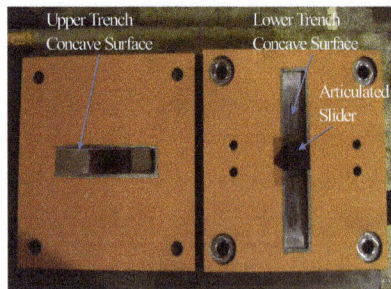

Fig. 9(a). Open-up view of trench friction pendulum system

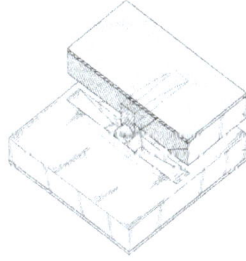

Fig. 9(b). A perspective view of trench friction pendulum system

In order to further enhance the functionality of the TFPS isolator (Tsai et al., 2010c), a base isolation system named the multiple trench friction pendulum system (MTFPS) with numerous intermediate sliding plates was proposed by Tsai et al. (2010d). As shown in Figures 10(a) and 10(b), the MTFPS isolator has multiple concave sliding interfaces that are composed of several sliding surfaces in each of two orthogonal directions, and an articulated slider located among trench concave sliding surfaces. The MTFPS represents more than one trench friction pendulum system connected in series in each direction. The friction coefficient, displacement capacity, and radius of curvature of each trench concave sliding surfaces in each direction can be different. The natural period and damping effect for a MTFPS isolator with several sliding surfaces can change continually during earthquakes. Therefore, a large number of possibilities of combinations are available for engineering designs. Such options are dependent on the needs of engineering practicing.

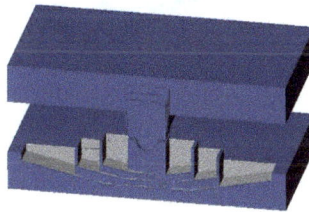

Fig. 10(a). Cross-sectional view of multiple trench friction pendulum system

Fig. 10(b). Perspective view of multiple trench friction pendulum system

As shown in Figs. 11(a) and 11(b), Tsai et al. (2007, 2008b) developed a direction-optimized friction pendulum system (DO-FPS) which consists of a spherical concave surface, a trench concave surface, as shown in Fig. 11(a), or a trajectory concave surface, as shown in Fig. 11(b), and an articulated slider. The DO-FPS isolator possesses important characteristics such as the natural period, displacement capacity and damping effect, which are functions of the directional angle of the sliding motion of the articulated slider during earthquakes. In order to improve the contact between the spherical and trench surfaces (or the trajectory concave surface), the slider consists of circular and square contact surfaces to match the spherical and trench surfaces, respectively. To further enhance the contact, it possesses a special articulation mechanism to accommodate any rotation in the isolator and maintain the stability of the isolated structures during earthquakes. In addition, the DO-FPS isolator can continually change the natural period and adjust the capacity of the bearing displacement and damping effect as a result of the change of the angle between the articulated slider and trench concave surface during earthquakes. This isolation system exhibits adaptive stiffness and adaptive damping by using different coefficients of friction and radii of curvature on the spherical and trench (or the trajectory) concave sliding surfaces to change the stiffness and damping to predictable values at specified and controllable angle of the sliding motion of the slider in the isolator although it is a completely passive device.

Fig. 11(a). Open-up view of first type direction optimized-friction pendulum system

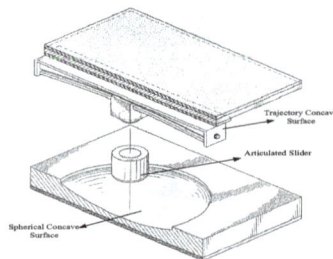

Fig. 11(b). Open-up view of second type direction optimized-friction pendulum system

As shown in Figures 12(a) and 12(b), Tsai et al. (2010e, 2011b) proposed and studied in theory and experiment on a base isolator that features variable natural period, damping effect and displacement capacity, named the multiple direction optimized-friction pendulum system (MDO-FPS). This device is mainly composed of several spherical concave sliding surfaces, several trench sliding concave surfaces and an articulated slider located among these spherical and trench concave sliding surfaces to make the isolation period changeable with the sliding direction from only the multiple trench sliding interfaces to the combinations of the multiple trench sliding interfaces and the multiple spherical sliding interfaces.

In addition, this bearing may have N number of sliding interfaces in the trench and spherical surfaces to possess adaptive features of stiffness and damping by using different coefficients of friction and radii of curvature on the trench and spherical concave surfaces leading to changeable stiffness and damping at specified displacement amplitudes. Therefore, the MDO-FPS isolator possesses important characteristics in natural period, damping effect and displacement capacity, which are functions of the direction of the sliding motion, coefficients of friction and radii of curvature on sliding interfaces, and sliding displacements.

The advantage of the isolator is able to change its natural period, damping effect and displacement capacity continually during earthquakes to avoid possibility of resonance induced by ground motions. This base isolator has more important features and flexibility than other types of base isolation devices for engineering practice. Practicing engineers will be able to optimize the isolator at various levels of earthquakes by adopting suitable parameters of friction coefficients and radii of curvature of the sliding interfaces.

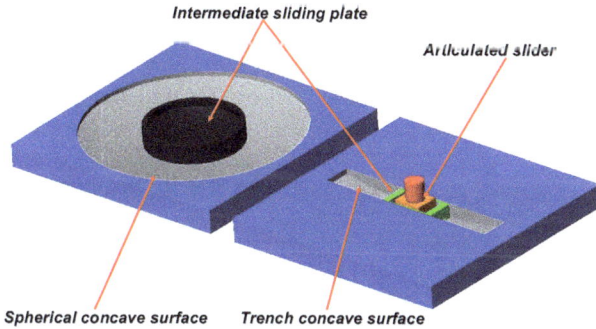

Fig. 12(a). Open-up view of MDO-FPS

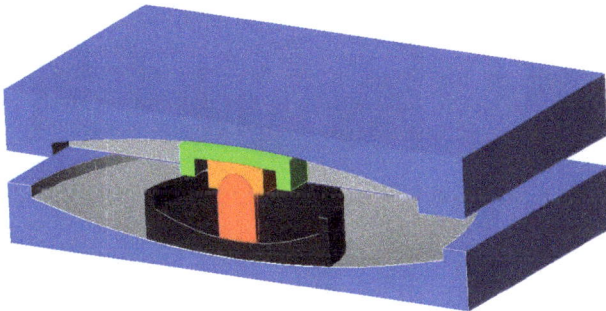

Fig. 12(b). Cross-sectional view of multiple direction optimized-friction pendulum system (MDO-FPS)

## 2. The static dynamics interchangeable – Ball pendulum system

The dynamics interchangeable–ball pendulum system (SDI-BPS) is schematized in Figs. 4(a) and 4(b) consisting of one upper concave surface (not necessary a spherical shape), one lower concave surface, several supporting steel balls to provide supports for long terms of service loadings and the frictional damping effect to the isolator at small displacements (see Case 1 of Fig. 13), several housing holes to lodge the supporting steel balls and one damped steel ball covered by damping materials to uphold the vertical loads resulting from the static and seismic loadings at large displacements (see Case 3 of Fig. 13) and supply additional damping to the bearing by deforming the damping material that could be a rubber material during earthquakes.

As shown in Case 1 of Fig. 13, almost all static loadings as a result of the weight of the equipment are sustained by the supporting steel balls and negligible loadings are taken by the damped steel ball while the system is under long terms of service loadings.

In the event of an earthquake, the static loadings and the dynamic loadings induced by the ground or floor accelerations are still supported by the supporting steel balls while the horizontally mobilized force is less than the total frictional force from the supporting steel balls, and the damped steel ball remains inactivated, similar to Case 1 of Fig. 13. The frictional force depends on the contact area and the coefficient of friction among the upper concave surface, the supporting steel balls and the housing holes located on the lower concave surface. This contact area and friction coefficient can be properly designed for the purpose of adjusting the frictional force and damping.

When the horizontal force exceeds the frictional force, the damped steel ball is activated and starts rolling on the concave surfaces. The vertical force resulting from the static and dynamic loadings is shared by the damped steel ball and the supporting steel balls. Simultaneously, the damping effect is provided by the supporting steel balls due to the frictional force and the damped steel ball as a result of the deformation of the damping material enveloping the damped steel ball under the condition of small isolator displacement, as shown in Case 2 of Fig. 13. The natural period of the isolated system is then dominated by the radii of curvature of the concave surfaces, which is equal to $2\pi\sqrt{\dfrac{R_1 + R_2}{g}}$ .

Where $R_1$ and $R_2$ are the radii of curvature of the upper and lower spherical concave surfaces, respectively; and $g$ is the gravity constant.

If the system is subjected to a large isolator displacement during an earthquake, the supporting steel balls will be detached from the upper concave surface, and the total vertical and horizontal loads will be supported by the damped steel ball only to result in more damping effect due to the larger deformation of the damping material, and no damping effect results from the frictional force caused by the supporting steel balls, as depicted in Case 3 of Fig. 13. Furthermore, the natural period of the isolated equipment is governed by the radii of curvature of the concave surfaces in this stage. The damping effect for the isolator is only provided by the deformation of the damping material covering the damped steel ball in the course of motions to reduce the size of the isolator as a result of smaller isolator displacements caused by earthquakes in comparison to a rolling isolation system with negligible damping.

As shown in Case 4 of Fig. 13, because the component of the gravity force from the equipment weight tangential to the concave surface provides the restoring force, the isolator will be rolling back to the original position without a significant residual displacement after earthquakes. Therefore, the damped steel ball is subjected to temporary loadings induced by earthquakes only, and the static loadings in the life span of service won't cause any permanent deformation to the damping material enveloping the damped steel ball.

In general, in the case of a service loading or a small earthquake, the static load is supported by the mechanism composed of the upper and lower concave surfaces and supporting steel balls with negligible supporting effect from the damped steel ball. On the other hand, in the events of medium and large earthquakes, the entire loads including static and dynamic loads are supported by the mechanism offered by the upper and lower concave surfaces and the damped steel ball while the isolation system is activated. These two mechanisms are interchangeable between the cases of static loading from the weight of equipment and seismic loading from the ground or floor acceleration.

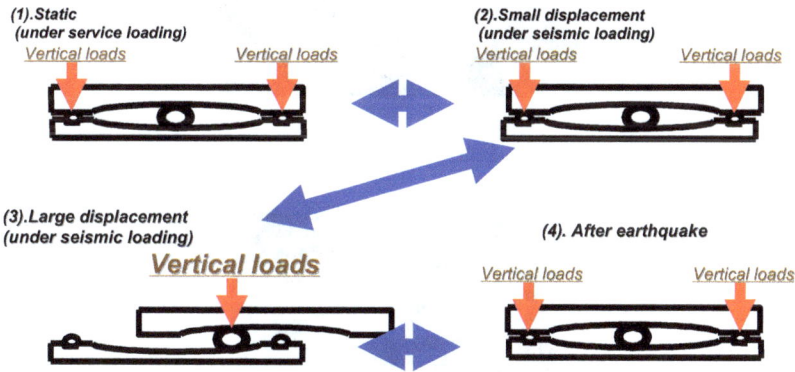

Fig. 13. Movements of a SDI-BPS isolator under service and seismic loadings

## 2.1 Characteristic of the SDI-BPS isolator

The test setup of the SDI-BPS isolator is depicted in Figures 14(a) and 14(b). In this test, the damped steel ball consisted of a steel rolling ball of 44.55mm in diameter covered with a thickness of 6.75mm damping material which was made of natural rubber material with hardness of 60 degrees in the IRHD standard (International Rubber Hardness Degree). The main purpose of this test was to investigate the mechanical behavior of the damped steel ball, therefore, supporting balls were removed during the component tests and all damping effect resulting from the system was provided by the damped steel ball. Figure 15 shows the relationship of the horizontal force to the horizontal displacement while the system was subjected to a vertical load of 4.56 $KN$ and a harmonic displacement of 50mm with a frequency of 0.3Hz. The enclosed area shown in the Fig. 15 provides a damping effect into the isolation system. The test result demonstrates that the deformations of the rubber material leaded to significant damping effects in the system with negligible deformation occurring in the steel material.

Fig. 14(a). Setup for component tests of the SDI-BPS isolator

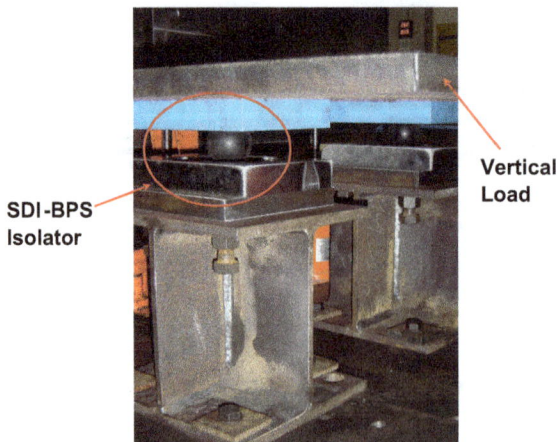

Fig. 14(b). Close view of component test of SDI-BPS isolator

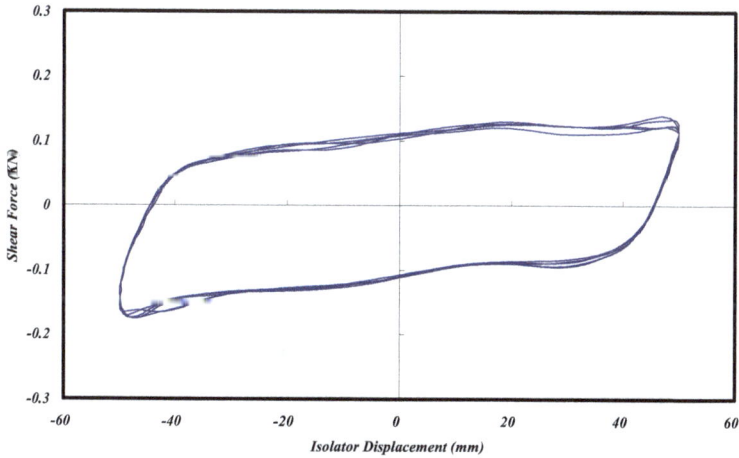

Fig. 15. Hysteresis loop of SDI-BPS isolator without supporting balls

## 2.2 Shaking table tests of motion sensitive equipment with SDI-BPS isolators

This section will investigate the performance of the SDI-BPS isolator installed in the motion sensitive equipment on seismic mitigation under tri-directional earthquakes through a series of shaking table tests of the vibration sensitive equipment isolated with the SDI-BPS isolator under tri-directional earthquakes. As shown in Figure 16, the tested vibration sensitive equipment with six inner layers was used to house the network server. The lengths in the two horizontal directions were 0.8 m and 0.6 m, respectively, and 1.98 m in height. The mass of the empty equipment was 110 kg. A mass of 108 kg at each layer from the first to the third layer and 54 kg each at the rest of the layers was added. In the case of the fixed base equipment, the natural frequency was 5.66 Hz. In the case of the isolated system, four SDI-BPS isolators with a radius of curvature of 1.0 m representing 2.84 seconds in natural period for the isolated equipment were installed beneath the equipment.

The input ground motions in these tests included the 1995 Kobe earthquake (Japan) and the 1999 Chi-Chi earthquake (recorded at TCU084 station, Taiwan). Figures 17(a) and 17(b) show the comparisons of the acceleration responses at the top layer of the equipment between the fixed base and SDI-BPS-isolated systems under tri-directional earthquakes. It is observed from these figures that the SDI-BPS isolator can effectively isolate the seismic energy trying to transmit into the vibration sensitive equipment during earthquakes. Figures 18(a) and 18(b) show the hysteresis loops of the SDI-BPS isolator under the various tri-directional earthquakes. These figures illustrate that the SDI-BPS isolator can provide damping to limit the bearing displacement, and accordingly, the bearing size was decreased. It also infers from these figures that a frictional type of damping was provided by the isolation system in small displacements and other type of damping was provided by the damped steel ball of the isolation system for large displacements, referring to the hysteresis loop in Figure 15, because the upper concave surface was lifted and away from the supporting steel balls without any contact. The isolator displacement history depicted in Figure 19 demonstrates that the response of equipment had been reduced by the isolation system with acceptable displacements in the isolators and that the isolator displacement approached zero in the end of the earthquake with negligible residual displacement.

Fig. 16. Vibration sensitive equipment isolated with SDI-BPS isolators

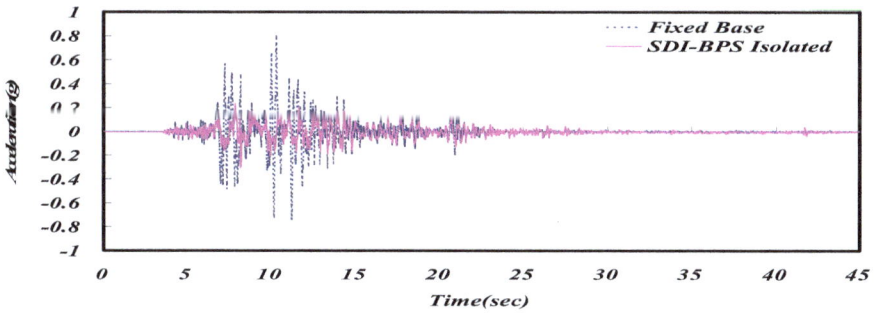

Fig. 17(a). Comparison of X-directional acceleration response at top layer between fixed-base and SDI-BPS-isolated systems under tri-directional Kobe earthquake (PGA = X0.388g + Y0.265g + Z0.116g)

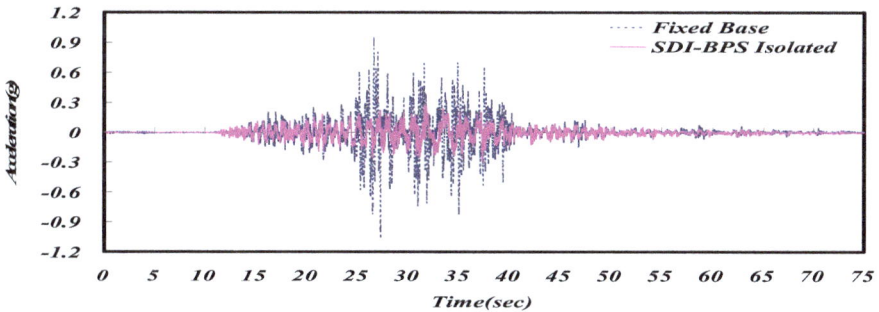

Fig. 17(b). Comparison of acceleration response in X-direction at top layer between fixed-base and SDI-BPS-isolated systems under tri-directional Chi-Chi (TCU084 Station) earthquake (PGA = X0.673g + Y0.271g + Z0.159g)

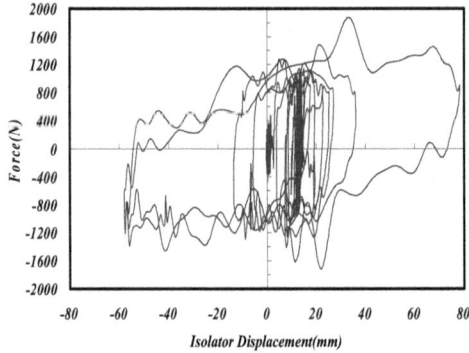

Fig. 18(a). Hysteresis loops of SDI-BPS isolator in X direction under tri-directional Kobe earthquake (PGA = X0.388g + Y0.265g + Z0.116g)

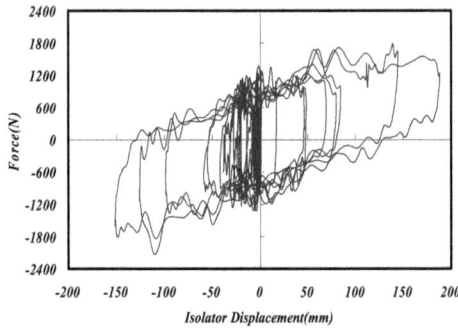

Fig. 18(b). Hysteresis loops of SDI-BPS isolator in X direction under tri-directional Chi-Chi (TCU084 Station) earthquake (PGA = X0.673g + Y0.271g + Z0.159g)

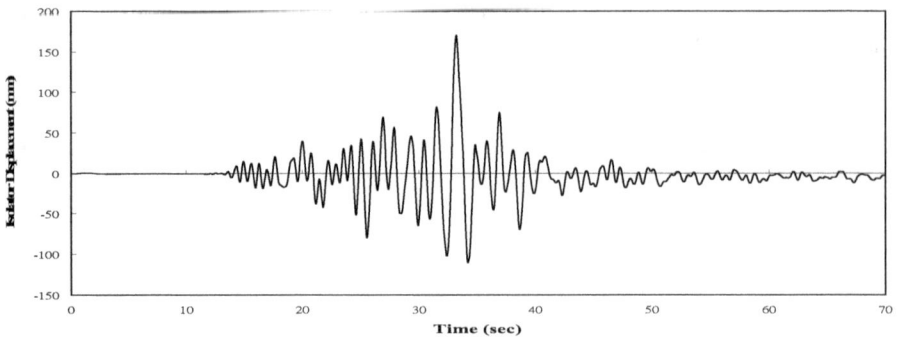

Fig. 19. Isolator displacement history in X-direction while isolated system subjected to tri-directional Chi-Chi (TCU084 Station) earthquake (PGA = X0.513g + Y0.170g +Z0.151g)

## 3. The direction-optimized friction pendulum system

A seismic isolation system, called the direction-optimized friction pendulum system (DO-FPS), consists of a concave trench, a plate with a spherically concave surface, and an articulated slider located in between, as shown in Figs. 20(a) and 20(b). The trench and the plate can usually possess different curvature radii, friction coefficients, and displacement capacities. As shown in Fig. 20(b), the concave trench is connected in series with the spherically concave plate in the X-direction, and this provides both the maximum natural period and the maximum displacement capacity. By contrast, in the Y-direction the isolator consists of just the spherically concave plate, and this provides both the minimum natural period and the minimum displacement capacity. The natural period and the displacement capacity vary with direction and are functions of the sliding angle between the articulated slider and the concave trench. In engineering practice, the DO-FPS isolator can be designed for differences in natural period and displacement capacity for the purpose of cost-effectiveness. During an earthquake, the DO-FPS isolator will adjust the natural period automatically to avoid the possibility of resonance. Following the earthquake, the concave trench and the spherically concave plate can offer a centering mechanism to return the isolated structure to its original position without significant residual displacements.

Fig. 20(a). Assembly of direction optimized friction pendulum system

Fig. 20(b). An open-up view of the DO-FPS isolator with trench and concave surface

### 3.1 Properties of the direction-optimized friction pendulum system

As shown in Figs. 21(a) and 21(b), an articulated slider is located between the concave trench and the spherically concave plate. The articulated slider includes two parts: the lower part at the trench side can have a rectangular cross section, while the upper part at the plate side can have a circular cross section. These two parts are joined with a hemispherical joint to produce an articulation mechanism in the middle of the slider. Hence, the articulated slider incorporates a rotational function to accommodate the relative rotation during its movement. As shown in Fig. 21(a), the slider is coated with Teflon composite on top, bottom, and two sides to reduce the frictional forces at its interfaces with not only the concave surfaces of the trench and plate but also the side walls of the trench.

Fig. 21(a). Assembly of articulated slider coated with Teflon composite

Fig. 21(b). Slider located between the concave trench and the spherical concave surface.

As shown in Fig. 21(b), the width of the articulated slider almost equals that of the trench; therefore, the sides of the slider and the walls of the trench almost remain in contact. These two sides of the articulated slider can thus provide additional damping in the direction parallel to the concave trench to help dissipate seismic energy.

### 3.2 Component tests of sliding interfaces

Figures 22(a) and 22(b) respectively show the schematic design and the test setup for the Teflon composite coated on the sliding surfaces. To examine the sliding characteristic of the sliding surface, the steel plates were coated with the Teflon composite and the high density chrome, respectively, to rub against each other. During the tests, the vertical pressures imposing at the interface are variable values.

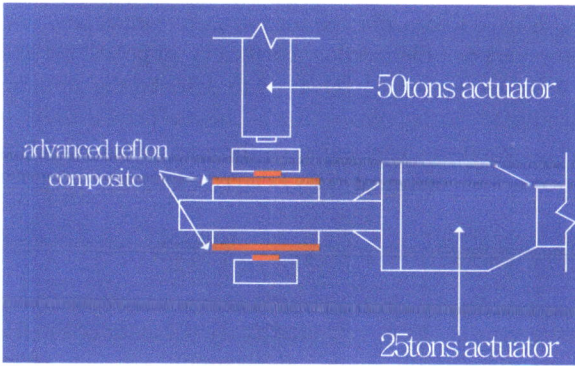

Fig. 22(a). The schematic design for the test of the Teflon composite

Fig. 22(b). Test setup for the Teflon composite

Figure 23 depicts a typically hysteretic loop when the vertical pressure was 98Mpa. Figure 24 demonstrate that the coefficients of friction are functions of the vertical pressure on the contact surface and the velocity of the sliding motion. In accordance with the analytical model proposed by Tsai (2005a), the total friction force acting on the sliding interface can be expressed as:

$$\mu(\dot{u}_b) = \frac{A}{\lambda_1 A + \lambda_2 P} \cdot \{1 + \alpha[1 - \exp(-\beta|\dot{u}_b|)]\} \times Coef \tag{2}$$

where $A$ represents the contact area at the sliding interface; $\lambda_1$ and $\lambda_2$ denote the parameters associated with the quasi-static friction force at the zero velocity; $P$ is the contact force normal to the sliding surface; and $\alpha$ is the amplification factor used to describe the increase of friction force with increasing the sliding velocity; and $\beta$ is the parameter which controls the variation of the friction coefficient with sliding velocity; and $\dot{u}_b$ is the siding velocity of the base isolator.

The term $A/_{\lambda_1 A + \lambda_2 P}$ is used to describe the friction coefficient at zero velocity. According to the experimental observations, the friction coefficient depends on the sliding velocity. Therefore, the term $1 + \alpha\left[1 - \exp(-\beta|\dot{u}_b|)\right]$ is used to describe the amplification factor of the friction coefficient relative to that at zero velocity. The coefficient, $Coef$, is a decay function representing the phenomenon of degradation of the friction force with the increase of the number of cyclic reversals. The coefficient of $Coef$ can be shown as:

$$Coef = (1 - \gamma_1) + \gamma_1 \cdot \exp(-\gamma_2 \cdot \int_0^t \frac{F_t - F_t^0}{F_t^0} \cdot du_b) \tag{3}$$

where $\gamma_1$ and $\gamma_2$ are parameters to describe the decay behavior of the friction coefficient at the Teflon interface associated with the energy accumulation in the history of the sliding motion; $F_t^0$ is the friction force when the sliding velocity is equal to zero; $F_t$ is the friction force at current time $t$; and $du_b$ is the displacement increment of the base isolator. The effect of the coefficient, $Coef$, may be neglected for the purpose of engineering practice.

$$\lambda_1 = 21.120 \ , \ \lambda_2 = 1.221 \times 10^{-7} (1 / Pa)$$

$$\alpha = 1.903 \ , \ \beta = 100.000 (\text{sec}/ m)$$

$$\gamma_1 = 0.1390 \ , \ \gamma_2 = 7.1537 (1 / m)$$

According to the model proposed by Al-Hussaini et al. (1994), the friction coefficient can be represented in the following form:

$$\mu(\dot{u}_b) = \mu_{max} - (\mu_{max} - \mu_{min})\exp(-\beta|\dot{u}_b|) \tag{4}$$

where $\mu_{min}$ and $\mu_{max}$ are the friction coefficients at zero sliding velocity and high sliding velocity, respectively.

Figure 25 shows the comparison between the theoretical and experimental results. It can be concluded that the numerical result obtained from the mathematical model has good agreement with the experimental results.

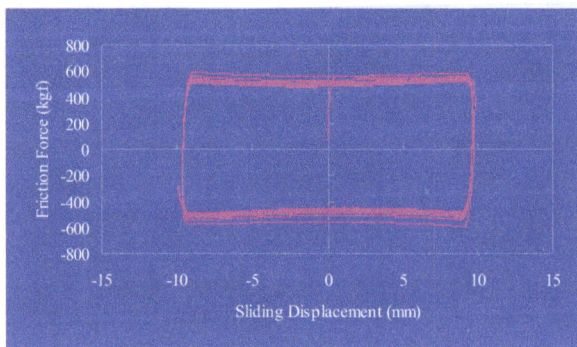

Fig. 23. Typical hysteretic loop for the Teflon composite under a vertical pressure of 98 Mpa and a Frequency of 1Hz

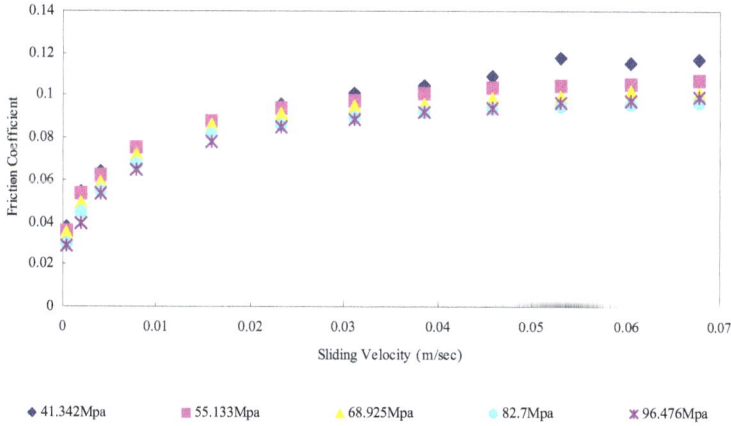

41.342Mpa          55.133Mpa          68.925Mpa          82.7Mpa          96.476Mpa

Fig. 24. Friction coefficients under various vertical pressures

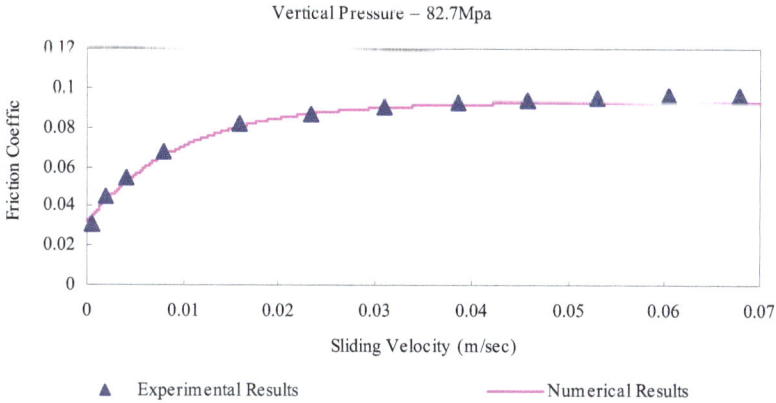

Vertical Pressure – 82.7Mpa

▲ Experimental Results          —— Numerical Results

Fig. 25. Friction coefficients under various sliding velocities

## 3.3 Equation of motion for a rigid mass isolated with DO-FPS isolators

As shown in Fig. 26, the time history of ground accelerations can be represented by the method of interpolation of excitation by assuming that the ground motion is a linear variation between time $t_{i-1}$ and $t_i$. If a rigid mass is isolated with DO-FPS isolators, as shown in Fig. 27, the equation of motion can be expressed as:

$$m\ddot{u}_b + c_b\dot{u}_b + k_b u_b = -\mu(\dot{u}_b)mg\,\mathrm{sgn}(\dot{u}_b) - m\ddot{u}_g^{i-1} - \frac{m\ddot{u}_g^{i} - m\ddot{u}_g^{i-1}}{\Delta t}\tau \qquad (5)$$

where $m$, $c_b$ and $k_b$ are the mass, damping coefficient and horizontal stiffness of the isolated mass, respectively; $u_b$ is the displacement of the base isolator relative to the ground; $\mu(\dot{u}_b)$ is the friction coefficient of the sliding surface, which is a function of sliding velocity; and $\ddot{u}_g^{i}$ is the ground acceleration at time $t_i$.

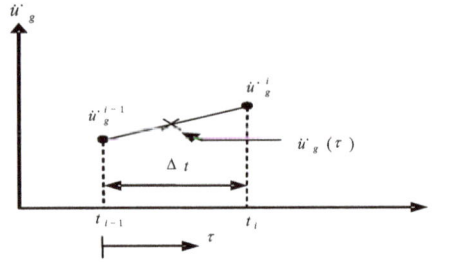

Fig. 26. Linear Interpolation for Ground Motions

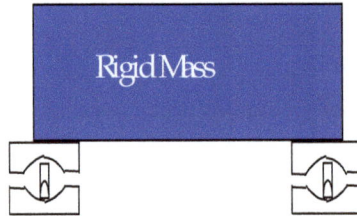

Fig. 27. A rigid mass isolated with DO-FPS isolator

The transient response of Eq. (5) can be given by:

$$u_c(\tau) = (A\cos\omega_d\tau + B\sin\omega_d\tau)\exp(-\xi\omega_n\tau) \tag{6}$$

where $\omega_n$ is the natural frequency; $\omega_d = \omega_n\sqrt{1-\xi^2}$ is the damped frequency; $\xi$ is the viscous damping ratio.

The particular solution of Eq. (5) between time $t_{i-1}$ and $t_i$ can be given by:

$$u_p(\tau) = \frac{1}{\omega_n^2}\left[-\mu(\dot{u}_b)g\,\mathrm{sgn}(\dot{u}_b) - \ddot{u}_g^{i-1} + \frac{2\xi}{\omega_n\Delta t}(\ddot{u}_g^i - \ddot{u}_g^{i-1})\right] - \frac{\ddot{u}_g^i - \ddot{u}_g^{i-1}}{\omega_n^2\Delta t}\tau \tag{7}$$

The sliding displacement of the base isolator between time $t_{i-1}$ and $t_i$ can be obtained from Eqs. (6) and (7) as:

$$u_b(\tau) = (A\cos\omega_d\tau + B\sin\omega_d\tau)\exp(-\xi\omega_n\tau)$$
$$+ \frac{1}{\omega_n^2}\left[-\mu(\dot{u}_b)g\,\mathrm{sgn}(\dot{u}_b) - \ddot{u}_g^{i-1} + \frac{2\xi}{\omega_n\Delta t}(\ddot{u}_g^i - \ddot{u}_g^{i-1})\right] - \frac{\ddot{u}_g^i - \ddot{u}_g^{i-1}}{\omega_n^2\Delta t}\tau \tag{8}$$

At the beginning of each time step, the sliding displacement is equal to that at the end of the previous time step, i.e. $u_b(0) = u_b^{i-1}$.

The coefficient $A$ of Equation (6) can be obtained as:

$$A = u_b^{i-1} - \frac{1}{\omega_n^2}\left[-\mu(\dot{u}_b)g\,\mathrm{sgn}(\dot{u}_b) - \ddot{u}_g^{i-1} + \frac{2\xi}{\omega_n\Delta t}(\ddot{u}_g^i - \ddot{u}_g^{i-1})\right] \tag{9}$$

The derivative of Eq. (8) respect to $\tau$ leads to:

$$\dot{u}_b(\tau) = (-\omega_d A \sin \omega_d \tau + \omega_d B \cos \omega_d \tau) \exp(-\xi \omega_n \tau)$$
$$-\xi \omega_n (A \cos \omega_d \tau + B \sin \omega_d \tau) \exp(-\xi \omega_n \tau) - \frac{\ddot{u}_g^i - \ddot{u}_g^{i-1}}{\omega_n^2 \Delta t} \tag{10}$$

Backsubstitution of $\dot{u}_b(0) = \dot{u}_b^{i-1}$ into Eq. (10) results in:

$$B = \frac{1}{\omega_d}(\dot{u}_b^{i-1} + \xi \omega_n A + \frac{\ddot{u}_g^i - \ddot{u}_g^{i-1}}{\omega_n^2 \Delta t}) \tag{11}$$

The sliding acceleration of the base isolator relative to the ground can be given by:

$$\ddot{u}_b(\tau) = (-\omega_d^2 A \cos \omega_d \tau - \omega_d^2 B \sin \omega_d \tau) \exp(-\xi \omega_n \tau)$$
$$-2\xi \omega_n (-\omega_d A \sin \omega_d \tau + \omega_d B \cos \omega_d \tau) \exp(-\xi \omega_n \tau) \tag{12}$$
$$+\xi^2 \omega_n^2 (A \cos \omega_d \tau + B \sin \omega_d \tau) \exp(-\xi \omega_n \tau)$$

### 3.3.1 Condition for non-sliding phase

The kinetic friction coefficient has been considered as the same as the static friction coefficient. Therefore, as the summation of the inertia and restoring forces imposing at the base raft is lower than the quasi-static friction force, i. e.:

$$\left| m(\ddot{u}_b + \ddot{u}_g) + c_b \dot{u}_b + k_b u_b \right| < \mu_{min} mg \tag{13}$$

Then the structure will behave as a conventional fixed base structure, and the sliding displacement, sliding velocity and relative acceleration are:

$$u_b = \text{constant} , \quad \dot{u}_b = \ddot{u}_b = 0 \tag{14}$$

### 3.3.2 Initiation of sliding phase

The base isolated rigid mass will behave as a fixed base structure unless the static friction force can be overcome. During the sliding phase, the equation given in the following should be satisfied:

$$\left| m(\ddot{u}_b + \ddot{u}_g) + c_b \dot{u}_b + k_b u_b \right| \ge \mu_{min} mg \tag{15}$$

Because the time increment adopted in the time history analysis (e.g. $\Delta t - 0.0005$ sec ) is quite smaller than that of the sampling time of the earthquake history, it is reasonable to assume that the direction of sliding at the current time step is the same as the previous time step. It should be noted that the direction of sliding remains unchanged during a particular sliding phase. At the end of each time step, the validity of inequality Eq. (15) should be checked. If the inequality is not satisfied at a particular time step, then the structure enter a non-sliding phase and behaves as a fixed base structure.

### 3.3.3 Simplified mathematical model for DO-FPS isolator

The simplified model based on the equilibrium at the slider of the DO-FPS isolator can be shown in the following. As shown in Fig. 28, horizontal forces $F_1$ and $F_2$ imposing at the concave trench and spherical sliding surfaces, respectively, can be expressed as:

$$F_1 = \frac{W}{R_1} u_1 + \mu W \operatorname{sgn}(\dot{u}_1) \tag{16}$$

and

$$F_2 = \frac{W}{R_2} u_2 + \mu W \operatorname{sgn}(\dot{u}_2) \tag{17}$$

where $W$ is the vertical load resulting from the superstructure ; $R_1$ and $R_2$ represent the radii of curvature of the concave trench and spherical sliding surfaces, respectively; $u_1$ and $u_2$ depict the horizontal sliding displacements of the slider relative to the centers of the concave trench and spherical sliding surfaces, respectively; $\mu$ represents the friction coefficient for the Teflon composite interface which depends on the sliding velocity; and $\dot{u}_1$ and $\dot{u}_2$ are the sliding velocities of the articulated slider.

Rearrangement of Eqs. (16) and (17) leads to:

$$u_1 = \frac{F_1 - \mu W \operatorname{sgn}(\dot{u}_1)}{W / R_1} \tag{18}$$

and

$$u_2 = \frac{F_2 - \mu W \operatorname{sgn}(\dot{u}_2)}{W / R_2} \tag{19}$$

With the aid of equilibrium at the articulated slider ($F = F_1 = F_2$), the total relative displacement between the centers of the concave trench and spherical sliding surfaces can be obtained as:

$$u_b = u_1 + u_2 = \frac{F - \mu W \operatorname{sgn}(\dot{u}_1)}{W / R_1} + \frac{F - \mu W \operatorname{sgn}(\dot{u}_2)}{W / R_2} \tag{20}$$

Rearrangement of Eq. (20) results in the base shear force:

$$F = \frac{W}{R_1 + R_2} u_b + \frac{R_1 \mu W \operatorname{sgn}(\dot{u}_1) + R_2 \mu W \operatorname{sgn}(\dot{u}_2)}{R_1 + R_2} = K_b u_b + F_f \tag{21}$$

where $K_b$ represents the horizontal stiffness of the DO-FPS isolator and can be expressed as:

$$K_b = \frac{W}{R_1 + R_2} \tag{22}$$

Hence, the isolation period of the DO-FPS isolated structure in the direction of the concave trench surface is as follows:

$$T_b = 2\pi \sqrt{\frac{m}{k_b}} = 2\pi \sqrt{\frac{R_1 + R_2}{g}} \tag{23}$$

Based on the exact solution, the relationship between the base shear force and sliding displacement can be obtained by using Eqs. (21) and (8).

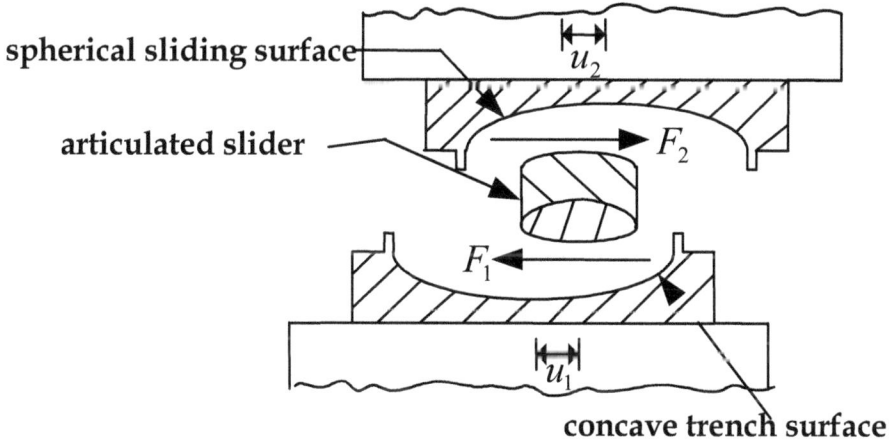

Fig. 28. The equilibrium of the forces on the spherical sliding and concave trench surfaces

### 3.3.4 Comparison of experimental and analytical results

In order to verify the accuracy of the exact solution derived in the previous sections, the shaking table tests of a rigid mass isolated with DO-FPS isolators were performed in Taiwan. The strong ground motions of the 1999 Chi-Chi (TCU084) have been given as earthquake loads during the shaking table tests. The DO-FPS isolator adopted in the shaking table tests has a concave trench and spherical sliding surfaces of 2.236m in radius of curvature. The fundamental period of the base isolated structure in the trench direction is 4.242sec.

The comparisons of the bearing displacement and velocity histories between the experimental and analytical results under the 1999 Chi-Chi earthquake (TCU084, EW component) of 1.211g in PGA are shown in Figs. 29(a) and 29(b), respectively. These figures tell us that the proposed solution can well simulate the sliding displacement and velocity of the nonlinear behavior of the device under strong ground motions. Figures 30(a) and 30(b) display the comparison between the recorded force-displacement loop and the calculated results for the DO-FPS base isolator. The results obtained from the numerical analysis are quite consistent with those from the experiments. The proposed algorithm can be given as a good tool for engineering professions to preliminarily design the displacement capacity and isolation period, etc. It should be noted that the method for the numerical analysis proposed in the previous section is only suitable for unidirectional loadings.

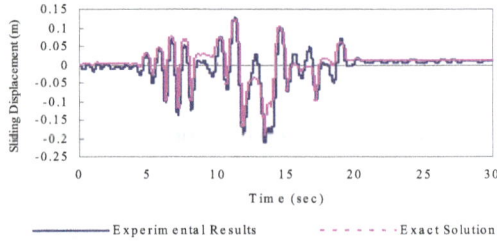

———— Experimental Results          - - - - - - Exact Solution

Fig. 29(a). Comparison of sliding displacement of DO-FPS isolator during 1999 Chi-Chi earthquake (TCU084, EW Component) of 1.211g in PGA

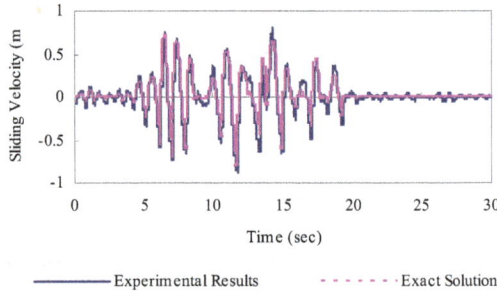

———— Experimental Results          - - - - - - Exact Solution

Fig. 29(b). Comparison of bearing sliding velocity between experimental and numerical results under 1999 Chi-Chi earthquake (TCU084, EW Component) of 1.211g in PGA

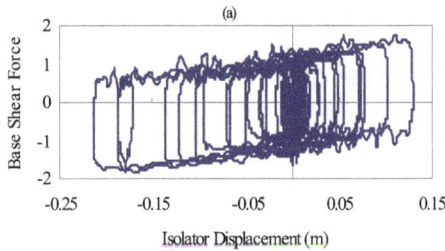

Fig. 30(a). Hysteresis loop of DO-FPS isolator under 1999 Chi-Chi earthquake (TCU084, EW Component) of 1.211g in PGA: (a) Experimental Results

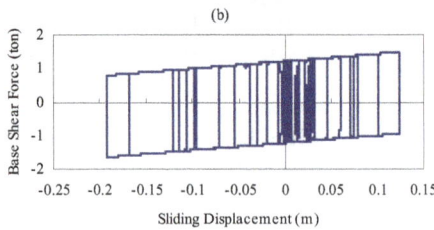

Fig. 30(b). Hysteresis loop of DO-FPS isolator under 1999 Chi-Chi earthquake (TCU084, EW Component) of 1.211g in PGA: (b) Exact Solution

### 3.4 Shaking table tests of equipment isolated with DO-FPS isolators

In order to examine the seismic behavior of motion sensitive equipment isolated with a direction-optimized friction pendulum system, a series of shaking table tests were carried out in the Department of Civil Engineering at Feng Chia University, as shown Fig. 31. In this full-scale experiment, a modem rack was adopted to simulate high-technology equipment such as server computers and workstations. The dimensions of the critical equipment were 0.8 m × 0.6 m × 1.98 m (length × width × height). Within the critical equipment were six levels, and lumped masses in the range from 50 kg to 100 kg were placed on these in 10 kg increments from top to bottom. The fundamental period of the critical equipment without isolators was measured in the shaking table test as 0.18 s.

In order to prove the benefit provided by the DO-FPS isolator, we used two extreme conditions with angles of 0° and 90°, respectively. The input ground motions included those of the earthquakes at El Centro (USA, 1940), Kobe (Japan, 1995), and Chi-Chi (station TCU084, Taiwan, 1999). Accelerometers and LVDTs were installed to measure the accelerations and displacements of the critical equipment plus DO-FPS isolators when subjected to the various ground motions.

Fig. 31. Critical equipment isolated with DO-FPS isolators

Figures 32(a) and 32(b) show comparisons of the roof acceleration responses of the critical equipment with and without DO-FPS isolators for angles of 0° and 90°, respectively, under the conditions of the El Centro earthquake (PGA of 0.4g). Figures 33(a) and 33(b) show analogous comparisons for the Kobe earthquake (PGA of 0.35g), while Figures 34(a) and 34(b) show those for the Chi-Chi earthquake (PGA of 0.3g). These results illustrate that the direction-optimized friction pendulum system effectively reduced the responses of the critical equipment by lengthening its fundamental period.

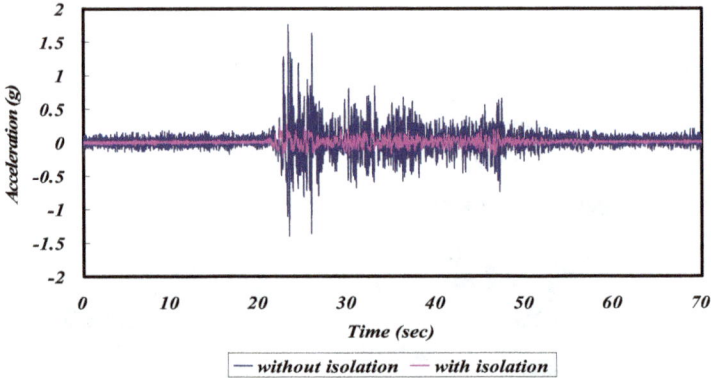

Fig. 32(a). Roof acceleration of equipment with and without DO-FPS isolators at angle of 90°under El Centro earthquake of 0.4g PGA

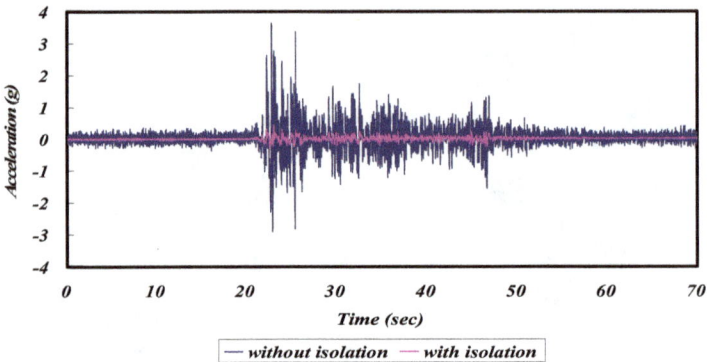

Fig. 32(b). Roof acceleration of equipment with and without DO-FPS isolators at angle of 0°under El Centro earthquake of 0.4g PGA

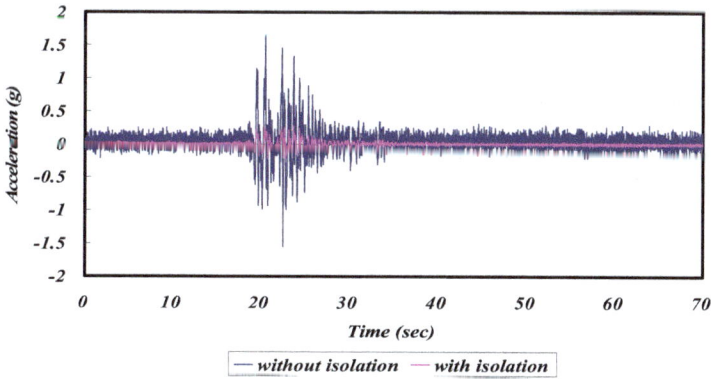

Fig. 33(a). Roof acceleration of equipment with and without DO-FPS isolators under at angle of 90° Kobe earthquake of 0.35g PGA

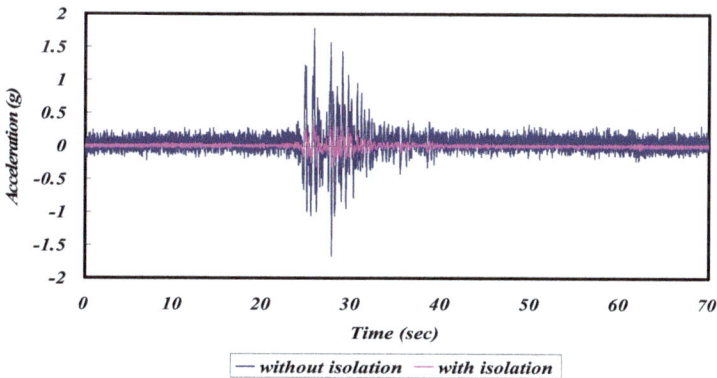

Fig. 33(b). Roof acceleration of equipment with and without DO-FPS isolators at angle of 0°under Kobe earthquake of 0.35g PGA

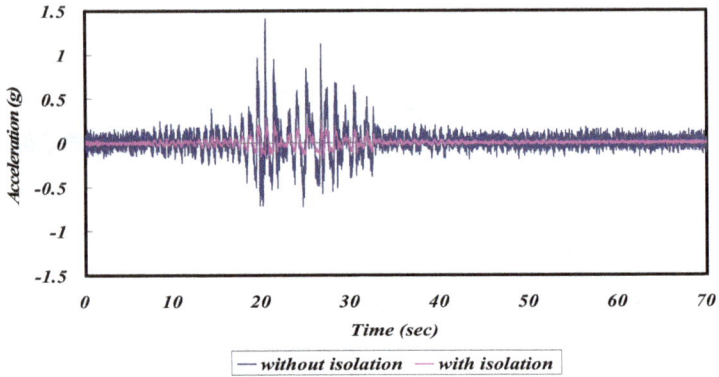

Fig. 34(a). Roof acceleration of equipment with and without DO-FPS isolators under at angle of 90° Chi-Chi earthquake of 0.3g PGA

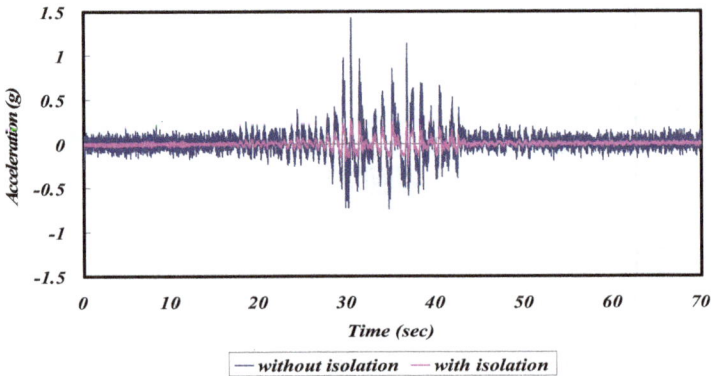

Fig. 34(b). Roof acceleration of equipment with and without DO-FPS isolators at angle of 0° under Chi-Chi earthquake of 0.3g PGA

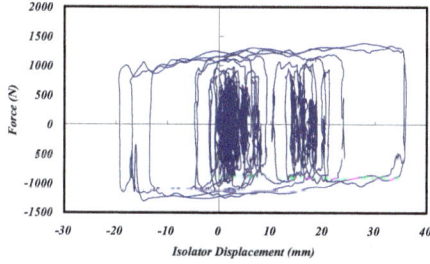

Fig. 35(a). Hysteresis loop of DO-FPS isolator at angle of 90° during 0.4 g El Centro ground motion

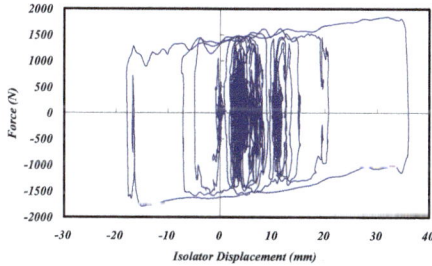

Fig. 35(b). Hysteresis loop of DO-FPS isolator at angle of 0° during 0.4 g El Centro ground motion

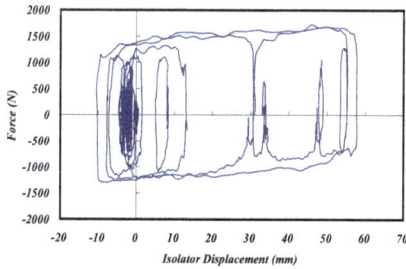

Fig. 36(a). Hysteresisloop of DOFPS isolator at angle of 90° during 0.35 g Kobe ground motion

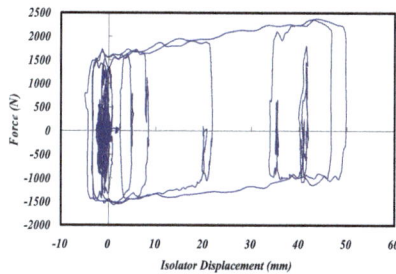

Fig. 36(b). Hysteresis loop of DO-FPS isolator at angle of 0° during 0.35 g Kobe ground motion

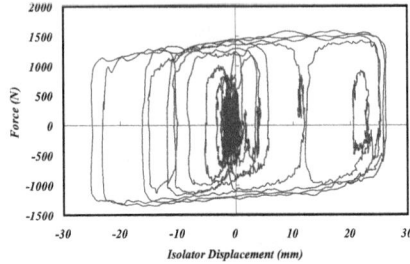

Fig. 37(a). Hysteresis loop of DOFPS isolator at angle of 90° during 0.3 g Chi-Chi ground motion

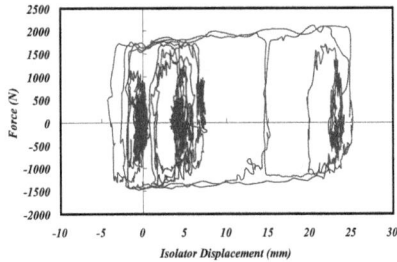

Fig. 37(b). Hysteresis loop of DOFPS isolator at angle of 0° during 0.3 g Chi-Chi ground motion

Figures 35(a) to 37(b) show the hysteresis loops of the DO-FPS isolator under the various earthquake conditions. The enclosed areas in these figures demonstrate that the DO-FPS isolator provided excellent damping of the entire system during the simulated earthquakes. Figures 38(a) to 40(b) show the roof displacement responses of the critical equipment. These figures illustrate that the DO-FPS isolator has a good mechanism for bringing the isolator back to its original position without significant displacements.

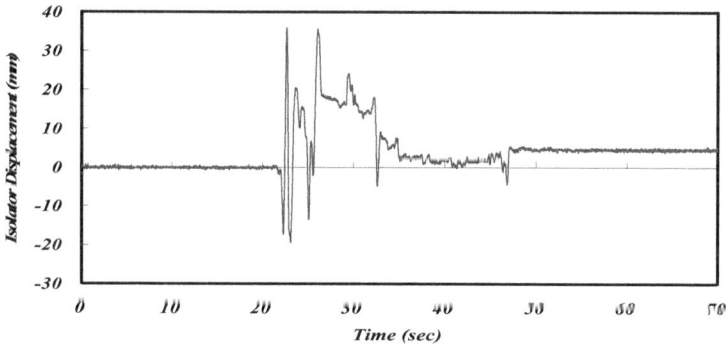

Fig. 38(a). Base isolator displacement of DOFPS isolator at angle of 90° during 0.4 g El Centro ground motion

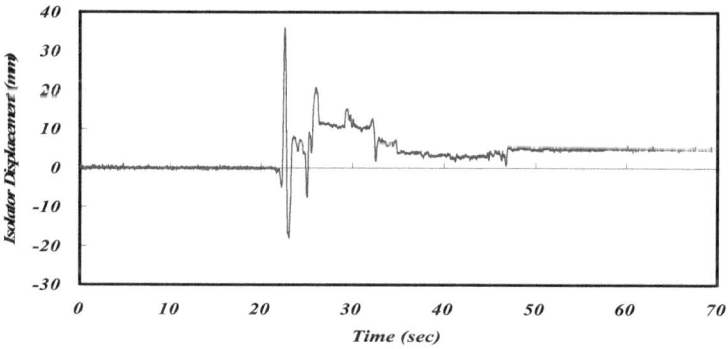

Fig. 38(b). Base isolator displacement of DO-FPS isolator at angle of 0° during 0.4 g El Centro ground motion

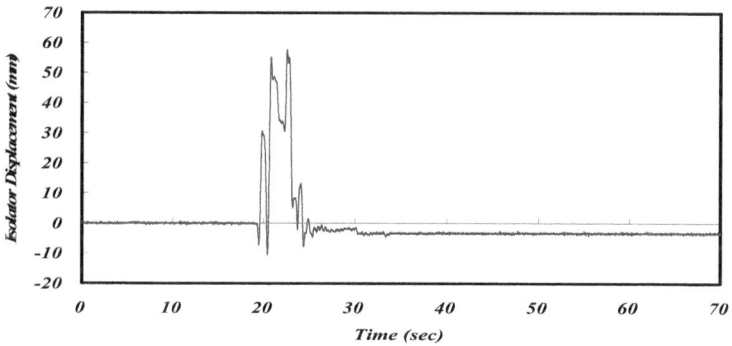

Fig. 39. Base isolator displacement of DO-FPS isolator at angle of 90° during 0.35 g Kobe ground motion

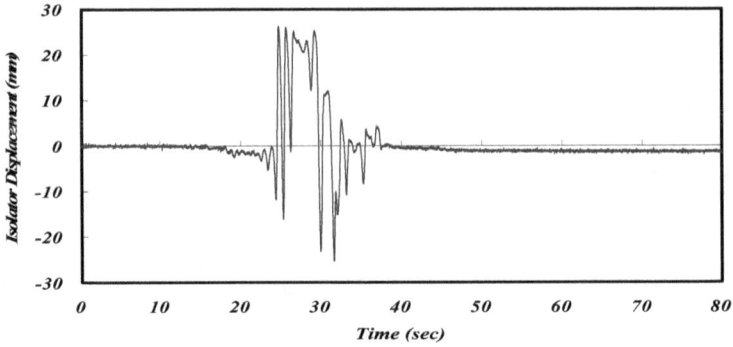

Fig. 40(a). Base isolator displacement of DO-FPS isolator at angle of 90° during 0.3 g Chi-Chi ground motion

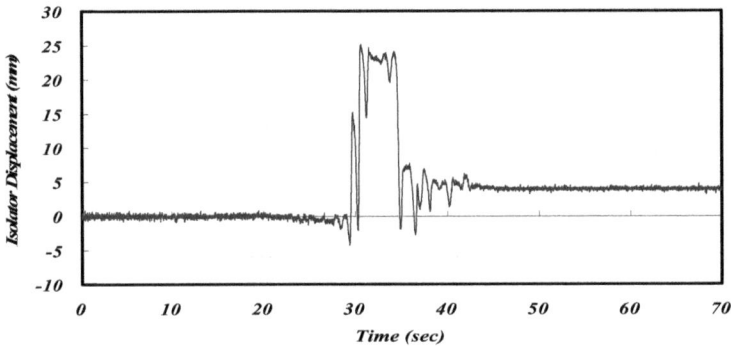

Fig. 40(b). Base isolator displacement of DO-FPS isolator at angle of 0° during 0.3 g Chi-Chi ground motion

As shown in Tables 1 and 2, the maximum roof accelerations of critical equipment have been reduced remarkably under various types of earthquakes. Therefore, the DO-FPS isolator can be recognized as an effective tool for upgrading the seismic resistance of high-technology facilities by isolating earthquake induced energy trying to impart into the equipment. This device also supplies significant damping for the isolation system through the frictional force resulting from the sliding motion of the slider on the sliding surface to reduce the isolator displacement and the bearing size.

## 4. Discussions

Earthquake proof technologies such as energy absorbing systems and base isolation systems have been accepted as powerful tools to protect structures and equipments from earthquake damage. At the same time, what we should bear in mind is that the design earthquakes utilized for estimating the performance of such technologies in safeguarding structures and equipments still leave room for future research. More recently, a new methodology based on damages indices to obtain design earthquake loads for seismic-resistant design of traditional building structures was proposed by Moustafa (2011). The prerequisite for examining the efficiencies of the earthquake proof technologies in protecting building, bridge, and lifeline structures and equipments is having rational design earthquake loads associated with the occurrence and their characteristics (e.g. time, location, magnitude, frequency content and duration, etc.). More research efforts in this subject are needed to make this possible.

| Max. Response | Roof Acceleration(g) | | | |
|---|---|---|---|---|
| Earthquake | Amplitude (PGA) | Fixed-Base Structure | Isolated Structure | Response Reduction |
| | 0.2g | 0.846 | 0.200 | 76.35% |
| El Centro | 0.3g | 1.492 | 0.239 | 83.96% |
| | 0.4g | 1.798 | 0.259 | 85.61% |
| | 0.2g | 0.589 | 0.187 | 68.26% |
| Kobe | 0.3g | 1.330 | 0.248 | 81.35% |
| | 0.35g | 1.687 | 0.274 | 83.75% |
| Chi-Chi | 0.2g | 0.605 | 0.177 | 70.79% |
| | 0.3g | 1.431 | 0.275 | 80.80% |

Table 1. Maximum roof accelerations and earthquake efficiency for the DO-FPS isolated critical equipment at angle of 90°

| Max. Response | Roof Acceleration(g) | | | |
|---|---|---|---|---|
| Earthquake | Amplitude (PGA) | Fixed-Base Structure | Isolated Structure | Response Reduction |
| | 0.2g | 0.827 | 0.285 | 65.53% |
| El Centro | 0.3g | 2.204 | 0.370 | 83.21% |
| | 0.4g | 3.723 | 0.472 | 87.31% |
| | 0.2g | 0.599 | 0.203 | 66.19% |
| Kobe | 0.3g | 1.345 | 0.293 | 70.39% |
| | 0.35g | 1.813 | 0.398 | 78.03% |
| Chi-Chi | 0.2g | 0.618 | 0.228 | 63.08% |
| | 0.3g | 1.451 | 0.282 | 80.56% |

Table 2. Maximum roof accelerations and earthquake efficiency for the DO-FPS isolated critical equipment at angle of 0°

## 5. Conclusions

The isolators presented in this chapter, which provide damping as a result of the deformed material or the frictional force between the sliding interfaces, have rectified the drawbacks of the rolling ball isolation system, such as little damping provided by the system, highly concentrated stress produced by the rolling ball or cylindrical rod due to the small contact area between the rolling ball (or cylindrical rod) and the concave surfaces, and scratches and damage to the concave surfaces caused by the ball or cylindrical rod motions during earthquakes. The presented isolators not only effectively lengthen the natural period of the vibration sensitive equipment but also provide significant damping to reduce the bearing displacement and size and the protection to the contact area between the damped steel ball (or articulated slider) and the concave surface to prevent any damage or scratch on the concave surfaces. Further, the advanced isolators possess a stable mechanical behavior during the life span of service. In addition, the isolators can isolate energy induced by earthquakes to ensure the safety and functionality of the vibration sensitive equipment located in a building. It can be concluded from these studies that the presented isolators in this chapter, including the rolling and sliding types of isolators, exhibit excellent features for preventing vibration sensitive equipment from earthquake damage.

## 6. References

Alhan, C. and Gavin, H. P. (2005). Reliability of Base Isolation for the Protection of Critical Equipment from Earthquake Hazards. *Engineering Structures*, Vol. 27, 1435-1449.

Al-Hussaini, T. M.; Zayas, V. A. and Constantinou, M. C. (1994). *Seismic Isolation of Multistory Frame Structures Using Spherical Sliding Isolation Systems.* NCEER Technical Report, NCEER-94-0007, National Center for Earthquake Engineering Research, State University of New York at Buffalo, NY.

Bakker, J. F. J. (1935). *Balance Block for Buildings.* US Patent No. 2014643.

Chung, L. L.; Yang, C. Y.; Chen, H. M. and Lu, L. Y. (2008). Dynamic Behavior of Nonlinear Rolling Isolation System. *Structural Control and Health Monitoring*, DOI: 10.1002/stc.305.

Cummings, F. D. (1930). *Building Construction (Quakeproof Building).* US Patent No. 1761659.

Eberhard, M. O.; Baldridge, S.; Marshall, J.; Mooney, W. and Rix, G. J. (2010). *The MW 7.0 Haiti Earthquake of January 12, 2010: USGS/EERI Advance Reconnaissance Team Report.* Report 2010–1048, U.S. Department of the Interior, U.S. Geological Survey.

Fan, Y. C.; Loh, C. S.; Yang, J. N. and Lin, P. Y. (2008). Experimental Performance Evaluation of an Equipment Isolation Using MR Dampers. *Earthquake Engineering and Structural Dynamics*, DOI: 10.1002/eqe.844.

Fathali, S. and Filiatrault, A. (2007). *Experimental Seismic Performance Evaluation of Isolation/Restraint Systems for Mechanical Equipment, Part 2: Light Equipment Study.* Technical Report MCEER-07-0022, University at Buffalo, The State University of New York.

Fenz, D. M. and Constantinou, M. C. (2006). Behavior of the Double Concave Friction Pendulum Bearing. *Earthquake Engineering and Structural Dynamics*, Vol. 35, No. 11, 1403–1424.

Fenz, D. M. and Constantinou, M. C. (2008a). Spherical Sliding Isolation Bearings with Adaptive Behavior-Theory. *Earthquake Engineering and Structural Dynamics*, Vol. 37, No. 2, 168-183.

Fenz, D. M. and Constantinou, M. C. (2008b). *Mechanical Behavior of Multi-spherical Sliding Bearings*. Technical Report MCEER-08-0007, Multidisciplinary Center for Earthquake Engineering Research, State University of New York at Buffalo, NY.

Kanamori, H. (1978). Quantification of Earthquakes. *Nature*, Vol. 271, No. 5644, 411-414.

Kasalanati, A.; Reinhorn, A. M.; Constantinou, M. C. and Sanders, D. (1997). Experimental Study of Ball-in-cone Isolation System. *Proceedings of the structures congress XV, ASCE*, Vol. 2. Portland, Oregon, 1191–1195.

Kemeny, Z. A. (1997). *Ball-in-cone Seismic Isolation Bearing*. US Patent No. 5599106.

Kim, J. K. (2004). *Directional Rolling Pendulum Seismic Isolation Systems and Roller Assembly Therefore*. US patent No. 6725612, Filing date: March 11, 2002..

Kim, Y. S. and Yun, C. B. (2007). Seismic Response Characteristics of Bridges Using Double Concave Friction Pendulum Bearings with Tri-linear Behavior. *Engineering Structures*, Vol. 29, No. 11, 3082-3093.

Lin, T. W. and Hone, C. C. (1993). Base Isolation by Free Rolling Rods under Basement. *Earthquake Engineering and Structural Dynamics*, Vol. 22, 261–73.

Morgan, T. A. and Mahin, S. A. (2008). The Optimization of Multi stage Friction Pendulum Isolators for Loss Mitigation Considering a Range of Seismic Hazard. In the *14th World Conference on Earthquake Engineering*, Beijing, China, Paper No. 11-0070.

Morgan, T. A. and Mahin, S. A. (2010). Achieving Reliable Seismic Performance Enhancement Using Multi-stage Friction Pendulum Isolators. *Earthquake Engineering and Structural Dynamics*, Vol. 39, 1443–1461.

Moustafa, A. (2011). Damage-Based Design Earthquake Loads for Single-Degree-Of-Freedom Inelastic Structures. *Journal of Structural Engineering, ASCE*, Vol. 137, No. 3, March, 2011, 456-467.

Naeim, F. and Kelly, J. M. (1999). *Design of Seismic Isolated Structures: from Theory to Practice*. John Wiley & Sons, ISBN: 0-471-14921-7, New York.

Penkuhn, A. L. K. (1967). *Three Point Foundation for Building Structures*. US Patent No. 3347002.

Rivin, E. I. (2003). *Passive Vibration Isolation*. The American Society of Mechanical Engineers, ISBN: 0-7918-0187-X.

Roussis, P. C. and Constantinou, M. C. (2005). *Experimental and Analytical Studies of Structures Seismically Isolated with an Uplift-restraint Isolation System*. Technical Report MCEER005-0001, Buffalo, NY.

Schär, F. (1910). *Foundation for Buildings*. US Patent No. 951028.

Skinner, R. I.; Robinson, W. H. and McVerry, G. H. (1993). *An Introduction to Seismic Isolation*. John Wiley & Sons, ISBN: 0 471 93433 X, New York.

Soong, T. T. and Dargush, G. F. (1997). *Passive Energy Dissipation Systems in Structural Engineering*. John Wiley & Sons, ISBN: 0-471-96821-8, New York.

Takewakin, I. (2009). *Building Control with Passive Dampers*. John Wiley & Sons, ISBN: 978-0-470-82491-7, Singapore.

Takewakin, I.; Murakami, S.; Fujita, K.; Yoshitomi, S. and Tsuji, M. (2011). The 2011 off the Pacific Coast of Tohoku Earthquake and Response of High-rise Buildings under Long-period Ground Motions. *Soil Dynamics and Earthquake Engineering,* doi:10.1016/j.soildyn. 2011.06.001.

Tsai, C. S. (2003). *Improved Structures of Base Isolation Systems.* Taiwan patent No. 207126, Publication Number: 00542278, Publication date: July 11, 2003, Application number: 091210175, Filing date: July 4, 2002 [in Chinese].

Tsai, C. S. (2004a). *Structure of an Anti-shock Device.* US Patent No. 6688051, Filing date: March 7, 2002.

Tsai, C. S. (2004b). *Structure of an Anti-shock Device.* US Patent No. 6820380, Filing date: Sept. 10, 2003.

Tsai, C. S. (2007). *Foundation Shock Eliminator.* US Patent No. 7237364, Filing Date: July 2, 2004.

Tsai, C. S. (2008a). *Foundation Shock Suppressor.* US Patent No. 7338035, Filing date: Dec. 9, 2004.

Tsai, C. S. (2008b). *Anti Shock Device.* US Patent No. 7409799, Filing date: Dec. 13, 2005.

Tsai, C. S.; Chiang, T. C. and Chen, B. J. (2003a). Finite Element Formulations and Theoretical Study for Variable Curvature Friction Pendulum System. *Engineering Structures,* Vol. 25, No. 14, 1719-1730.

Tsai, C. S.; Chiang, T. C. and Chen, B. J. (2003b). Seismic Behavior of MFPS Isolated Structure under Near-fault Earthquakes and Strong Ground Motions with Long Predominant Periods. In the *2003 ASME Pressure Vessels and Piping Conference, Seismic Engineering,* Cleveland, Ohio, U. S. A., J. C. Chen, Editor, Vol. 466, 73-79.

Tsai, C. S.; Chiang, T. C. and Chen, B. J. (2005a). Experimental Evaluation Piecewise Exact Solution for Predicting Seismic Responses of Spherical Sliding Type Isolated Structures. *Earthquake Engineering and Structural Dynamics,* Vol. 34, No. 9, 1027–1046.

Tsai, C. S.; Cheng, C. K.; Chen, M. J. and Yu, S. H. (2005b). Experimental Study of MFPS-isolated Sensitive Equipment. in the *2005 ASME Pressure Vessels and Piping Conference, Seismic Engineering,* C. S. Tsai, editor, Denver, Colorado, U.S.A., 8:11-18.

Tsai, C. S.; Lin, Y. C.; Chen, W. S.; Chen, M. J. and Tsou, C. P. (2006a). The Material Behavior and Isolation Benefits of Ball Pendulum System. in the *2006 ASME Pressure Vessels and Piping Conference, Seismic Engineering,* James F. Cory, editor, Vancouver, BC, Canada, Paper No. PVP2006-ICPVT11-93252.

Tsai, C. S.; Chen, W. S.; Yu, S. H. and Yang, C. T. (2006b). Shaking Table Tests of Critical Equipment with Simple Isolators. in the *2006 ASME Pressure Vessels and Piping Conference, Seismic Engineering,* James F. Cory, Editor, Vancouver, BC, Canada., July 23-27, No. PVP2006-ICPVT11-93250.

Tsai, C. S.; Chen, W. S.; Chiang, T. C. and Chen, B. J. (2006c). Component and Shaking Table Tests for Full-scale Multiple Friction Pendulum System. *Earthquake Engineering and Structural Dynamics,* Vol. 35, No. 11, 1653–1675.

Tsai, C. S.; Chen, W. S.; Chiang, T. C. and Lin, Y. C. (2007). Application of Direction Optimized-Friction Pendulum System to Seismic Mitigation of Sensitive Equipment. in the *2007 ASME Pressure Vessels and Piping Conference, Seismic Engineering*, Artin Dermenjian, editor, San Antonio, Texas, U.S.A., Paper No. PVP2007-26552.

Tsai, C. S.; Chen, W. S.; Lin, Y. C.; Tsou, C. P.; Chen, C. C. and Lin, C. L. (2008a). Shaking Table Tests of Motion Sensitive Equipment Isolated with Static Dynamics Interchangeable-Ball Pendulum System. In: the *14th World Conference on Earthquake Engineering*, Beijing, China, Paper No. 11-0010.

Tsai, C. S.; Lu, P. C.; Chen, W. S.; Chiang, T. C.; Yang, C. T. and Lin, Y. C. (2008b). Finite Element Formulation and Shaking Table Tests of Direction-Optimized Friction Pendulum System. *Engineering Structures*, Vol. 30, No. 9, 2321-2329.

Tsai, C. S.; Lin, Y. C. and Su, H. C. (2010a). Characterization and Modeling of Multiple Friction Pendulum Isolation System with Numerous Sliding Interfaces. *Earthquake Engineering and Structural Dynamics*, Vol. 39, No. 13, 1463–1491.

Tsai, C. S.; Lin, Y. C. and Su, H. C. (2010b). Seismic Responses of a Building Isolated with Multiple Friction Pendulum System Subjected to Multi-directional Excitations. In the *2010 ASME Pressure Vessels and Piping Conference, Seismic Engineering*, O'Brien, Cheryl C. (ed.), Bellevue, Washington, U. S. A., Paper No. PVP2010-25587.

Tsai, C. S.; Chen, W. S. and Chiang, T. C. (2010c). Experimental and Numerical Studies of Trench Friction Pendulum System. *Structural Engineering and Mechanics, An International Journal*, Vol. 34, No. 2.

Tsai, C. S. and Lin, Y. C. (2010d). Mechanical Characteristics and Modeling of Multiple Trench Friction Pendulum System with Multi-intermediate Sliding Plates. *International Journal of Aerospace and Mechanical Engineering*, Vol. 4, No. 1.

Tsai, C. S.; Lin, Y. C. and Su, H. C. (2010e). Characteristic and Modeling of multiple Direction Optimized-Friction Pendulum System with Numerous Sliding Interfaces Subjected to Multi-directional Excitations. In the *2010 ASME Pressure Vessels and Piping Conference, Seismic Engineering*, O'Brien, Cheryl C. (ed.), Bellevue, Washington, U. S. A., Paper No. PVP2010-25598.

Tsai, C. S.; Wang, Y. M. and Su, H. C. (2011a). Experimental Study of a Full Scale Building Isolated with Multiple Friction Pendulum System with Multiple Sliding Interfaces. In the *2011 ASME Pressure Vessels and Piping Conference, Seismic Engineering*, O'Brien, Cheryl C. (ed.), Baltimore, Maryland, U. S. A., Paper No. PVP2011-57352.

Tsai, C. S.; Hsueh, C. I. and Su, H. C. (2011b). Experimental Investigation on Performance of Multiple Direction-Optimized Friction Pendulum System with Multiple Sliding Interfaces in Mitigating Structural Responses during Earthquakes. In the *2011 ASME Pressure Vessels and Piping Conference, Seismic Engineering*, O'Brien, Cheryl C. (ed.), Baltimore, Maryland, U. S. A., Paper No. PVP2011-57598.

Tsai, M. H.; Wu, S. Y.; Chang, K. C. and Lee, G. C. (2007). Shaking Table Tests of a Scaled Bridge Model with Rolling-type Seismic Isolation Bearings. *Engineering Structures*, Vol. 29, 694-702.

Touaillon, J. (1870). *Improvement in Buildings*. US Letters Patent No. 99973.

Wu, C. J. (1989). *Anti-earthquake Structure Insulating the Kinetic Energy of Earthquake from Buildings*. US Patent No. 4881350.

Zayas, V. A. (1987). *Earthquake Protective Column Support*. US Patent No. 4644714.

Zayas, V. A.; Low, S. S. and Mahin, S. A. (1987). *The FPS Earthquake Resisting System Report*. EERC Technical Report, UBC/EERC-87/01, University of California, Berkeley, CA.

# 5

# Seismic Damage Estimation in Buried Pipelines Due to Future Earthquakes – The Case of the Mexico City Water System

Omar A. Pineda-Porras[1] and Mario Ordaz[2]

[1]Energy and Infrastructure Analysis (D-4), Los Alamos National Laboratory, Los Alamos,
[2]Instituto de Ingeniería, Ciudad Universitaria, Universidad Nacional
Autónoma de México (UNAM)
[1]USA
[2]Mexico

## 1. Introduction

Since the mid-70s, there have been advances in the development of models to better understand how earthquakes affect buried pipelines. These natural events can cause damage due to two phenomena: seismic wave propagation and permanent ground deformation. The combined effect of both phenomena in pipeline damage estimation is a subject still complex to address, especially if the objective is to estimate damage due to future earthquakes. In this chapter, the damage assessment methods only consider the impact of seismic wave propagation. The effects of permanent ground deformation phenomena, like ground subsidence, landslides, and ground rupture, are omitted.

The exceptional damage caused by the 1985 Michoacan earthquake in Mexico City has encouraged researchers to develop sophisticated tools to estimate ground motion in the Valley of Mexico from Pacific coastal earthquakes, including the important site effects largely observed in the city. These tools have helped to better understand how earthquakes affect buildings and other structures like pipeline systems. The most remarkably case of pipeline damage caused by the 1985 seismic event is the extensive damage suffered by the Mexico City Water System (MCWS) that left almost 3.5 million people without water, and caused water service disruptions over a period of two months. The 1985 MCWS damage scenario has been extensively analyzed for developing models to better understand how seismic wave propagation affects buried pipelines; some of those models are employed in the future damage prediction methods described in this manuscript.

Fragility functions are typically the tools most used to assess seismic damage in buried pipelines. These functions relate pipeline damage with seismic intensity. Pipeline damage is generally expressed as a linear pipe repair density. Seismic intensity is usually quantified through a seismic parameter. There are many seismic parameters used as arguments of fragility functions; the most important of these are described in Section 2. Section 3 describes the most important fragility functions proposed until now, including the two employed in the seismic damage estimation for the MCWS presented in Section 4. Finally, Section 5 contains a summary of the most important conclusions of this work.

## 2. Seismic parameters related to damage in buried pipelines

An historical revision of all the seismic parameters employed to represent seismic intensity in fragility functions is summarized in this section. The seismic parameters described in detail are Mercalli modified intensity (MMI), peak ground acceleration (PGA), peak ground velocity (PGV), maximum ground strain, and a recently proposed composite parameter (Section 2.5). Other parameters used as fragility function arguments are not included here because there is not enough evidence of their relationship with pipeline damage; among them are permanent ground displacement, Arias intensity, spectral acceleration, and spectral intensity.

### 2.1 Mercalli modified intensity
Though it is a parameter of subjective nature, MMI was used as damage indicator for pipelines in the 80s and 90s (Eguchi 1983 and 1991; Ballantyne et al. 1990; and, O'Rourke T. et al. 1998). A likely reason for the development of MMI-based fragility relations in the past was the extended use of that parameter to describe damage to aboveground structures. Lately, the installation of seismic stations and the availability of seismic records have made it easier to estimate parameters like PGA and PGV, which are better related to buried pipeline damage.

### 2.2 Peak Ground Acceleration (PGA)
PGA was largely employed as a damage indicator for pipelines during 25 years, from the study of Katayama et al. (1975), to the PGA-based fragility function of Isoyama et al. (2000). Though it has been largely demonstrated that PGV is related more closely to pipeline damage than PGA, as it is further explained in the following paragraphs, there are several reasons to explain why PGA, instead of PGV, was used to create some fragility functions before 2000. Most seismic stations record time histories of acceleration instead of velocity; then, PGA can be directly obtained from seismic records without involving the integration process needed for computing PGV. Most attenuations laws provide estimates of PGA (before 2000, PGV attenuation laws were limited); thus, for practical purposes, PGA was the ideal parameter for analyzing pipeline damage, and therefore, creating pipeline fragility relations.

### 2.3 Peak Ground Velocity (PGV)
PGV is by far the most widely used seismic parameter for pipeline seismic fragility functions. Generally, PGV shows good correlation with pipeline damage, although some studies (Sections 2.4 and 2.5) have shown that, for pipeline located in soft soils, there are some complications, mainly due to the assumptions related to PGV's use as a damage indicator.
PGV is better related to pipeline damage than PGA mainly due to two reasons: 1) PGV is related to ground strain –the main cause of pipeline damage due to seismic wave propagation (Section 2.3.2)–; and, 2) PGA is more related to inertia forces –forces that do not affect buried structures like pipelines–. Many studies have empirically demonstrated that PGV is better pipeline damage predictor than PGA (O'Rourke T. et al. 1998; Isoyama et al. 2000; and, Pineda 2002).

PGV has been extensively used as damage indicator for pipelines considering two assumptions: 1) PGV is directly related with maximum ground strain ($\varepsilon_g$); and 2) transient ground strain is the main cause of pipeline damage due to seismic wave propagation. The relationship between PGV and $\varepsilon_g$ can be analyzed in Equation 1 (Newmark 1967), where C is seismic wave velocity. From Equation 1, PGV is directly related to $\varepsilon_g$ only if C is constant. Since $\varepsilon_g$ is non-dimensional, PGV and C must be expressed with the same velocity units.

$$\varepsilon_g = \frac{PGV}{C} \tag{1}$$

### 2.4 Maximum ground strain ($\varepsilon_g$)
Because transient ground strain is the assumed main cause of pipeline damage due to seismic wave propagation, $\varepsilon_g$ is straightforwardly the optimum parameter for analyzing the relationship between pipeline damage and seismic intensity. Rigorously, $\varepsilon_g$ can be estimated from displacement time histories D(t) (Equation 2). In Equation 2, x is a space variable, $\varepsilon(t)$ is ground strain time history, and max represents the maximum of the expression between absolute value brackets | |.

$$\varepsilon_g = max|\varepsilon(t)| = max\left|\frac{\partial D(t)}{\partial x}\right| \tag{?}$$

There are three major problems for estimating $\varepsilon_g$ through Equation 2. First, $\varepsilon_g$ is generally obtained through the double integration of acceleration time histories; this process causes loss of information due to the involved mathematical operations. Procedures like correction of base line, filtering and tapering could generate ambiguous results if the parameters used in those operations are modified. Second, the derivation process of $\varepsilon_g$ with respect to a space variable (x) implies that the seismic records used in the analysis need to be referenced to an absolute time scale; this is a very significant limitation because only ground motion information from seismic arrays using the same time reference, and preferably located in the place of interest (e.g., the zone covered by a pipeline system), would be useful. The third and probably the most important problem is the high cost involved in installing and operating seismic arrays covering large extensions (e.g., area covered by a pipeline network).

In order to avoid the above-mentioned problems of Equation 2, Equation 1 has been used to obtain conservative estimates of $\varepsilon_g$. PGV can be easily obtained from seismic records or other sources (e.g., attenuation laws); on the contrary, C is far from being easy to obtain, which complicates the estimation of $\varepsilon_g$. We include two examples to show the complexity of estimating C for the purpose of estimating $\varepsilon_g$ with Equation 1.

In the first example, the $\varepsilon_g$-based fragility relation proposed by O'Rourke and Deyoe (2004) was computed by assuming C values of 500 m/sec for Rayleigh waves (surface waves), and 3,000 m/sec for S-waves (body waves). Later, the study of Paolucci and Smerzini (2008) suggests that the apparent propagation velocity of S-waves is closer to 1,000 m/sec. O'Rourke M. (2009) employed the new suggested C value for S-waves and proposed a new version of the 2004 fragility relation. Changing C from 3,000 m/sec to 1,000 m/sec in Equation 1 implies that $\varepsilon_g$ increases with a factor of three.

The second example deals with the estimation of $\varepsilon_g$ in soft soil zones. Singh et al. (1997) analyzed ground strains at the Roma micro-array in Mexico City for four earthquakes. They concluded that Equation 1 could be used to estimate $\varepsilon_g$ by using a phase velocity (Rayleigh waves) of 600 m/sec instead of the value of C at the natural period of lakebed sites (estimated as 1,500 m/sec). Singh et al. (1997) indicate that the discrepancy in the value of C could be due to local heterogeneities within the array. .

PGV is a more convenient parameter than $\varepsilon_g$ for analyzing pipeline damage due to seismic wave propagation for three reasons. First, PGV is easier to estimate than $\varepsilon_g$. Second, many studies have proved that PGV is well correlated with pipeline damage. Third, theoretically, there is a direct relationship between PGV and pipeline damage considering two assumptions already mentioned in this section. Notwithstanding these three points, there is evidence of a case in which PGV is not the best parameter for relating pipeline damage with seismic intensity: the particular case of Mexico City.

### 2.5 The novel composite parameter PGV²/PGA

As it will be described in Section 3.4, Pineda and Ordaz (2007 and 2009) demonstrated that $PGV^2/PGA$ is better correlated to pipeline damage than PGV alone for soft soils; a plausible explanation for this is the strong relationship of $PGV^2/PGA$ to permanent ground displacement (PGD), which is a ground motion parameter related to very-low frequency contents. Though in the past O'Rourke (1998) demonstrated that PGV is better pipeline damage predictor than PGD, studies that focused exclusively on the relationship between pipeline damage and PGD (or $PGV^2/PGA$) for soft soil sites have not been done yet. Finally, two things must be noted: 1) Pineda and Ordaz (2007 and 2009) employed $PGV^2/PGA$ instead of PGD due to the rigorous theoretical relationship between both parameters; the availability of detailed PGA and PGV maps for the 1985 Michoacan event (see Pineda 2006 for details about those maps); and the lack of information on ground motion to produce reliable PGD maps for the 1985 earthquake (Pineda 2006). Second, Pineda and Ordaz (2007 and 2009) define soft soils as those soils with natural periods equal to or higher than 1.0 sec.

## 3. Seismic fragility functions for buried pipelines

Buried pipeline seismic fragility functions relate pipeline damage rates with different levels of seismic intensity. Damage rates are usually defined as the number of pipe repairs per unit length of pipeline (e.g., number of repair per kilometer, rep/km). Seismic intensity can be quantified through a diverse group of ground motion parameters computed from seismic records (Section 2). Though there are many studies focused on computing analytical pipeline fragilities (Hindy and Novak 1979; O'Rourke M. and El Hmadi 1988; and Mavridis and Pitilakis 1996). This study only includes empirical pipeline fragility functions computed from pipeline damage documented after earthquakes. For the sake of brevity, equations on the fragility relations are not included here; but can be obtain from those studies in the reference section of this chapter.

### 3.1 Early fragility functions for buried pipelines

As noted previously, empirical correlation between buried pipeline damage and ground motion intensity parameters has been studied since the mid-70s. Katayama et al. (1975) employed pipeline damage scenarios from six earthquakes; four in Japan (Kanto, 9/01/1923; Fukui, 6/28/1948; Niigata, 6/16/1964; and, Tokachi-oki, 5/16/1968), one in Nicaragua (Managua, 12/23/1972), and one in the United States (San Fernando, 2/09/1971) to compute fragility functions for segmented cast iron (CI) and asbestos cement (AC) pipelines in terms of PGA. Katayama et al. (1975) included fragility functions for poor, average, and good soil conditions.

Early in the 80s, Isoyama and Katayama (1982) employed the 1971 San Fernando earthquake damage scenario to compute a PGA-based fragility function. The same damage data and information on other three pipeline damage scenarios (Santa Rosa, 10/01/1969; Nicaragua, 12/23/1972; and Imperial Valley, 10/15/1979) was used by Eguchi (1983 and 1991) to compute a set of fragility functions in terms of MMI for the following pipeline types: welded steel gas welded joints (WSGWJ), AC, concrete (C), PVC, CI, welded steel caulked joints (WSCJ), welded steel arc-welded joints–Grades A & B steel (WSAWJ A&B), polyethylene (PE), ductile iron (DI), and welded steel arc-welded joints Grade X steel (WSAWJ X). Eguchi (1983 and 1991) concluded that AC and concrete pipes are more vulnerable than PVC pipes; and PVC pipes are more vulnerable than CI pipes and welded steel pipes with caulked joints. DI pipes experienced on average about 10 times fewer repairs per unit length than the worst performing pipes; and finally, the repair rate of X -grade steel pipes with arc-welded joints was approximately ten times smaller than that of DI pipes.

In the late 80s, Barenberg (1988) proposed the first documented PGV-based fragility function for buried CI pipelines employing damage data from three U.S. earthquakes (Puget Sound, 4/29/1965; Santa Rosa, 10/01/1969; and, San Fernando, 2/09/1971). The fragility function of Barenberg (1988) suggests that a doubling of PGV will lead to an increase in the pipeline damage rate by a factor of about 4.5.

Early in the 90s, Ballantyne et al. (1990) expanded the pipeline damage data of Barenberg (1988) with damage information from other three U.S. earthquakes (Puget Sound, 4/29/1949; Coalinga, 5/02/1983; and Whittier Narrows, 10/01/1987) and proposed new fragility functions by using MMI as a measure of seismic intensity.

Three PGA-based fragility functions were also published in the early 90s. The Technical Council on Lifeline Earthquake Engineering (TCLEE) of the American Society of Civil Engineers (ASCE) published a comprehensive study on seismic loss estimation for water systems (ASCE-TCLEE 1991) in which PGA-based fragility relations were computed from a reanalysis of the damage data of Katayama et al. (1975) and the 1983 Coalinga pipeline damage scenario. Hamada (1991) proposed another PGA-based fragility function by analyzing the damage scenarios of earthquakes from United States (San Fernando, 2/09/1971) and Japan (Miyagiken-oki, 6/12/1978; and Nihonkai-chubu, 5/26/1983). O'Rourke T. et al. (1991) related pipeline damage with PGA employing damage scenarios from seven earthquakes: six from U.S. (San Francisco, 4/18/1906; Puget Sound, 4/29/1965; Santa Rosa, 10/01/1969; San Fernando, 2/09/1971; Imperial Valley, 10/15/1979; Coalinga, 5/02/1983; and Loma Prieta, 10/18/1989), and one from Japan (Miyagiken-oki, 6/12/1978).

### 3.2 Fragility functions in terms of PGV

A notable change in the literature on seismic fragility functions for pipelines is observed from 1993, when PGV began to be the preferred seismic parameter for pipeline fragility relations, and PGA and MMI were no longer used for developing new fragility functions (with some exemptions described later in this section).

O'Rourke and Ayala (1993) proposed a new pipeline fragility function in terms of PGV by using the damage data points of Barenberg (1988) and damage information from three earthquakes, one from the United States (Coalinga, 5/02/1983), and two from Mexico

(Michoacan, 9/19/1985; and, Tlahuac, 4/25/1989). The damage data employed for computing the fragility function are related to pipelines made of AC, CI, concrete, and prestressed concrete cylinder pipes. The fragility relation of O'Rourke and Ayala (1993) was later incorporated into the U.S. Federal Emergency Management Agency's loss assessment methodology HAZUS-MH (FEMA 1999). This fragility function can be used for damage prediction of brittle pipelines. For ductile pipelines, the fragility relation must be multiplied by a suggested factor of 0.3 (FEMA 1999).

Eidinger et al. (1995) and Eidinger (1998) reanalyzed the pipeline damage data of O'Rourke and Ayala (1993), along with information from the 1989 Loma Prieta pipeline damage scenario, in order to propose a set of fragility functions in terms of PGV considering pipe material, joint type, and soil corrosiveness. Edinger's fragility functions estimated damage for CI, welded steel (WS), AC, concrete, PVC, and DI pipes.

Hwang and Lin (1997) computed a pipeline fragility function in terms of PGA by analyzing pipeline damage data obtained from six previous studies (Katayama et al. 1975; Eguchi 1991; ASCE-TCLEE 1991; O'Rourke T. et al. 1991; Hamada 1991; and Kitaura and Miyajima 1996).

O'Rourke T. et al. (1998) employed a GIS-based methodology to investigate factors affecting the water supply service of the Los Angeles Department of Water and Power and the Metropolitan Water District, after the 1994 Northridge earthquake. Analyses of the relationship between damage rate and seismic intensity employed seven seismic parameters: MMI, PGA, PGV, PGD, Arias Intensity, Spectral Acceleration, and Spectral Intensity (SI). Pipeline fragility relations in terms of MMI, SI, PGA, and PGV, are also included in the study. O'Rourke T. et al. (1998) concluded that PGV relates to the pipeline damage better than any other parameter and proposed PGV-based fragilities for steel, CI, DI, and AC pipelines. Later, O'Rourke T. and Jeon (1999) developed a fragility relationship (for CI pipes) for scaled velocity, a parameter based on PGV but normalized for the effects of pipe diameter.

Isoyama et al. (2000) computed fragility functions in terms of PGA and PGV by analyzing the pipeline damage scenario left by the 1995 Hyogoken-nanbu earthquake. A multivariate analysis computed empirical correction factors to account for pipe material, pipe diameter, ground topography, and liquefaction, in the fragility relation.

The American Lifeline Alliance (ALA), a public-private partnership between the FEMA and the ASCE, published a set of algorithms to compute the probability of damage from earthquake effects to several components of water supply systems (ALA 2001). For buried pipelines, the PGV-based fragility function published by the ALA was computed from a set of 81 damage rate-PGV data points from 12 seismic damage scenarios. Similar to the fragility functions of Eidinger et al. (1995) and Eidinger (1998), the ALA's fragility relation provides factor to account for pipe material, joint type, and soil corrosiveness.

Jeon and O'Rourke T. (2005) reanalyzed the pipeline damage data from a previous study (O'Rourke T. et al. 1998). They compared the correlation between CI pipeline damage rates from the 1994 Northridge earthquake and PGV computed in different ways (geometric mean PGV, maximum PGV, and maximum vector magnitude of PGV). Their results show that maximum PGV, computed as the peak recorded value, is better correlated with pipeline damage. Jeon and O'Rourke T. (2005) also provide fragility functions for WSJ Steel, CI, DI, and AC, pipelines.

## 3.3 Fragility functions in terms of maximum ground strain

O'Rourke M. and Deyoe (2004) analyzed the differences of the fragility relations published by O'Rourke M. and Ayala (1993) and O'Rourke T. and Jeon (1999). The analysis identified some reasons for the differences, including the wave type that dominated each seismic scenario, the presence of corrosion in some pipes, and the low statistical reliability of some data points. By removing doubtful data points and classifying the remaining data points according to the presumably dominating wave type, O'Rourke M. and Deyoe computed PGV-based pipeline fragility functions for surface waves (Rayleigh) by assuming phase velocity of 500 m/sec, and body waves (S-waves) by assuming apparent velocity of 3,000 m/sec. They also proposed a fragility function in terms of $\varepsilon_g$. The new $\varepsilon_g$-based fragility function also considers the effect of permanent ground deformation because O'Rourke and Deyoe (2004) included repair rate-$\varepsilon_g$ data points from the 1994 Northridge earthquake (Sano et al. 1999) and from Japan (Hamada and Akioka 1997). A recent modification to the $\varepsilon_g$-based fragility relation (O'Rourke M. 2009) uses an apparent velocity of 1000 m/sec for S-waves; they base that assumption on a study by Paolucci and Smerzini (2008).

## 3.4 Seismic fragility functions based on the 1985 MCWS damage scenario

In the literature, there are two observed tendencies for developing fragility functions: the use of damage scenarios for several pipelines systems and earthquakes; and the use of damage scenarios for only a specific pipeline system and earthquake. On the one hand, while the first tendency provides general pipeline fragilities, characterized by its wide applicability due to the typical mixture of pipeline types and other factors (ALA, 2001), it is also usually related to high uncertainty levels. On the other hand, using information about a specific pipeline system and well-studied damage scenarios, the uncertainty could be controlled because the number of unknown variables related to pipeline damage (e.g., variables related to earthquake environment, soil properties, and pipeline conditions), for that particular study case, is lower than if several systems and events are included in the analysis. In order to reduce the uncertainty in the damage estimation for the MCWS, in this study only two fragility functions (Pineda and Ordaz, 2003 and 2007) are used to estimate the damage in the pipeline system.

Pineda and Ordaz (2003) proposed a fragility formulation for buried pipelines employing the 1985 damage scenario published by Ayala and O'Rourke M. (1989), and the detailed maps for Mexico City proposed by Pineda and Ordaz (2004). In order to analyze the variability of the relationship between damage rates (rep/km) and PGV for several ranges of seismic intensity, nine scenarios were used to generate a PGV-based fragility function. In this study, the functional form chosen for the fragility relation is the cumulative normal function because it better fit the cloud of damage rate-PGV data points when compared with the resulting fit employing other function types (linear, power, etc.). This fragility function is employed in the damage assessment of Section 4.2.

In a subsequent study, Pineda and Ordaz (2007) proposed $PGV^2/PGA$ as a new seismic parameter for buried pipelines. From a theoretical development, they found that $\lambda_{pr}$, defined by Equation 3, interrelate the peak ground response parameters PGA, PGV, and PGD. $\lambda_{pr}$ has three relevant characteristics: it is non-dimensional; it is always greater than or equal to 1.0; and it can be a measure of spectra bandwidth.

$$\lambda_{pr} = \frac{PGA \cdot PGD}{PGV^2} \tag{3}$$

By isolating PGD in Equation 3, it is found that PGD and PGV²/PGA have a direct relationship through $\lambda_{pr}$ (Equation 4). Based on the assumption that ground strain, the main cause for the pipeline damage, is related to PGD; pipeline damage is then also related to PGV²/PGA and $\lambda_{pr}$. Because $\lambda_{pr}$ varies in a delimited range for the places where the MCWS is located in Mexico City, it was implicitly included in the fragility formulation. This PGV²/PGA-based fragility function is used in the damage assessment discussed in Section 4.3.

$$PGD = \lambda_{pr}(PGV^2/PGA) \tag{4}$$

Recently, Pineda and Ordaz (2009) computed fragility relations for 48-inch segmented pipelines considering the effects of ground subsidence, a phenomenon largely observed in the Valley of Mexico. They analyzed the relationship between pipeline damage and seismic intensity (measured in terms of PGV²/PGA) for two levels of differential ground subsidence (DGS). The proposed fragility functions fall above and below a previous fragility relation for 48-inch pipelines that does not explicitly consider the effects of DGS in the damage (Pineda 2006).

## 4. Seismic damage estimation in the MCWS due to future earthquakes

The main objective of this section is to describe a new approach for estimating pipeline damage at the MCWS due to future earthquakes nucleated at the subduction zones of the Pacific Mexican coast. This new study is based on a fragility function in terms of PGV²/PGA (a seismic parameter described in Section 2.5) that has shown better correlation with buried pipeline damage (Pineda and Ordaz, 2007) than PGV alone. Pineda and Ordaz (2003) previously employed this parameter to predict future damage at the MCWS.

### 4.1 Description of the MCWS and its 1985 damage scenario

The MCWS pipeline network is formed by more than 600 km of pipes, with diameters ranging from 20- to 72-inches, and made of several pipe materials: asbestos-cement, concrete, cast iron, and steel. This pipeline system is particularly vulnerable to earthquakes because it is mainly located in the lake zone of the Valley of Mexico where deep clay deposits cause great dynamic amplification of seismic waves.

The MCWS was severely affected by the 1985 earthquake, causing extensive damage that left around 40% of the 8.5 million people living in the Federal District without water service. A comprehensive report on the damage was published by Ayala and O'Rourke M. (1989). They observed that two-thirds of the damage was located at pipe joints; and that the observed failure types include lateral crushing of pipes, and crushing and unplugging of joints. The authors state that the main apparent cause for the extensive damage in the pipeline network was the propagation of surface waves (Rayleigh waves). These greatly amplified waves caused that most of the damage were localized in sites with natural period higher than 2 s, and with clay deposits between 30-m and 70-m deep. The authors also indicate that the second most important factor that likely influenced the great damage in the pipeline system was the largely observed ground settlement typical of the lake zone, a phenomenon that could have reduced the pipeline capacity to withstand wave propagation.

## 4.2 General aspects of the estimation of future damage in the MCWS

The MCWS 1985 damage scenario has been used to compute fragility functions to better understand the relationship between pipeline damage and seismic intensity (Section 3.4). Two of those fragility functions, the one in terms of PGV (Pineda and Ordaz, 2003), and another in terms of PGV²/PGA (Pineda and Ordaz, 2007) are used in the future damage estimation described in Sections 4.3 and 4.4, respectively.

In general, the procedure to estimate the expected number of pipe repairs for a set of earthquake scenarios is the same, independently of the seismic parameter of interest (PGV or PGV²/PGA). The main steps are 1) A set of postulated earthquake scenarios is calculated with the Program Z (Ordaz et al. 1996) for magnitudes (m) between 6.6 and 8.4, and for focal distances (R) between 250 and 450 km. 2) The MCWS network is divided in segments of 100 m or shorter, and the seismic parameter value (PGV or PGV²/PGA) is calculated at each segment mid-point for each earthquake scenario. 3) The expected number of pipe repairs for each earthquake scenario is then computed by adding all the damage rate-length products for the whole pipeline network. The summary of these calculations is shown in Sections 4.3 and 4.4.

## 4.3 Future damage estimation using PGV

Based on the 1985 MCWS pipeline damage scenario, Pineda and Ordaz (2003) proposed a fragility function to estimate damage to buried pipelines. In Figure 1 the fragility function is showed along with the data points employed in its calculation. The 1985 damage scenario was divided in zones depending on the PGV values; a total of nine different PGV intervals (I1 to I9 in Figure 1) were used to generate the data points for the relationship damage rate-PGV. More details about the calculation of this fragility curve are found at the 2003 paper. In Equation 5, RR is the expected number of pipe repairs per kilometer of pipeline, $\Phi$ represents the cumulative normal function defined by Equation 6, where $\mu$ and $\sigma$ are parameters related to the mean and standard deviation of the damage rate-PGV relationship. In Equations 5 and 6, PGV is in cm/sec.

$$RR = \begin{cases} 0 & \text{if} \quad 0 < PGV < 5.35 \text{ cm/sec} \\ 0.1172 + 0.7281 \cdot \Phi(PGV; 51.8964, 19.7811) & \text{if} \quad 5.35 \leq PGV < 95 \text{ cm/sec} \\ 0.00137 \cdot PGV + 0.70458 & \text{if} \quad PGV \geq 95 \text{ cm/sec} \end{cases} \quad (5)$$

$$\Phi(PGV; \mu, \sigma) = \int_{-\infty}^{PGV} \frac{1}{\sqrt{2\pi}\sigma} e^{-(1/2)[(v-\mu)/\sigma]^2} dv \quad (6)$$

A total of 75 earthquakes scenarios were employed to estimate the MCWS damage due to future seismic events nucleated at the subduction zones of the Pacific Mexican coast. These scenarios represent events with magnitudes between 6.6 and 8.4, and focal distances between 250 and 450 km. The number of pipe repairs for each scenario was rounded to the nearest integer (Table 1). These results can also be observed in Figure 2. It is observed that for magnitudes higher than 7.6, there is an exponential tendency in the variation of the number of repairs (NR), and magnitude. In fact, after fitting a double exponential function in terms of m and R, an equation to simplify the estimation of NR was obtained (Equation 7). NR, then, can be estimated with Equation 7 (where e is the exponential parameter) for earthquake scenarios with magnitudes equal or higher than 7.6, and focal distances between

250 km and 450 km. The fitting of Equation 7 is a good representation of the damage estimates from Table 1 because the R-square parameter was 0.996. The results for magnitudes lower than 7.6 did not have any clear tendency on the m and R domains, so it was no possible to find a similar function to the one shown in Equation 7.

$$NR = 0.00075 \cdot e^{1.6295m} \cdot e^{-0.00344R} \tag{7}$$

Equation 7 shows how NR is related to m and R; however, for a better damage prediction, the figures from Table 1 must have preference over calculations with the equation due to the added error caused by the fitting.

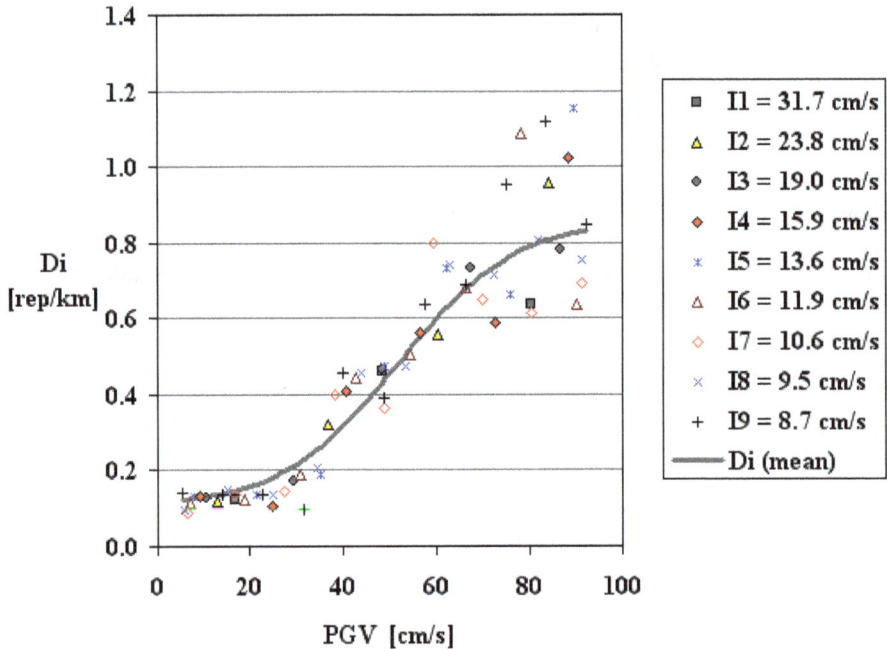

Fig. 1. Fragility Function for the MCWS in terms of PGV (Pineda and Ordaz, 2003). Di is damage rate in number of pipe repairs per kilometer of pipeline (it is also called RR in this manuscript)

| MAGNITUDE | FOCAL DISTANCE | | | | |
|:---:|:---:|:---:|:---:|:---:|:---:|
| | R = 250 km | R = 300 km | R = 350 km | R = 400 km | R = 450 km |
| 6.6 | 1 | 0 | 0 | 0 | 0 |
| 6.7 | 4 | 1 | 0 | 0 | 0 |
| 6.8 | 10 | 3 | 1 | 0 | 0 |
| 6.9 | 18 | 7 | 2 | 1 | 0 |
| 7.0 | 27 | 15 | 7 | 2 | 1 |
| 7.1 | 38 | 24 | 13 | 7 | 3 |
| 7.2 | 47 | 35 | 23 | 14 | 7 |
| 7.3 | 54 | 42 | 33 | 23 | 11 |
| 7.4 | 59 | 50 | 40 | 33 | 23 |
| 7.5 | 66 | 57 | 47 | 39 | 32 |
| 7.6 | 76 | 63 | 54 | 45 | 38 |
| 7.8 | 102 | 89 | 70 | 60 | 51 |
| 8.0 | 141 | 115 | 98 | 87 | 68 |
| 8.2 | 208 | 165 | 134 | 113 | 99 |
| 8.4 | 289 | 240 | 198 | 163 | 136 |

Table 1. Expected number of pipe repairs in the MCWS due to postulated earthquake
scenarios (PGV-based model; Pineda and Ordaz, 2003)

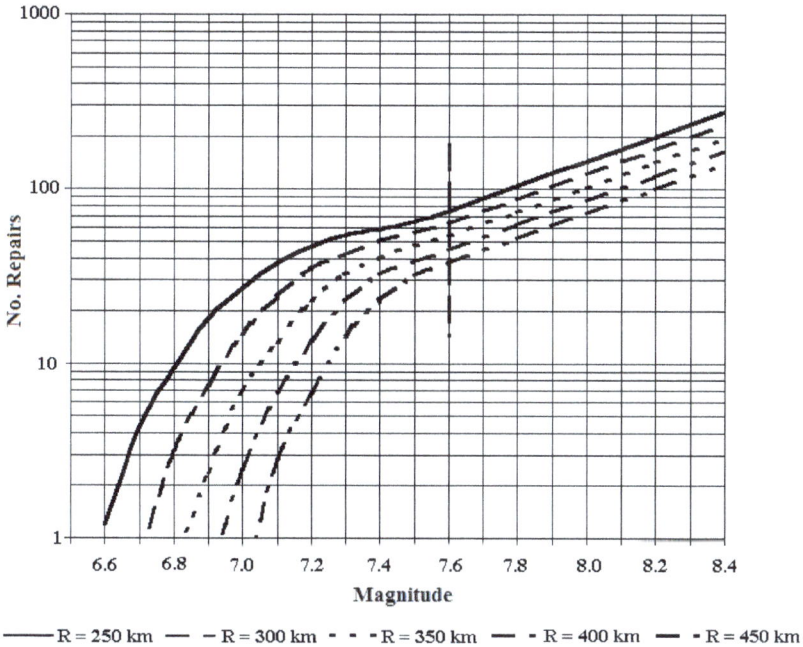

Fig. 2. Seismic damage prediction model for the MCWS based on a PGV-based fragility
function (Pineda and Ordaz, 2003)

## 4.4 Future damage estimation using the composite parameter PGV²/PGA

The PGV²/PGA-based fragility function proposed by Pineda and Ordaz (2007) for the MCWS (Figure 3) has three parts where the damage rate can be— zero, constant, or linearly dependent of PGV²/PGA (Equation 8). The no-damage zone is defined for seismic intensity levels not associated to pipeline damage in the 1985 damage scenario. A likely explanation of the constant-damage zone is the presumably about-to-fail precondition of some pipe segments previously to the 1985 event. The PGV²/PGA map employed by Pineda and Ordaz (2007) to generate the fragility function is shown in Figure 3.

$$RR = \begin{cases} 0 & \text{if} & PGV^2/PGA < 1.8 \text{ cm} \\ 0.122 & \text{if} & 1.8 \leq PGV^2/PGA < 8.72 \text{ cm} \\ 0.032(PGV^2/PGA) - 0.157 & \text{if} & PGV^2/PGA \geq 8.72 \text{ cm} \end{cases} \quad (8)$$

In Figure 4, the PGV²/PGA-based fragility function is showed along with the data points employed in its calculation. In a similar way to the employed to calculate the PGV-based fragility function (Section 4.3), the 1985 damage scenario was divided in zones depending on the PGV²/PGA values; a total of nine different PGV²/PGA intervals (I1 to I9 in Figure 3) were used to generate the data points for the relationship damage rate- PGV²/PGA. The 2007 paper contains more details about the calculation of this fragility curve.

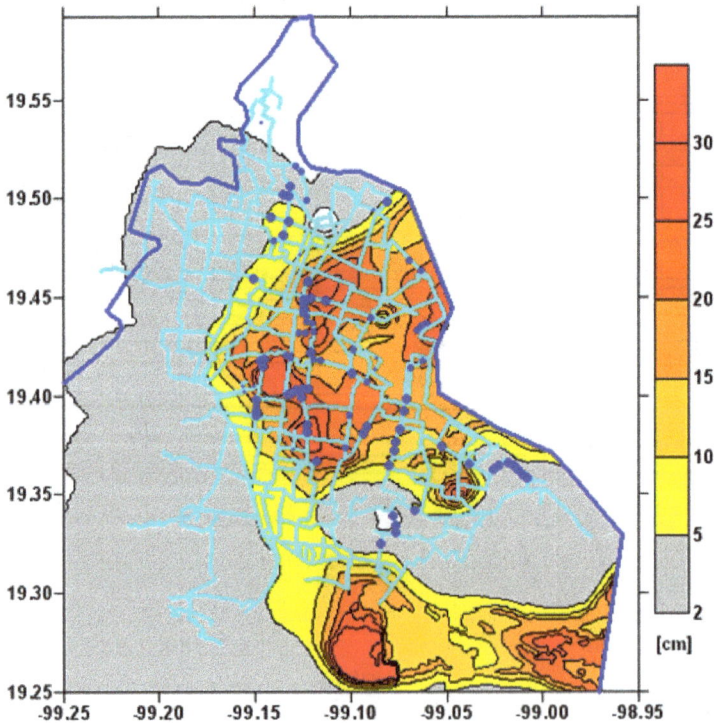

Fig. 3. Mexico City water system, PGV²/PGA map, and damage sites after the 1985 Michoacan earthquake

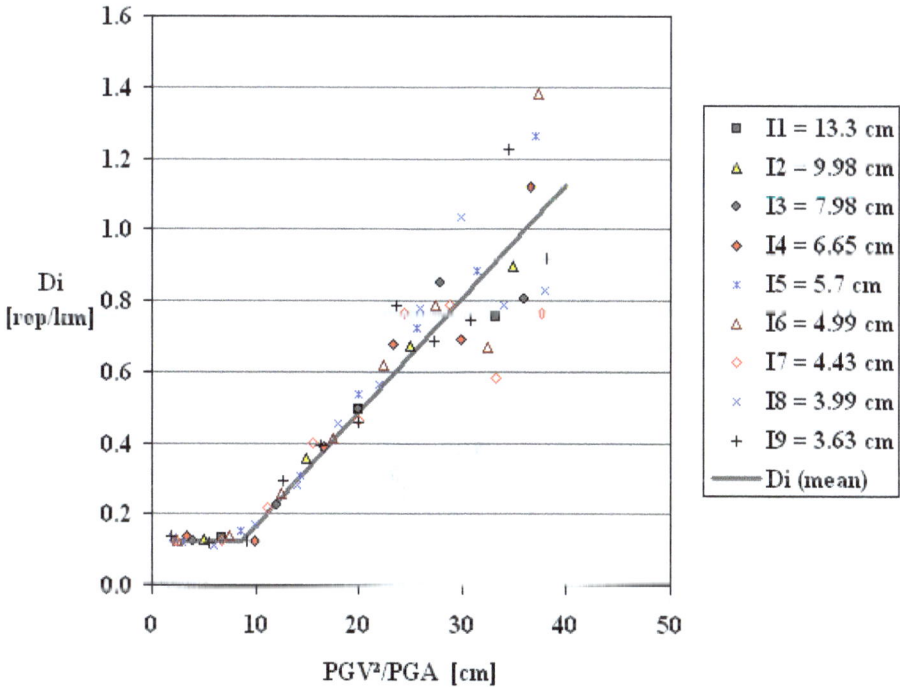

Fig. 4. Fragility Function for the MCWS in terms of $PGV^2/PGA$ (Pineda and Ordaz, 2007). Di is damage rate (rep/km); it is also called RR through this manuscript

The damage estimation for the MCWS employing the fragility function in terms of $PGV^2/PGA$ was done with 80 earthquake scenarios calculated with the Program Z (Ordaz et al. 1996) for seismic events nucleated at the Mexican Pacific subduction zones with magnitudes between 6.6 and 8.4, and focal distances between 250 and 450 km, as it was done with the model described in Section 4.3. Figure 5 shows an example of postulated seismic scenario for a magnitude of 8.4 and focal distance of 250 km. This event is much stronger than the 1985 Michoacan event (m=8.1; R=295 km) because it has a larger magnitude and a shorter focal distance.

The expected number of pipe repairs for the 80 earthquake scenarios are shown in Table 2 and plotted in Figure 6. In this prediction model, an exponential relationship between the number of pipe repairs and magnitude is observed for magnitudes higher than 7.8 for all focal distances; this relationship can be represented by the exponential function shown in Equation 9, where a and b are fitting parameters that vary depending on the focal distance values (Table 3). The fit of Equation 9 for all five focal distances is good because the R-square parameters have values close to one. Figure 7 shows the fitted curves with Equation 9 using the parameters from Table 3. These curves can be compared with the original data points from Table 2. Because there is no a clear tendency in the variation of NR with respect to m and R for magnitudes lower than 7.8, a fitted curve could not be found.

$$NR = a \cdot e^{b \cdot m} \tag{9}$$

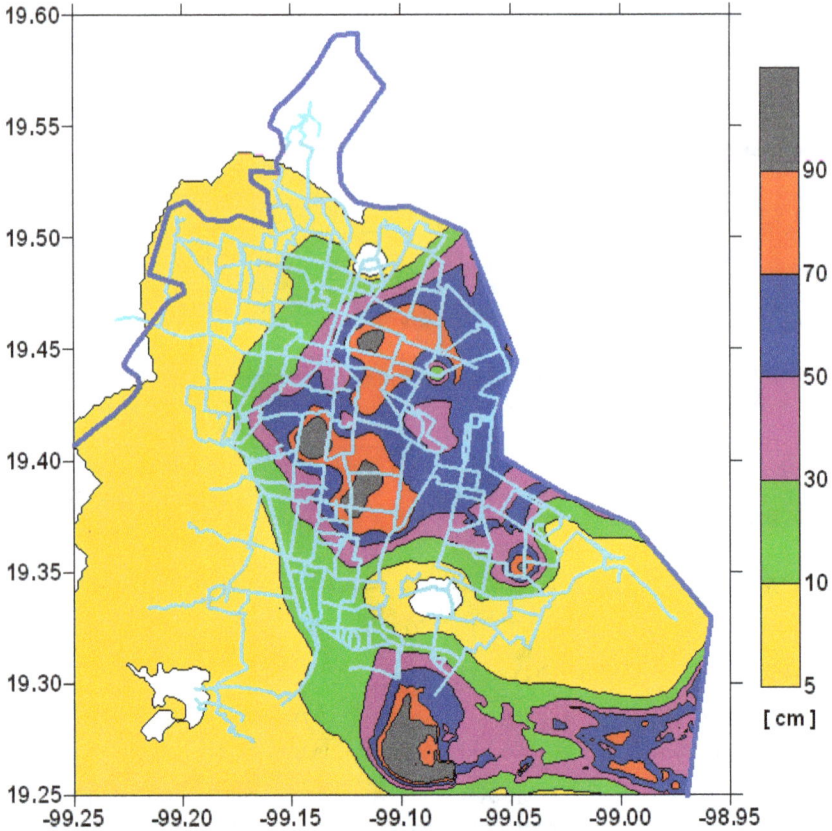

Fig. 5. Earthquake scenario map for a postulated event (m=8.4; R=250km) and the MCWS

## 4.5 Comparison of results between the PGV and PGV²/PGA models

A comparison between both prediction models reveals that the PGV²/PGA-based model (Section 4.4) predicts a lower number of pipe repairs than the PGV-based model (Section 4.3), for magnitudes around 8.0 and lower; and for higher magnitudes, the PGV²/PGV-based model predicts a higher number of pipe repairs (Figure 8). These results mean that the 2003 model overestimates the damage in the MCWS for earthquakes with magnitudes up to 8.0, and underestimates damage figures for stronger earthquakes. This conclusion is based on the fact that the proposed model uses a parameter better related to buried pipeline damage.

One important advantage of the PGV²/PGA model is the linear relationship between RR and PGV²/PGA in the fragility function. This simple functional form makes it easier to assess the damage for very strong earthquakes (events stronger than the 1985 quake). On the contrary, the PGV model could be unreliable for earthquakes stronger than the 1985 quake because the PGV-based fragility function assumes a linear relationship RR-PGV for PGV values higher than 95 cm/sec, something that could not be demonstrated in the 2003 study.

| MAGNITUDE | FOCAL DISTANCE | | | | |
|:---:|:---:|:---:|:---:|:---:|:---:|
| | R = 250 km | R = 300 km | R = 350 km | R = 400 km | R = 450 km |
| 6.6 | 0 | 0 | 0 | 0 | 0 |
| 6.7 | 0 | 0 | 0 | 0 | 0 |
| 6.8 | 2 | 0 | 0 | 0 | 0 |
| 6.9 | 5 | 1 | 0 | 0 | 0 |
| 7.0 | 9 | 4 | 1 | 0 | 0 |
| 7.1 | 14 | 8 | 4 | 1 | 0 |
| 7.2 | 19 | 13 | 8 | 4 | 1 |
| 7.3 | 25 | 18 | 13 | 8 | 4 |
| 7.4 | 32 | 24 | 18 | 13 | 8 |
| 7.5 | 39 | 35 | 29 | 23 | 18 |
| 7.6 | 46 | 39 | 35 | 32 | 28 |
| 7.7 | 56 | 47 | 39 | 36 | 32 |
| 7.8 | 74 | 58 | 48 | 40 | 36 |
| 8.0 | 147 | 124 | 98 | 73 | 59 |
| 8.2 | 261 | 216 | 183 | 158 | 138 |
| 8.4 | 474 | 390 | 328 | 279 | 241 |

Table 2. Expected number of pipe repairs in the MCWS due to postulated earthquake
scenarios ($PGV^2$/PGA-based model)

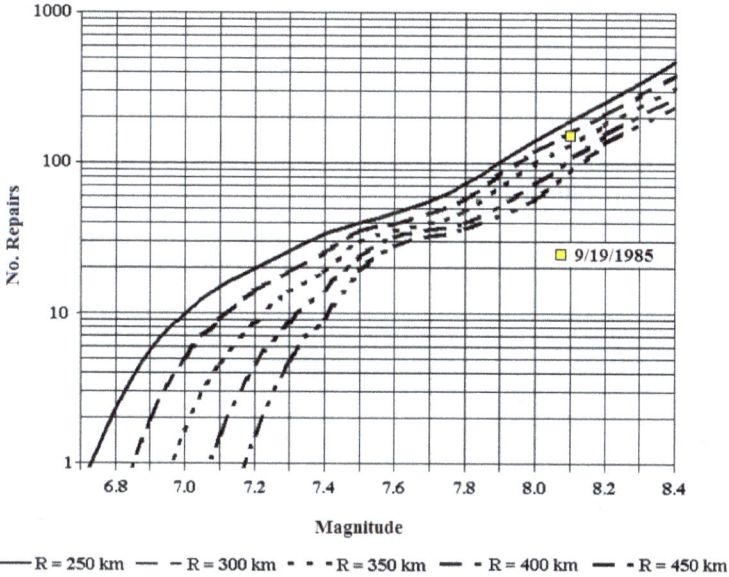

Fig. 6. Seismic damage prediction model for the MCWS based on a $PGV^2$/PGA-based
fragility function. The square marker corresponds to the 1985 Michoacan earthquake case
(m=8.1; R=295 km)

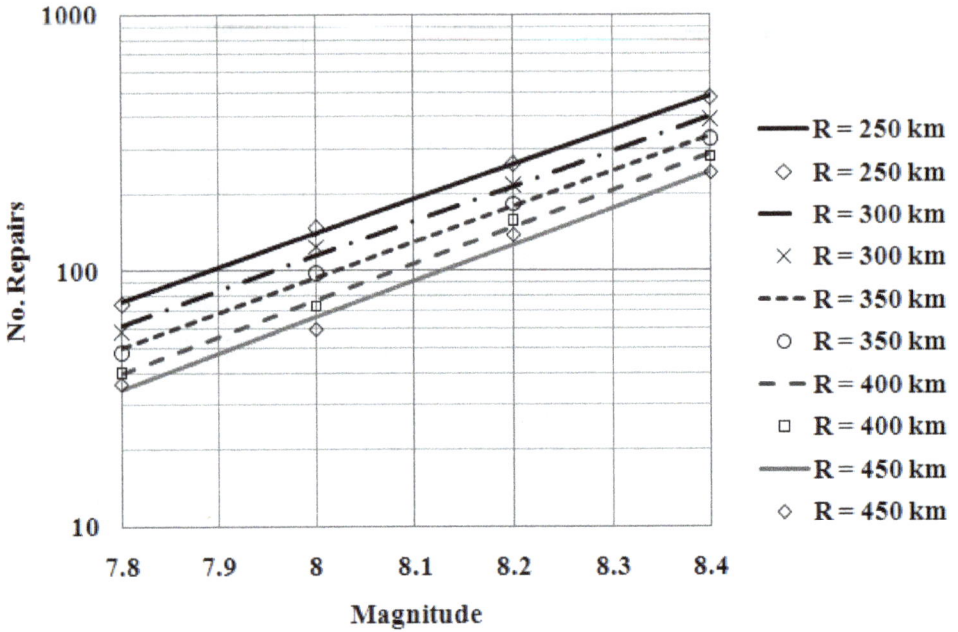

Fig. 7. Fitted curves for Equation 9

| | FOCAL DISTANCE | | | | |
|---|---|---|---|---|---|
| | R = 250 km | R = 300 km | R = 350 km | R = 400 km | R = 450 km |
| a | 2.972E-09 | 1.448E-09 | 8.453E-10 | 2.916E-10 | 2.927E-10 |
| b | 3.073 | 3.1369 | 3.18 | 3.287 | 3.268 |
| $R^2$ | 0.998 | 0.995 | 0.998 | 0.997 | 0.99 |

Table 3. Parameters a and b for Equation 9, and R-square parameters

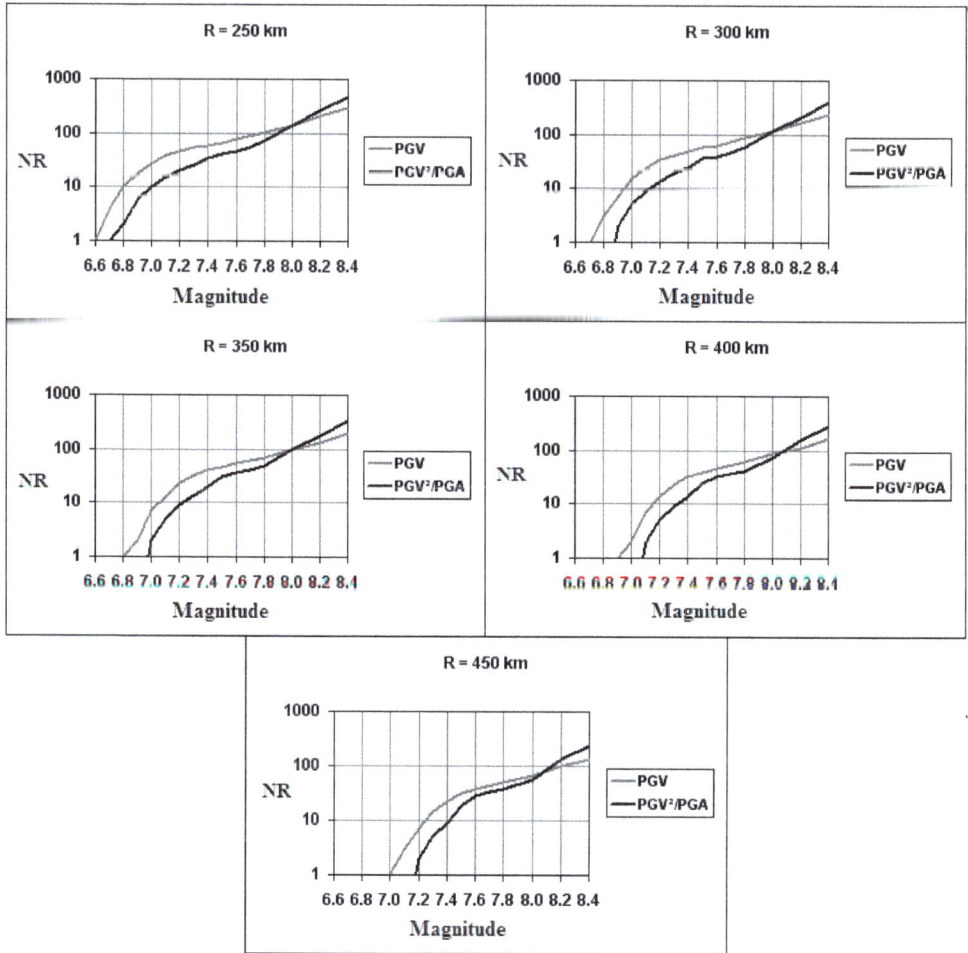

Fig. 8. Comparison of results between the PGV and PGV²/PGA models

## 5. Conclusions

One important challenge in the field of earthquake engineering for pipelines is the damage estimation due to future events. In this chapter, the reader will find a comprehensive state-of-the-art revision of the seismic parameters that have been employed as damage indicators for pipelines, and the most important seismic fragility functions proposed until now. In order to show a case of damage estimation due to future earthquake, we describe two prediction models for the Mexico City Water System (MCWS).

The results of this research reveal that a previous damage estimation study for the MCWS, based on a PGV fragility function, overestimated the expected number of pipe repairs caused by earthquakes with magnitudes around 8.0 to 8.1 and lower, and underestimated the damage for stronger earthquakes. This new study employs a recently proposed fragility function for the MCWS in terms of $PGV^2/PGA$: A composite parameter in terms of PGA, and PGV. Because a previous study (Pineda and Ordaz, 2007) demonstrates that $PGV^2/PGA$ is better related to pipeline damage in Mexico City than PGV alone, the results of the future damage estimation for the MCWS showen here are believed to be more reliable than those obtained with the PGV-based fragility function (Pineda and Ordaz, 2003).

## 6. Acknowledgements

This research has been partially funded by the Energy and Infrastructure Analysis (D-4) group at Los Alamos National Laboratory and the Institute of Engineering of the National Autonomous University of Mexico (UNAM).

## 7. References

American Lifelines Alliance, ALA (2001). *Seismic Fragility Formulations for Water Systems*, American Society of Civil Engineers (ASCE) and Federal Emergency Management Agency (FEMA). www.americanlifelinesalliance.org

ASCE-TCLEE (1991). *Seismic Loss Estimation for a Hypothetical Water System*. Technical Council on Lifeline Earthquake Engineering (TCLEE) of the American Society of Civil Engineers (ASCE), Monograph No.2, C.E. Taylor (eds).

Ayala, G. & O'Rourke, M. (1989). *Effects of the 1985 Michoacan Earthquake on Water Systems and other Buried Lifelines in Mexico*, Multidisciplinary Center for Earthquake Engineering Research, Technical Report NCEER-89-0009, New York.

Ballantyne, D. B.; Berg, E.; Kennedy, J.; Reneau, R. & Wu, D. (1990). *Earthquake Loss Estimation Modeling for the Seattle Water System*, Report to U.S. Geological Survey under Grant 14-08-0001-G1526, Technical Report, Kennedy/Jenks/Chilton, Federal Way, WA.

Barenberg, M.E. (1988). Correlation of Pipeline Damage with Ground Motions. *Journal of Geotechnical Engineering*, ASCE, June, 114 (6), 706-711.

Eidinger, J. (1998). *Water Distribution System – The Loma Prieta, Californa, Earthquake of October 17, 1989 - Lifelines*, USGS, Professional Paper 1552-A, Anshel J. Schiff (ed.), U.S. Government Printing Office, Washington, A63-A78.

Eidinger, J.; Maison, B.; Lee, D. & Lau, B. (1995). East Bay Municipal District Water Distribution Damage in 1989 Loma Prieta Earthquake. *Proceedings of the Fourth U.S. Conference on Lifeline Earthquake Engineering*, ASCE-TCLEE, Monograph No. 6, 240-247.

Eguchi, R. T. (1983). Seismic Vulnerability Models for Underground Pipes. *Proceedings of Earthquake Behavior and Safety of Oil and Gas Storage Facilities, Buried Pipelines and Equipment*, ASME, PVP-77, New York, June, 368-373.

Eguchi, R. T. (1991). Seismic Hazard Input for Lifeline Systems. *Structural Safety*, 10, 193-198.

FEMA (1999). *Earthquake Loss Estimation Methodology HAZUS-MH – Technical Manual*. FEMA, Washington DC, http://www.fema.gov/hazus.

Hamada, M. (1991). Estimation of Earthquake Damage to Lifeline Systems in Japan. *Proceedings of the Third Japan-U.S. Workshop on Earthquake Resistant Design of Lifeline Facilities and Countermeasures for Soil Liquefaction*, San Francisco, CA, December 17-19, 1990; Technical Report NCEER-91-0001, NCEER, State University of New York at Buffalo, Buffalo, NY, 5-22.

Hamada, M. & Akioka, Y. (1997). Liquefaction Induced Ground Strain and Damage to Buried Pipes. *Proceedings of Japan Society of Civil Engineers Earthquake Engineering Symposium*, Vol. 1, pp. 1353–1356 (in Japanese).

Hindy, A. & Novak, M. (1979). Earthquake Response of Underground Pipelines. *Earthquake Engineering and Structural Dynamics*, 106, 451–476.

Hwang, H. & Lin, H. (1997). *GIS-based Evaluation of Seismic Performance of Water Delivery Systems.* Technical Report, CERI, University of Memphis, Memphis, TN.

Isoyama, R.; Ishida, E.; Yune, K. & Shirozu, T. (2000). Seismic Damage Estimation Procedure for Water Supply Pipelines. *Proceedings of the Twelfth World Conference on Earthquake Engineering*, CD-ROM Paper No. 1762, 8pp.

Isoyama, R. & Katayama, T. (1982). *Reliability Evaluation of Water Supply Systems during Earthquakes.* Report of the Institute of Industrial Science, University of Tokyo, 30 (1) (Serial No. 194).

Jeon, S. S. & O'Rourke, T. D. (2005). Northridge Earthquake Effects on Pipelines and Residential Buildings. *Bulletin of the Seismological Society of America*, 95 (1), 294-318.

Katayama, T.; Kubo, K. & Sato, N. (1975). Earthquake Damage to Water and Gas Distribution Systems. *Proceedings of the U.S. National Conference on Earthquake Engineering*, EERI, Oakland, CA, 396-405.

Kitaura, M. & Miyajima, M. (1996). *Damage to Water Supply Pipelines. Special Issue of Soils and Foundations.* Japanese Geotechnical Society, Japan, January, 325-333.

Mavridis, G. & Pitilakis, K. (1996). Axial and Transverse Seismic Analysis of Buried Pipelines. *Proceedings of the Eleventh World Conference on Earthquake Engineering*, Acapulco, Mexico, 81-88.

Newmark, N. M. (1967). Problems in Wave Propagation in Soil and Rocks. *Proceedings of the International Symposium on Wave Propagation and Dynamic Properties of Earth Materials*, University of New Mexico Press.

Ordaz, M.; Perez Rocha, L.E.; Reinoso, E.; Montoya, C. & Arboleda, J. (1996-2002) *Program Z*, Instituto de Ingeniería, National Autonomous University of Mexico (UNAM).

O'Rourke, M. J. (2009). Analytical Fragility Relation for Buried Segmented Pipe. *Proceedings of the TCLEE 2009: Lifeline Earthquake Engineering in a Multihazard Environment.* Oakland, CA, 771-780.

O'Rourke, M. J. & Ayala, G. (1993). Pipeline Damage due to Wave Propagation. *Journal of Geotechnical Engineering*, ASCE, 119 (9), 1490-1498.

O'Rourke, M. J. & Deyoe, E. (2004). Seismic Damage to Segmented Buried Pipe. *Earthquake Spectra*, (20) 4, 1167–1183.

O'Rourke, M. J. & El Hmadi, K. (1988). Analysis of Continuous Buried Pipelines for Seismic Wave Effects. *Earthquake Engineering and Structural Dynamics*, 16, 917-929.

O'Rourke, T. D. & Jeon, S. S. (1999). Factors Affecting the Earthquake Damage of Water Distribution Systems. *Proceedings of the Fifth U.S. Conference on Lifeline Earthquake Engineering*, Seattle, WA, ASCE, Reston, VA, 379-388.

O'Rourke, T. D.; Stewart, H. E.; Gowdy, T. E. & Pease, J. W. (1991). Lifeline and Geotechnical Aspects of the 1989 Loma Prieta Earthquake. *Proceedings of the Second International Conference on Recent Advances in Geotechnical Earthquake Engineering and Soil Dynamics*, St. Louis, MO, 1601-1612.

O'Rourke, T. D.; Toprak, S. & Sano, Y. (1998). Factors Affecting Water Supply Damage Caused by the Northridge Earthquake. *Proceedings of the Sixth U.S. National Conference on Earthquake Engineering*.

Paolucci, R. & Smerzini, C. (2008). Earthquake-induced Transient Ground Strains from Dense Seismic Networks. *Earthquake Spectra*, 24 (2), 453-470.

Pineda, O. (2002). Estimación de Daño Sísmico en la Red Primaria de Distribución de Agua Potable del Distrito Federal. *Master of Engineering Thesis*, Institute of Engineering, National Autonomous University of Mexico (UNAM), Mexico City. (In Spanish)

Pineda, O. & Ordaz, M. (2003). Seismic Vulnerability Function for High-diameter Buried Pipelines: Mexico City's Primary Water System Case. *2003 ASCE International Conference on Pipeline Engineering and Construction*, American Society of Civil Engineers, Baltimore, USA.

Pineda, O. & Ordaz, M. (2004). Mapas de Velocidad Máxima del Suelo para la Ciudad de México. *Revista de Ingeniería Sísmica*, Mexican Society of Earthquake Engineering; (71) 37-62. (*In Spanish*)

Pineda, O. (2006). Estimación de Daño Sísmico en Tuberías Enterradas. *Doctorate of Engineering Thesis*, Institute of Engineering, National Autonomous University of Mexico (UNAM), Mexico City. (*In Spanish*)

Pineda, O. & Ordaz, M. (2007). A New Seismic Intensity Parameter to Estimate Damage in Buried Pipelines due to Seismic Wave Propagation. *Journal of Earthquake Engineering*, (11) 773-786.

Pineda, O. & Ordaz, M. (2010). Seismic Fragility Formulation for Segmented Buried Pipeline Systems Including the Impact of Differential Ground Subsidence. *Journal of Pipeline Systems Engineering and Practice*, ASCE, Vol. 1, No. 4, pp. 141-146.

Sano, Y.; O'Rourke, T. & Hamada, M. (1999). GIS Evaluation of Northridge Earthquake Ground Deformation and Water Supply Damage. *Proceedings of Fifth U.S. Conference on Lifeline Earthquake Engineering*, TCLEE Monograph No.16, ASCE, pp. 832–839.

Singh, S. K.; Santoyo, M.; Bodin, P. & Gomberg, J. (1997). Dynamic Deformations of Shallow Sediments in the Valley of Mexico. Part II: Single-station Estimates. *Bulletin of the Seismological Society of America*, 87, 540-550.

# 6

# Masonry and Earthquakes: Material Properties, Experimental Testing and Design Approaches

Thomas Zimmermann and Alfred Strauss
*University of Natural Resources and Life Sciences Vienna*
*Institute for Structural Engineering*
*Austria*

## 1. Introduction

Earthquakes are natural disasters which can occur without warning and they can affect large areas. Earthquakes are often accompanied by aftershocks which may cause additional damage to an already damaged structure or lead to failure. Consequences of earthquakes, such as rock falls, fires, explosions etc., can be very large in the affected areas. An example is the 1906 earthquake in San Francisco. Thereby many lives were killed by the ensuing fire. But not only past events have had fatal consequences, also events of recent years caused countless deaths and consequential damages e.g. the 2010 earthquake in Haiti with a death toll of more than 250,000 people (Eberhard et al., 2011) or the 2011 earthquake in Fukushima, Japan with high consequential damages e.g. to the nuclear power plant (Takewaki et al., 2011) or both earthquakes in Christchurch, New Zealand with high damage on cultural heritage (Ingham & Griffith, 2011; Ingham et al., 2011).

## 2. Masonry

### 2.1 Historical overview in masonry construction

Masonry is one of the oldest types of construction. Taking account structural-physical properties and the quite easy construction process, this construction system is used until today. The building material brick was easy to produce. In ancient times, clay was put into a model, the surface was smoothed and then the brick was exposed to air for drying. In later times the raw material was extruded and baked in a kiln. Masonry construction methods were already well known in ancient times, about 5000 BC bricks were used in Mesopotamia. A well-known example for the usage of masonry in these times is the tower of Babel, which was built around 600 BC. Clay bricks were used as building material and bitumen as mortar. Even then masonry offered a faster and cheaper alternative to natural stones in the construction process.

In contrast to natural stone construction methods, manufacturing of regular building stones was a revolution and enables systematic construction methods. Monumental structures, for instance the Pantheon in Rome were built by using masonry. Masonry construction methods do not offer the possibility of build plane top panels and lintels, therefore the construction of arches was developed and enhanced during that period (Maier, 2002).

Masonry always has to be constructed in bond to guarantee an adequate bearing capacity. Depending on the time period, different bond types were used, additionally it has to be distinguished between pure and mixed masonry. Especially in the Early Middle Ages, mixed masonry was constructed. The facings of the wall were built with bricks but the core was filled up with quarry stones and lower quality bricks. Later on, masonry was used through the whole thickness of a structure, with continuous horizontal joints in each row, independent from the thickness of the wall.

By placing bricks in an adequate way, a cross bonding through the whole thickness of the wall was achieved. As a result of this the decisive bond of the bricks depends on the thickness of the wall. Walls with a thickness of the width of a brick usually were built in a stretching bond and walls with a thickness of the length of a brick in a heading bond. Further masonry was built with different bond types e.g. Markish or Wendish, Gothic or Monastic bond. Thereby in every brick layer one header brick is followed by a few stretcher bricks.

In the 16th century, structured bonds like cross bond and block bond were used, in which a stretcher course is followed by a header course. In the 17th century, the Dutch or Flemish bond has been used. Thereby a header course is followed by a mixed layer of headers and stretchers. The common bond methods are depictured in Fig. 1.

The bearing capacity of a masonry bond is reached by avoiding vertical joints which go through multiple layers of construction stones. An additional capacity can be reached by using anchorages, which set up a force-fit connection of opposite walls and especially of masonry and wooden floor slabs. In the rebuilding period after World War II masonry construction became important again.

Developing of reinforced concrete and enhancing of the corresponding construction methods showed the limits of masonry construction methods. Nevertheless masonry is still used in restoration of historic buildings and in housing construction. It becomes more important again during the last years, because of the increasing requirements to thermal insulation, almost special large-sized honeycomb bricks are used.

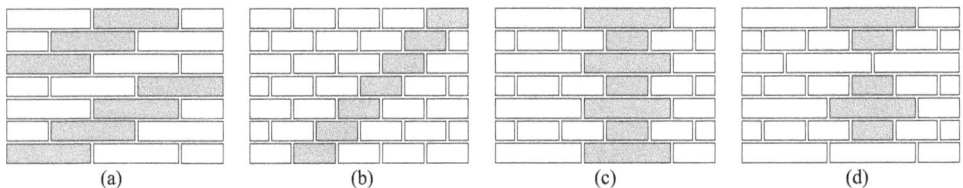

(a)                     (b)                     (c)                     (d)

Fig. 1. Various bonds of masonry (a) stretching bond, (b) heading bond, (c) block bond, (d) cross bond; (Bargmann, 1993)

### 2.1.1 Bricks

In contrast to natural stones, which have to be cut to the appropriate form and size, artificial construction stones as bricks are shaped by the models they are made with. The requirements to construction stones and mortar are defined in various codes. These codes define basic materials, definitions, dimensions, classes of raw density and strength and the assurance of quality.

Nowadays research objectives in brick design should offer the following properties: (a) increasing of efficiency in production process of construction stones; (b) increasing of the bricklayer's capacity; (c) increasing of the bearing capacity; (d) increasing of the building physics.

An important change in the production process was made with the industrial revolution, when the process was automatized. The change from the manual handling to automatically extruding machines yields in constant shapes. Moreover it ensures an alignment of the clay minerals due to the high pressure. This orientation of the clay minerals is the texture. It causes the anisotropic material behaviour depending on the direction of the brick. During the time various shapes and dimensions of bricks for different purposes occurred.

### 2.1.2 Mortar

Mortar is used to build up compensating layers to connect not accurately fitting bricks to an accurate and usable masonry. The most important characteristics of masonry mortar are its processability, its plasticity, water retention capacity, compressive strength and the bond strength between bricks and mortar. The components of mortar are binding agent, additives and water. Throughout additives and admixtures certain properties can be achieved.

The classification of mortars is based on the type of the binding agent, the production process and the type. Binding agents can be distinguished in non-hydraulic and hydraulic binders.

Non-hydraulic binders become hardened only on air and must not be under permanent moisture and water influence. Non-hydraulic binders are loam, common lime, gypsum, anhydrite, magnesia and fireclay.

Hydraulic binders become hardened both on air and under water and are resistant against permanent moisture. Hydraulic binders are cement, special types of lime and mixed binders. Mortars can be produced on construction side or in plant.

Lime was used as binding agent for masonry verifiable in 1000 BC (Hilsdorf, 1965) and it is up to now the determining binding agent for masonry mortar. Already in ancient times it was well known that admixtures like volcanic ashes or brick powder have a positive influence on strength and moisture resistance of the mortar (Grimm, 1989). Organic components enhance or modify manufacturing and hardening process, especially casein has an improving influence on water demand and water retention capacity (Conrad, 1990).

Extensive analyses of historic mortars have shown information about their composition, mixture and characteristics, which are quite different to modern mortars (Schäfer & Hilsdorf, 1990; Wisser & Knöfel, 1987). Additional components like tuff and puzzolan can be added to lime mortar to make the mortar hardening under water.

Table 1 gives an overview of different mortar types. Nowadays the mortar layer becomes thinner and thinner because the surface of modern bricks is very flat ensured by the production process. Further the thermal properties of mortar are improved continuously.

### 2.2 Material properties: Bricks, mortar, masonry
### 2.2.1 Bricks

Older structures often built with bricks, which are not conforming to today's ones. A comparison of old bricks in terms of material properties and characteristics can be found in (Egermann & Mayer, 1987). The following types were tested: (a) usual common brick (MZ); (b) extruded brick (SM); (c) hand-smoothed brick (HM); (d) historical brick 1796 (QU); (e) historical brick 1884 (BE).

The bricks SM and HM are made from the same raw material as the usual common brick MZ, but they were baked with 800 °C instead of 1000 °C. The historical brick QU from 1796 has been made manually, and the historical brick BE was formed by a screw extruder.

| Loam-mortar | made out of moistured loam with added chaffed components, hardening through drying; application: indoor rooms, well protected outdoor walls, clay floor in agricultures |
|---|---|
| Gypsum-mortar | made out of gypsum with sand and lime; application: gypsum, sand, lime as wall and ceiling plaster |
| Lime-mortar | lime-sand mortar is the usual mortar in structural engineering, made out of slaked-lime, sand and water |
| Lime-cement-mortar | extended cement-mortar is made by add cement to lime-mortar; application: if lime-mortar cannot be used as a result of the type of the bricks, the strength of the mortar and the expected moisture (lime-sand bricks, floating bricks, loaded piles and arches, weather sides and outside wall plaster |
| Cement-mortar | made out of cement and additional sand; application: for heavily used constructions and for structural members (piles, arches) which are exposed to moisture (foundations) |
| Raw-cement-mortar | cement, sand and additional components: fly-ash, gravel, stones |
| Floor-mortar | terrazzo-mortar, magnesia-mortar, xylolite, pavement |

Table 1. Mortar types (Bargmann, 1993)

| Specimen | Density | | Compressive strength in load direction | | Splitting tensile strength | | Elastic modulus in load direction | | Poisson's ratio in load direction | |
|---|---|---|---|---|---|---|---|---|---|---|
| | [g/cm³] | | [MPa] | | [MPa] | | [MPa] | | [-] | |
| | mean | cov | mean | cov | mean | cov | mean | cov | mean | cov |
| MZ | 1.83 | 1.6 | 43.0 | 18.1 | 3.94 | 32.4 | 22669 | 5.5 | 0.19 | 6.4 |
| SM | 1.90 | 1.2 | 31.3 | 16.3 | 3.76 | 18.0 | 11867 | 18.7 | 0.13 | 12.5 |
| HM | 1.82 | 1.4 | 15.6 | 22.2 | 1.82 | 21.7 | 5716 | 21.2 | 0.10 | 33.5 |
| QU | 1.65 | 4.3 | 9.5 | 56.2 | 0.52 | 37.4 | 2726 | 11.7 | 0.14 | 19.1 |
| BE | 1.49 | 72.3 | 13.9 | 38.5 | 2.42 | 40.1 | 8379 | 35.2 | 0.21 | 21.9 |

Table 2. Properties of historical bricks (Egermann & Mayer 1987)

As it can be seen in Table 2, the strength properties of hand-smoothed (HM) and historical bricks (QU, BE) are by trend lower than from machine-made bricks (MZ, SM). The scattering increases due to the manual production process. The consequent enhancement of the oxidation technique mainly led to a reduction of the spreading in the mechanic parameters. On the other hand the shaping methods influence other important parameters which are decisive for research of the structures. As discussed above, the orientation of clay minerals have a significant importance on the compressive strength and can be influenced by the production process. Taking account the compressive strengths of different brick types the results of the experimental study shows that in reference to common bricks (MZ) compressive strength of the extruded bricks (SM) is 73 % and the hand-smoothed (HM) is just 37 %. The compressive strengths of historical bricks (QU, BE) is below the strength of the common bricks.

The bearing capacity of masonry is essentially influenced by the splitting tensile strength of bricks. Common (MZ) and extruded bricks (SM) have approximately the same values for splitting tensile strength, whereas the value of the hand-smoothed bricks is just at 50 %.
If a masonry element is loaded by a centric compressive load, the failure results from cracks caused by tensile stresses. This tensile stress state develops due to different lateral Poisson ratios of bricks and mortar. Additionally, the production process has an important influence on stiffness. In reference to common bricks (MZ) the elastic modulus of the extruded bricks (SM) is 50 % and the hand-smoothed bricks (HM) have a value of about 25 %. As written above, the orientation of the clay minerals causes a relationship of the strength and deformation parameters to the considered direction of loading. Particularly common bricks (MZ) and extruded bricks (SM) show this behaviour. Hand-smoothed and historical bricks (HM, QU, BE) do not show this behaviour according to the considered direction. Therefore it can be concluded, that shaping under high pressure induces an anisotropic material behaviour. In addition to the strength values of construction stones given in Table 3, research results on historical Viennese bricks from the 19th century are given in (Furthmüller & Adam, 2009; Pech, 2010; Zimmermann & Strauss, 2010a). Properties of contemporary brick types, mortar and masonry can be found in (Schubert & Brameshuber, 2011).

| Reference | Compressive strength [MPa] | | Tensile strength [MPa] | | Elastic modulus [MPa] | | Fracture energy [Nmm/mm²] | | Density kg/m³ | |
|---|---|---|---|---|---|---|---|---|---|---|
| | mean | cov | mean | cov | mean | cov | mean | cov | mean | cov |
| (a) | 29.5 | 34.6 | 2.1 | 33.3 | 12055 | 25.6 | 0.056 | 3.6 | 1510 | 4.0 |
| (b) | 22.5 | 26.6 | - | - | - | - | - | - | - | - |
| (c) | 19.3 | 39.7 | - | - | 13489 | 52.6 | - | - | 1467 | 6.6 |

Table 3. Material parameters from historical Viennese bricks from the 19th century (a): (Furthmüller & Adam, 2009), (b): (Pech, 2010), (c): (Zimmermann & Strauss, 2010a)

### 2.2.2 Mortar

The percentage of binding agent in the hardened state in comparison to the additives is 1:2 up to 1:3 in reference to weight. The whole hydraulic part of the binding agent, including puzzolanic additives is 10-25 %. The configuration of the mortar has a high influence on the material characteristics. For compressive bearing capacity of masonry the influence factors are the compressive and tensile strength of the mortar and additionally its deformation behaviour. For shear and flexural capacity, initial shear strength and tensile bond strength are the decisive parameters. It is quite difficult to get adequate test specimens from masonry. Therefore these characteristics mostly have to be estimated (Schubert & Brameshuber, 2011).
If masonry is under compressive load, (see Sec. 2.2.3), tensile strength of bricks is decisive for the capacity. The modulus of lateral elongation of the mortar is crucial for the tensile stress state inside the bricks. Compressive strength of mortar is influenced especially by binding agent, the percentage of the components and porosity. Pure lime-mortars have compressive strength in a range from 1.0 to 2.0 MPa. If the hydraulic fraction increases, then also the compressive strength increases, which can be up to 5.0 MPa and more for lime-cement-mortar. If the hydraulic fraction increases, the elastic modulus increases and ductility decreases. Bonding characteristics between mortar and bricks can be specified

with the following parameters: (a) shear strength $f_{vk}$; (b) tensile bond strength $f_{hz}$; (c) coefficient of friction μ. Bonding characteristics are mainly influenced by the mortar type and its components. The type of bricks and further the moisture characteristics has an influence.

### 2.2.3 Masonry

Masonry is defined as a composite material. The bearing behaviour under compressive, tensile, flexural and shear load is different to homogenous materials like concrete or steel. The composite material itself consists of the singular bricks, horizontal and vertical joints. Depending on the scale, masonry can be seen as (a) inhomogenous or (b) homogenous construction material. In case (a) the characterisation has to be made separately for bricks, for mortar and their interaction, in case (b) the characterisation can be done with global parameters on the smeared masonry element; see Fig. 2.

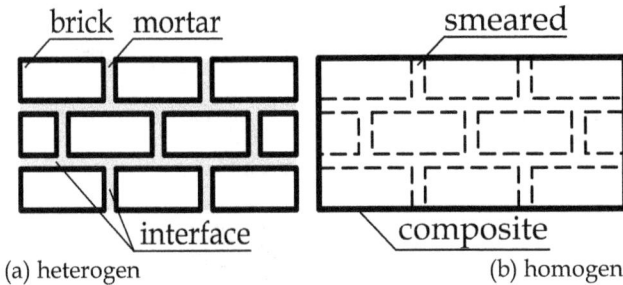

Fig. 2. Characterisation of masonry depending on the scale

#### 2.2.3.1 Compressive strength

Compressive strength of masonry is much higher than its tensile or flexural strength. As a result masonry is mainly used for structures under compressive load and the bearing capacity of masonry is described generally by the parameter compressive strength $f_k$. Taking into account the different components of masonry, the strengths are different. The compressive strength of the bricks $f_b$ is much higher than the compressive strength of the mortar $f_m$, therefore the failure mechanism of masonry under compressive load can be characterised.

Compressive strength of masonry can be seen as a function of the strength of the bricks and the strength of the mortar. The formula was detected empirically with the influences of the singular strengths on the overall strength of masonry and shows a nonlinear behaviour between the strengths of bricks and mortar. According to EC 6 the characteristic compressive strength of masonry $f_k$, which is defined as the 5 %-fractile value, can be determined from the mean values of the compressive strength of bricks and mortar. These mean values can be determined by execute tests according to EN 1052-1 or from the following equation:

$$f_k = K \cdot f_b^{\alpha} \cdot f_m^{\beta} \tag{1}$$

The factor $K$ and the exponents α and β have to be taken from EC 6, e.g. for old bricks $K$ = 0.60 the exponents α = 0.65 and β = 0.25. The relation of eq (1) is just valid for compressive loads perpendicular to the horizontal joints. If the compressive load is applied in the direction of horizontal joints, the strength has to be reduced as a result of the influence of the vertical joints. Researches from (Glitzka, 1988) quantify this reduction as follows:

$$f_{k||} = 0.75 f_k \tag{2}$$

If masonry is under compressive loads, deformations occur in parallel as well as in perpendicular direction in accordance to the load direction. As a result from the differences in material behaviour of bricks and mortar, horizontal stress occurs and causes failure of masonry. There are also differences in the lateral deformations of the masonry components, which can be observed, if masonry is loaded until the uniaxial compressive strength of mortar is almost reached. In this stress state the lateral elongation of mortar is much larger than those of the bricks, but the lateral deformations of the mortar are restricted by the bricks. Hence in the mortar occurs a spatial compressive stress state, whereas the brick is loaded by compression and tension.

The compressive strength of masonry under compressive loads is mostly determined by lateral tensile strength of bricks and the elastic modulus of mortar. However the elastic modulus can be identified in the tests quite difficulty. Therefore, in most applications the compressive strength of the bricks $f_b$ is taken as the initial parameter for defining the overall compressive strength of masonry. Some correlation factors between lateral tensile strength $f_{bt}$ and compressive strength of bricks are listed in the literature (Schubert & Brameshuber, 2011). For common bricks, this correlation is:

$$f_{bt} = 0.026 f_b \tag{3}$$

Tensile stresses inside of bricks cause cracks and fracture of bricks. To follow the progression of the cracks up to failure, the interaction phases between bricks and mortar have to be considered.

If masonry is put under compressive load and vertical cracks appear, the limit of lateral tensile strength has been exceeded, and on the appearing cracks, lateral tensile stress is reduced. If the load increases, the proximate cross-sections carry the lateral tensile stress up to an exceeding of the next maximum possible tensile stress state. After a few formations of cracks, the structure fails. The function of the critical stress state is depicted as the enveloping line of the fracture and shows the local reachable limit state. If masonry is under a marginal vertical compressive loading, the tensile stress state inside the bricks is below the enveloping curve of fracture. Therefore vertical stress can be increased, until the brick fails in tension or the mortar fails in compression.

The material performance of masonry under compressive load is defined in EC 6 by means of the parable-rectangle-diagram, see Fig. 3. The limit strain is defined with $\varepsilon_{m1}$ and $\varepsilon_{mu}$.

Another important parameter for defining the material behaviour of masonry is the elastic modulus. According to EC 6 the short time elastic modulus can be determined as the secant modulus from tests according to EN 1052-1 or calculated directly from compressive strength:

$$E = 1000 f_k \tag{4}$$

From experimental tests, see also Sec. 2.3, for masonry made of solid bricks with the old-Austrian shape type (height 6.5-7.5 cm) independent from the mortar type, a deviation from the recommended values for the elastic modulus was found out.

$$E = 300 f_k \qquad (5)$$

The reason for the lower elastic modulus for masonry made from solid bricks can be seen in the higher percentage of horizontal joints per altitude compared to masonry made out of new honeycomb bricks. Commonly used honeycomb bricks have a height of 25 cm and therefore just 4 horizontal joints per meter altitude difference in relation to 13 joints (masonry made of solid bricks), which is a multiplying factor of 3.3 for the joints. The same ratio can be found in the correlation factors for elastic modulus and compressive strength.

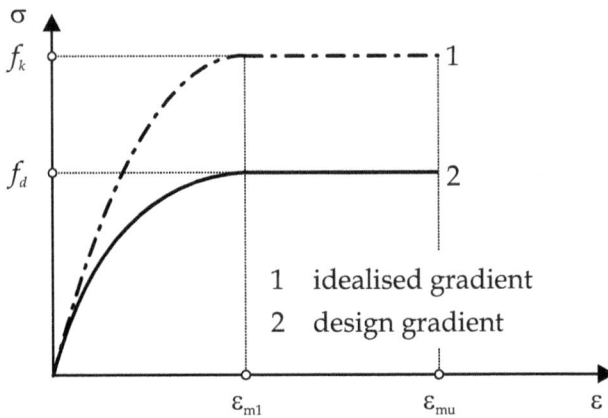

Fig. 3. Stress-Strain-Curve of masonry under compressive load

### 2.2.3.2 Tensile strength

Generally in masonry walls there is no constant tensile load over the whole cross section perpendicular to the horizontal joint. Besides the dead load, in most cases, masonry has to bear vertical loads. Anyway if there occur a tensile load, two mechanisms of failure can be distinguished: (a) failure of bond between mortar and brick, which is influenced mainly by tensile bond strength $f_{hz}$ between these components, and (b) failure of bricks, if tensile bond strength between mortar and bricks is larger than the tensile strength of bricks.

According to the codes of masonry construction, a planned tensile load perpendicular to the horizontal joints has to be avoided, because the resistance and capacity perpendicular to the horizontal joints has a large scatter.

EC 6 defines boundary conditions, when tensile strength perpendicular to the horizontal joints may be considered. In these cases, the failure of the structural member must not cause a failure of the whole structure. Tensile load in the direction of horizontal joints results from a load in the direction of the wall. To bear these tensile stresses, the walls have to be built in bond, whereat it is convenient to overpressure the occurring tensile stresses with perpendicular compressive stresses.

Tension bearing capacity of masonry in the parallel direction to the horizontal joints is defined mainly by the characteristics of the mortar. To define the deformation behaviour of masonry walls parallel and perpendicular to the horizontal joints extensive research is documented in (Bakes, 1983). However in this analysis the friction between the mortar of the vertical joint and the brick was neglected. Results show until fracture an almost linear material behaviour. Fig. 4 shows different failure modes caused by tension.

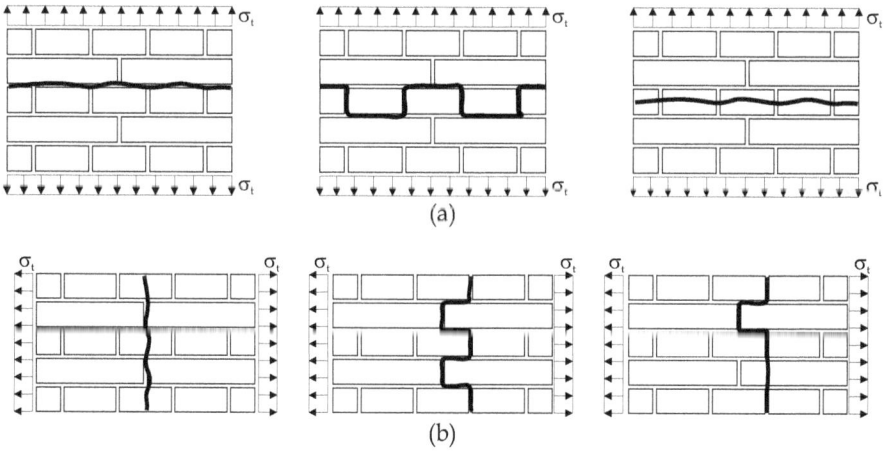

Fig. 4. Masonry under tension load, (a) tension perpendicular to horizontal interface (b) tension parallel to horizontal interface

### 2.2.3.3 Flexural behaviour

In contrast to the pure tensile load perpendicular to the horizontal joints, which can be excluded in nearly all load cases, flexural loading is a load case which is quite common. If a wall is loaded by wind or earth pressure perpendicular to its surface, then flexural stresses in the perpendicular and the parallel direction of the horizontal joints occur (Fig. 5). Designing principles assume no tensile or flexural stresses perpendicular to the horizontal joints. According to EC 6, the gap in the horizontal joints is just acceptable until to the half of the cross section.

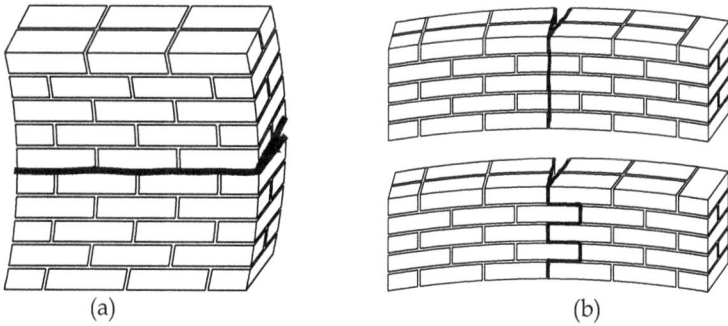

Fig. 5. Flexural load (a) parallel to the horizontal joints $f_{xk1}$ and (b) perpendicular to the horizontal joints $f_{xk2}$

### 2.2.3.4 Shear strength

The global bearing behaviour of structures made of masonry is influenced by loads acting in horizontal direction like wind and earthquake loads. For the load transfer vertical shear walls are required, which are loaded with shear forces in the wall direction. Therefore the behaviour of the shear wall is decisive for the bearing capacity of the whole structure. If an element is cut from the shear wall normal stresses are acting in vertical direction and shear stresses are acting at all four edges. This theory assumes that the vertical joints between the bricks nearby the fracture state do not transfer shear stresses, and consequently the vertical joints are neglected because of the low value of the stresses. Additionally the shrinking process in the mortar reduces the bond between mortar interface and brick. Further due to the low compression state in vertical direction no significant friction forces can be developed in the vertical joints.

Through the combination of normal and shear forces in the direction of the wall, a two-axial loading is induced. A plane stress state develops in the direction of the shear wall. The theory of Mann & Müller is defined for a stretcher bond with an overlapping of the stretcher of a half length of the bricks and a ratio of width and length of the bricks of 1:2. The shear stresses inside the horizontal joints induce a torque, which is compensated by the equilibrium on every single brick by a pair of forces. Assuming a linear distribution of stresses over the half length of the brick the stress state can be calculated as follows:

$$\sigma_{x1,2} = \sigma_x \pm \tau \cdot \frac{Q_x}{\Delta y} \quad \text{with } Q_x = 2 \cdot \tau \cdot \Delta x \tag{6}$$

The fracture depends on the ratio of the different loads and the material parameters and can be distinguished into four different failure modes (Mann & Müller, 1978): (a) failure of masonry due to compression; (b) rocking (gap in the horizontal joints at the bottom part of the wall); (c) friction failure of the horizontal joints; (d) tension failure of the bricks.

Failure of masonry due to compression (line $a$ in Fig. 6) appears, if the maximal compressive stress $\sigma_{x1}$ becomes higher than the compressive strength $f_k$ of masonry.

$$\tau = \left( f_k - \sigma_x \right) \cdot \frac{\Delta y}{2 \Delta x} \tag{7}$$

Rocking (b) appears, if the minimal compressive stress $\sigma_{x2}$ becomes zero. It is assumed that the horizontal joints cannot bear tensile stresses (eq. (8), $b_1$ Fig. 6). If a tensile strength is considered ($f_{hz}$ = tensile bond strength), the fracture condition can be formulated as stated in eq. (9) ($b_2$ Fig. 6).

$$\tau = \sigma_x \cdot \frac{\Delta y}{2 \Delta x} \tag{8}$$

$$\tau = \left( \sigma_x + f_{hz} \right) \cdot \frac{\Delta y}{2 \Delta x} \tag{9}$$

Friction failure (line $c$ in Fig. 6) appears, if in the area of the horizontal joints with minimal compressive stress the friction resistance is exceeded. The fracture condition can be defined with Mohr-Coulomb's law.

$$\tau = f_{vko} + \mu \cdot \sigma_x \tag{10}$$

The compressive stresses $\sigma_{x1,2}$ and the shear loads as shown in eq. (6) induce a principal stress state inside the bricks, which results in the fourth failure mode, tensile failure of bricks ($d$ Fig. 6). The principal stress state induces the fracture of the bricks, and therefore the tensile strength of bricks $f_{bt}$ becomes decisive.

$$\tau = \frac{f_{bt}}{2.3} \cdot \sqrt{1 + \frac{\sigma_x}{f_{bt}}} \qquad (11)$$

The discussed failure modes can be pictured as one curve in the $\sigma\tau$-diagram (Fig. 6). The curve encloses the area, in which no fracture and failure occurs. Stress states which are outside of the curve, lead to one of the four failure modes due to the discussed criteria.

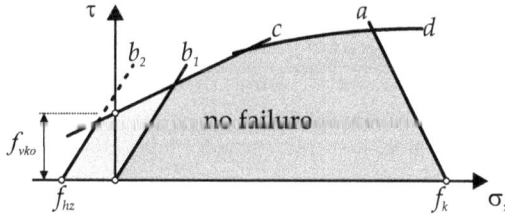

Fig. 6. Curve of the failure mode in the $\sigma\tau$-diagram

### 2.3 Experimental testing

From the 1950ies up to now, extensive research has been done with respect to the correlation of the strengths of used materials and the bearing capacity of structures, whereas most of interest was spent on uniaxial compressive strength. The research led to different calculation models and during the last centuries to a harmonisation of the European codes and standardised design concepts.

For definition of the design parameters of bricks and mortar, testing values are required. Therefore it was a need to standardise the testing methods and applications. On the other hand, efforts were put on resistance against horizontal loads, i.e. shear or dynamic loads and mostly also in combination with bending. As a result of a higher sensitivity in respect to earthquake it was necessary to adapt the codes. This adaption causes quite complex calculation methods and the need of additional material parameters. Various testing methods were discussed for determining the needed material parameters. A simple approach, which has been included into the European codes, is due to basic parameters, i.e. the compressive strengths of bricks and mortar. For defining the horizontal resistance, this approach needs more information, which can be given by the coefficient of friction $\mu$, the initial shear strength $f_{vko}$ and the tensile strength of the bricks $f_{bt}$. Therefore standardised values are available, which can be taken for the calculation and design or derived from the basic parameters, correlation factors are listed in (Schubert & Brameshuber, 2011).

For existing structures, it is required to get information about the basic material parameters in a non-destructive way as possible. There are a few non-destructive testing methods available, but they do not offer a direct conclusion on the existing strengths. One possibility is the sampling of wall-like specimens. This gives a deep insight into the strength of the tested component. However, by sampling test specimens of the building the structural integrity may get disturbed and in the extreme case the stability against collapse decreases.

Moreover, there are remarkable costs and therefore in most of the cases the number of test samples is not sufficient enough for a detailed static assessment of the existing structure. Especially in older structures the scatter of material parameters are considerably high and then a large number of samples is required to make serious assessments.

## 2.3.1 Testing methods of static parameters

In-situ testing methods of mortar and bricks have been enhanced to avoid disturbing existing structures. Additionally, the results were calibrated on wall specimens and on model structures and there are lots of standardised testing methods for determining parameters like brick size, shape, density etc.

- testing of masonry compressive strength by direct methods:
    - testing in laboratory of adequate, new constructed masonry specimens according to EN 1052-1 – Determination of compressive strength.
    - in-situ testing on structures, e.g. Flat-Jack-Test (Pech & Zach, 2009).
    - sampling of test specimens and testing in laboratory, according to EN 1052-1 – Determination of compressive strength.
    - sampling of representative test specimens for laboratory testing
- testing of shear strength of masonry:
    - adequate testings on masonry specimens, no standardised methods available
    - definitions and specifications of storage, size of test specimens, load application, boundary conditions can be found in the Mauerwerk Kalender
- testing of flexural strength of masonry:
    - general:
        - different test setups in literature, no valid results, many influencing parameters cannot be evaluated yet
        - values are not necessarily required for the assessment of existing structures
    - test method according to EN 1052-2 – Determination of flexural strength:
        - flexural in the plane of the wall-panel
        - test setup for determining the flexural strength parallel and perpendicular to the horizontal joints
    - test method according to EN 1052-5 – Determination of bond strength by the bondwrench method:
        - for torque application e.g. at the top of the wall, pure torque load
        - there cannot be estimated a common relationship between the test results of tensile bond strength and shear strength
- testing of tension strength of masonry:
    - There is no standardised test method available. In the ESECMaSE-Projekt the "Direct tension Test" was proposed as test method.
- testing of compressive strength of the components (indirect testing of masonry) and calculation of the overall compressive strength of masonry:
    - testing of strength of the bricks according to EN 772-1 – Determination of compressive strength. Compressive strength of the bricks is determined perpendicular to the horizontal joints, which is sufficient precise. The possibility of a reduced bearing capacity due to diagonal principal compressive stresses is neglected.
    - testing of strength on small sized test specimens
    - testing of strength by rebound hammer (Pech & Zach, 2009).
    - testing of mortar strength on a standardised prism, according to EN 1015-11 – Determination of flexural and compressive strength of hardened mortar
    - testing of mortar strength by stamp compression test (Pech & Zach, 2009).

- testing of mortar strength by determination of penetration resistance and the related deformation
- testing of splitting tensile strength of bricks:
  - There is no standardised test method available. Testing can be performed analogous to concrete, where load application is done with stripes of a width of 10 mm. The test method can also be applied on drill cores.
- testing of flexural strength of bricks:
  - testing method according to RILEM-Recommendations TC 76 – flexural strength of units (de Vekey, 1988)
- testing of centric brick tensile strength:
  - There is no standardised test method available.
- testing of coefficient of friction and initial shear strength between brick and mortar:
  - method according to EN 1052-3 – Determination of initial shear strength.
- additional physical analysis on bricks:
  - EN 772-13 – Determination of net and gross dry density of masonry units.
  - EN 772-16 – Determination of dimensions.

For estimating the strength of the bricks and as basic value for the calculative determination of the masonry compressive strength from the values from the components (e.g. according to national part of EC 6) the bricks have to be tested according to EN 772. As an alternative for existing structures, the compressive strength can be determined in a non destructive way by rebound and penetration methods.

Estimating of compressive strength of the bricks on existing structures has to follow EN 1998-3 for the required minimum number of testing samples. Existing structures are defined as those objects, which were built before the actual standards for masonry were valid, therefore these objects cannot ensure the required quality. The required amount of tests for a sufficient result of an existing structure with homogenous material and a knowledge class KL3 it is necessary to perform one test series per 1000 m² total floor area or two test serials per structure. For a knowledge class KL2 50 % of the required tests for KL3 have to be done. The definition of the knowledge classes is standardised in EN 1998-3 and depends on the geometry, constructional details and the materials. According to EN 1998-3 a test series is defined by the following parameters: (a) at least three test specimens (masonry) or (b) at least three test locations for testing strength of the components by taking specimens of the bricks and mortar for compressive strength tests or (c) at least six test locations for testing strength of the components by rebound and penetration methods.

After determining the test locations, the bricks usually are taken from masonry by means of a masonry saw or they are chiselled out. The test locations have to be documented. In order to minimize the size of the disturbed masonry, instead of the five to six full bricks according to EN 772-1, four to six half bricks can be taken instead. Although the size of the test specimens generally should not be decisive for compressive strength, especially the inhomogeneities in historical bricks can be the reason for high scatter and unsafe test results. For average determination of a test location there have to be taken five values, divergent from EN 772-1. In case of non-destructive testing with the rebound hammer there have to be taken 10 individual test results for each test location for determining the compressive strength.

In addition to performing the tests, they should be documented. The documentation should include the following parts: (a) object/structure; (b) date of testing; (c) situation of the test locations (identification on building plan); (d) testing method and standardisation to normative values; (e) characteristic masonry strength for each test location, test serial, type of masonry; (f) compressive strengths of bricks and mortar for each test location, test serial, type of masonry if the components were tested individually; (g) type of construction stones according to EN 771-1.

### 2.3.2 Testing methods of dynamic parameters

By investigation of the behaviour of masonry walls under cyclic load, essential information of the load-displacement-curve can be determined. The load-displacement-behaviour depends on the decisive failure mode. By means of pseudo-dynamic experiments, the dynamic behaviour of structures can be determined depending on time.

The experimental research of the cyclic behaviour of masonry walls has been investigated during the last years in lots of research projects, e.g. ESECMaSE and SEISMID. Usually, shear wall tests were performed, in which under constant vertical loads a cyclic horizontal load is applied in a quasi-static way. The defined boundary condition on the top of the wall is either a fixed support or a cantilever arm, which has the possibility of free rotation. This experimental setup allows the determination of special dynamic parameters, e.g. energy dissipation and hysteretic damping. The obtained test data can be used for further relevant parameters for seismic design concepts, like behaviour factor, stiffness and stiffness degradation. This topic is discussed in (Knox & Ingham, 2011; Tomazevic et al., 1996a; Zimmermann et al., 2010a; Zimmermann et al., 2010b). A further possibility for experimental testing is the performance of shake table tests on a vibrating table. In contrast to the shear walls discussed above, precise acceleration spectra can be taken for loading. The direct analysis of whole structures (walls, slabs, openings and floors) allows a better consideration of the load bearing behaviour. Experimental work is reported e.g. in (Benedetti et al., 1998; Tomazevic et al., 1996b; Tomazevic, 2007).

## 3. Seismic loads

Seismic loads are considered in EC 8. Part 1 specifies the basics, the loads from seismic impacts and structural design concepts in seismic influenced regions. The code specifications cover the design concepts by requirements on geometry, design by verification of the load bearing capacity and considerations for construction details. The other parts of EC 8 include specifications for bridges (2), existing structures (3), silos, tanks and pipelines (4), foundations and retaining walls (5) and towers, pylons and chimneys (6). Each structure has to be designed considering the unfavourable limit states, including and not including seismic loads. Therefore, depending on the used material, the corresponding EC is the basis for design, considering the specifications of EC 8.

### 3.1 Seismic zones

The exposure to seismic loads is specified by one single parameter, which is the reference-top level ground acceleration $a_{gR}$ for foundation class A. In Austria, the value of this ground acceleration is from 0.18 up to 1.34 m/s². This reference-top level ground acceleration has to be multiplied with the coefficient of importance $\gamma_I$ to obtain the ground acceleration $a_g$.

$$a_g = \gamma_I \cdot a_{gR} \tag{12}$$

The reference-top level ground acceleration at a coefficient of importance of $\gamma_I = 1.0$ corresponds to a reference exceeding probability of $P_{NCR} = 10\ \%$ within 50 years or a reference recurrence period of $T_{NCR} = 475$ years.

## 3.2 Categories of importance

Failure of a structure, its impacts on human life, public safety and social an economic effects are defined by means of the categories of importance and the related coefficients of importance $\gamma_I$. In case of structural design, structures with a low importance for public safety are category I (e.g. structures with an agricultural using, $\gamma_I = 0.8$), common structures are category II, structures where gatherings have to be considered (e.g. school buildings) are category III and the most important structures are classified in category IV (e.g. hospitals) both categories have a range of $\gamma_I = 1.0$ to 1.4.

## 3.3 Foundation classes

Foundation has an applicable influence on the exposure of a structure to seismic loads. Foundation can be divided in classes A – E, $S_1$ and $S_2$, according to EC 8. The classification of the local foundation should be done considering the shear wave velocity $v_{s30}$, if the value is known; otherwise the classification should be done with the number of blows of the Standard-Penetration-Test, $N_{SPT}$. Additional investigations to the required static analysis are necessary, if the local foundation is classified to $S_1$ or $S_2$, or if the structure has the category of importance III or IV.

## 3.4 Response spectrum

The dynamic impact of an earthquake on a structure is generally characterised by a horizontal response spectrum. Thereby on the abscissa it is plotted the natural oscillation time $T$ and on the ordinate the maximal amplitude of the response acceleration of a planar single degree of freedom, which has a constant natural oscillation time over the duration of the seismic impacts. The response spectrum is timely independent and depicts the smoothing and enclosing distribution of many earthquakes.

### 3.4.1 Horizontal-elastic response spectrum

The horizontal seismic impact can be described by means of two orthogonal components, which are independent from each other and can be characterised by the same response spectrum. The horizontal component of the elastic response spectrum $S_a(T)$ is defined at 5 % viscous damping by four groups. In general there are two modes of spectra, type 1 and type 2. Type 1 is assumed for larger magnitudes of surface waves, $M_S > 5.5$ and type 2 for smaller magnitudes, $M_S \leq 5.5$. The parameters of the elastic response spectrum are listed in EC 8. In Fig. 7 the characteristics of the response spectrum are depicted for the different foundation classes.

The elastic acceleration spectrum $S_a(T)$, depending on the settling time $T$, can directly be transformed into an elastic displacement response spectrum $S_{De}(T)$ by following condition:

$$S_{De}(T) = S_a(T) \cdot \left[ \frac{T}{2\pi} \right]^2 \tag{13}$$

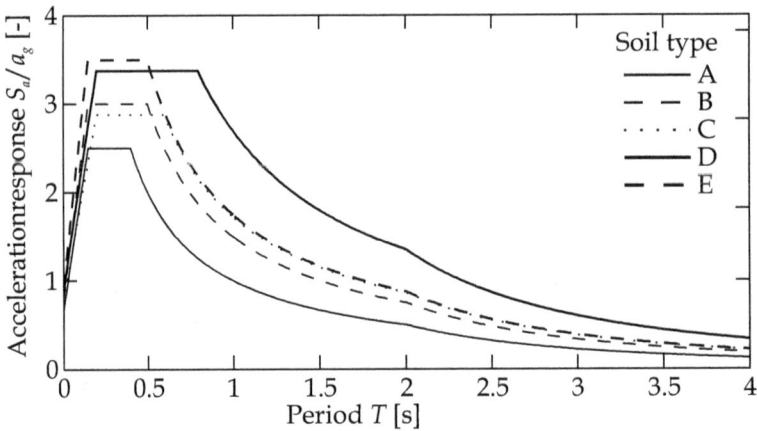

Fig. 7. Elastic response spectra of type 1 for foundation classes A – E with 5 % viscous damping

The correlation of spectral acceleration and displacement for a single degree of freedom with a predefined natural period is described by the ADR-spectrum (Acceleration Displacement Response), whereby the abscissa shows the time of displacement response $S_{De}(T)$ and the ordinate the acceleration response $S_a(T)$. For periods exceeding 4.0 s, the elastic acceleration response spectrum of type 1 can be obtained from the elastic displacement response spectrum by inverting eq. (13), according to EC 8, Annex A.

### 3.4.2 Vertical-elastic response spectrum
The vertical component of seismic impact can be neglected in Austria and is therefore listed only for ensuring completeness. It can be described by the vertical elastic response spectrum $S_{ve}(T)$. Just as the horizontal spectrum, the vertical spectrum can be divided in four groups, whereby the maximal values are considerably smaller. In general the dominant impact is the horizontal response spectrum.

### 3.4.3 Design value of ground displacement
The design value of ground displacement $d_g$ can be estimated in accordance with the design value of the ground acceleration as follows:

$$d_g = 0.025 \cdot a_g \cdot S \cdot T_C \cdot T_D \qquad (14)$$

### 3.4.4 Design spectrum
Observations of earthquake impacts have shown that structures can reduce seismic impacts by nonlinear reactions. To use the advantages of a linear calculation, the energy dissipation is considered by a reduced response spectrum (= design spectrum). The ratio of the estimated maximal exposure to the real appearing lower exposure is defined by the behaviour factor $q$. This coefficient is an approximation of the ratio of the seismic forces which would act on the structure if the response at 5 % viscous damping would be completely elastic and of those forces which can be used for a barely satisfactory linear design. Table 4 gives an overview of the behaviour factor $q$.

| Material | Range | Relevant code |
|---|---|---|
| concrete | 1.50 – 5.85 | EC 8 Table 5.1 |
| steel | ≤ 1.50 – 6.50 | EC 8 Table 6.1 resp. Table 6.2 |
| bond between concrete and steel | ≤ 1.50 – 6.50 | EC 8 Table 6.2 resp. Table 7.2 |
| wood | 1.50 – 5.00 | EC 8 Table 8.1 |
| masonry | 1.50 – 3.00 | EC 8 Table 9.1 |

Table 4. Behaviour factor $q$ according to EC 8

The four groups of the horizontal design spectrum can be defined as follows:

$$0 \le T \le T_B : S_d(T) = a_g \cdot S \cdot \left[ \frac{2}{3} + \frac{T}{T_B} \cdot \left( \frac{2.5}{q} - \frac{2}{3} \right) \right]$$

$$T_B \le T \le T_C : S_d(T) = a_g \cdot S \cdot \frac{2.5}{q}$$

$$T_C \le T \le T_D : S_d(T) \begin{cases} = a_g \cdot S \cdot \frac{2.5}{q} \cdot \left[ \frac{T_C}{T} \right] \\ \ge \beta \cdot a_g \quad with \ \beta = 0.2 \end{cases} \tag{15}$$

$$T_D \le T \le 4s : S_d(T) \begin{cases} = a_g \cdot S \cdot \frac{2.5}{q} \cdot \left[ \frac{T_C T_D}{T^2} \right] \\ \ge \beta \cdot a_g \quad with \ \beta = 0.2 \end{cases}$$

in which $S_d(T)$ = elastic design response spectrum $a_g$ = design ground acceleration according eq. (12), $T$ = settling time, $S$ = soil parameter, $T_B$, $T_C$, $T_D$ = period, $q$ = behaviour factor and $\beta$ = lower bound of spectrum, recommended value is 0.2.

If in eq. (15) the vertical component $a_{vg}$ is used instead of the design ground acceleration $a_g$ and $S = 1$, all four groups of the vertical design spectrum can be defined. The coefficient of behaviour should not exceed 1.5 for masonry.

### 3.4.5 Alternative Interpretation of seismic impacts

Alternatively, seismic impacts can be considered by means of natural or simulated time dependent distributions of acceleration. In spatial models of a structure there have to be considered three simultaneous time dependent distributions of acceleration, whereby the same distribution may not be used for both horizontal directions. Simulated distributions of acceleration have to be generated in a way, that the response spectra describe the elastic response spectra of Sec. 3.4.1 and Sec. 3.4.2 for 5 % viscous damping. The duration of the distributions of acceleration has to be consistent to the characteristics of the earthquake, which refers to $a_g$ and to the earthquake's magnitude. If there is no further information available, the minimum duration of the stationary part of the distribution of acceleration should be considered with 10 s. Moreover it should be used a minimum of three distributions of acceleration, whose average for the zero period yields at least to a value of $a_g S$ for the considered location. No ordinal value in the range of $0.2 T_1$ up to $2.0 T_1$ shall be smaller than 90 % of the corresponding value of the design spectrum, whereby $T_1$ is the natural period of the structure in the related direction.

### 3.5 Seismic design

Seismic loads are inertial loadings, which act on the mass points and they are the result of multiplying mass with acceleration. Hence the loading is not applied external on the structure, but is produced by ground movements and deformations in the structure itself. Therefore seismic loading depends on both the place of location and on the structure itself. In earthquakes prone areas, the seismic aspects have to be considered already in the conceptual state of designing structures, therewith both the requirements on structural safety and minimizing failure effects can be fulfilled with bearable costs, compare with EC 8 Part 2.1, (Bachmann, 2002a, 2002b).

### 3.5.1 General principles

Generally, structures have to be designed from the constructive aspect as easy as possible for ensuring a definite and direct load transfer, avoiding uncertainties in modelling and increasing the safety of the structure. In buildings, the floor slabs have a decisive significance. In the plane of the slab the inertial forces are bounded and transferred to the vertical members. Therefore the slabs should have a sufficient stiffness in their plane and the behaviour of horizontal panels. Particularly in case of mixed structures, large openings and changes of the stiffness, the behaviour of the panel and the interaction between horizontal and vertical structural members should be ensured.

For avoiding non-uniform torsional loading on bearing structural members it has to be ensured that stiffness is distributed preferably constant around the structure and a sufficient torsional stiffness is warranted. In addition to constructive aspects, an adequate foundation and connective elements to the superstructure are required for ensuring a homogenous seismic loading of the structure and the load transfer to the ground. Structures with load-bearing walls with diverging values of length and stiffness should have a box-shaped or cellular foundation; separate parts of the foundation should be connected with a ground slab or a flexible foundation beam.

### 3.5.2 Criteria of regularity

For purposes of seismic design of construction it has to be distinguished between regular and non-regular structures. This differentiation influences the calculation model, the calculation method and the behaviour factor, compare Table 5.

| Regularity | | Acceptable simplifications | | Behaviour factor |
|---|---|---|---|---|
| ground plan | vertical section | model | linear-elastic calculation | |
| Yes | Yes | plane | simplified | reference value |
| Yes | No | plane | modal | reduced value |
| No | Yes | spatial* | simplified | reference value |
| No | No | spatial | modal | reduced value |

* according EC 8 Part 4.3.3.1(8) under special circumstances an planar model can be used in each of the both directions

Table 5. Seismic design according to constructive regularity

A structure has to fulfil the following requirements that it can be classified as regularly in respect to the **ground plan**: The distribution of the horizontal stiffness and mass in accordance to two perpendicular axes should be symmetric. The shape of the ground plan has to be compact and should be enhanced by a polygon line, offsets and niches must not contain more than 5 % of the floor area. Comparing to the stiffness of the horizontal members, the slabs have to assure a sufficient stiffness in their plane to ensure the load transfer. Slenderness ratio $\lambda$ has to fulfil $L_{max}/L_{min} \leq 4$, in which $L_{max}$ and $L_{min}$ are the maximum and minimum perpendicular dimensions of the structure.

For each floor and in each direction of calculation the effective excentricity has to be in x-direction $e_{0x} \leq 0.30\ r_x$ and $r_x \geq l_s$ and in y-direction $e_{0y} \leq 0.30\ r_y$ and $r_y \geq l_s$ respectively, in which $e_o$ is the distance between centre of stiffness and centre of mass, $r$ is the radius of torsion and $l_s$ the radius of inertia of the floor mass. In case of one floor the radius of torsion is defined as the square root from the torsion stiffness in reference to the horizontal stiffness. The radius of inertia is defined as the square root of the polar moment of inertia in reference to the centre of mass. In case of several floors, the centre of stiffness and the radius of torsion can be determined just approximately. Simplifying it can be regarded as regular, if load-bearing structural members range from the foundation up to the top edge of the structure and if the bending lines of the stiffening systems under horizontal loading are not different. An approximate approach for the calculation is defined in the national part of EC 8, Annex B.

That a structure can be determined as regularly in respect to the **vertical section**, it has to conform to the requirements below. All horizontally operating stiffening systems have to range from the foundation up to the top edge of the structure, respectively up to the adequate height of structural members. Horizontal stiffness and the mass of the respective floors have to be constant over their height or decrease steadily from the bottom to the top. In case of frame structures, the ratio of the real strength of proximate floors to the required strength from the calculation should not diverge too much. If there are offsets, the conditions from Fig. 8 must be considered.

The offsets have to be designed symmetrically and may not exceed more than 20 % of the previous dimensions. In case of a single offset within the lower 15 % of the total height, the offset may not exceed 50 % of the ground plan dimension. Then the continuous part below should be able to bear at least 75 % of the total horizontal shear load. In the event of asymmetric offsets, in each vertical section the sum of offsets may not exceed 30 % of the dimensions of the ground plan and each offset may not be larger than 10 % of the previous dimension, see Fig. 8. In general, the requirements of EC 8 with regard to regularity in ground plan and vertical section can be summarized, that a compact construction type with symmetric distributed mass and stiffness has a positive influence on the seismic loadbearing capacity, compare Fig. 9.

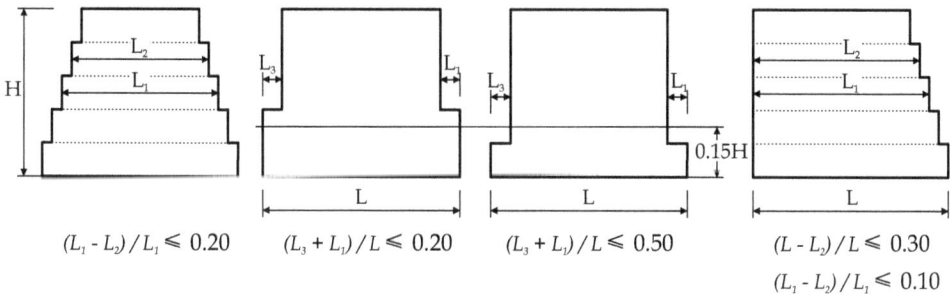

$$(L_1 - L_2)/L_1 \leq 0.20 \qquad (L_3 + L_1)/L \leq 0.20 \qquad (L_3 + L_1)/L \leq 0.50 \qquad (L - L_2)/L \leq 0.30$$

$$(L_1 - L_2)/L_1 \leq 0.10$$

Fig. 8. Criteria of regularity of structures in reference to the vertical section, according to EC 8

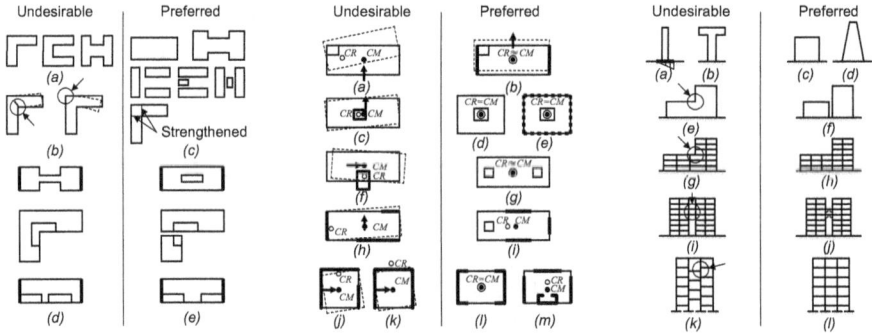

CM=centre of mass, CR=centre of stiffness

(a) horizontal setup        (b) distribution of mass and        (c) vertical setup
                            horizontal stiffness,

Fig. 9. Comparison of disadvantageous and favourable structural characteristics, from (Paulay & Priestley, 1996)

### 3.5.3 Coefficients of combination

In the seismic load case, variable loads have to be considered in a reduced way. Their value is defined by the factors of combination $\psi_{E,i} = \varphi.\psi_{2,i}$. Thereby $\psi_{2,i}$ is the value, which takes account of the quasi constant part, as recommended in Annex A1, EC 0. The factor $\varphi$ defines the probability of a simultaneous occurrence of the variable loads in each floor. In Austria it is recommended to use $\varphi = 1.0$, or the values from Table 6.

The masses, which are required for calculating the inertial forces, which have to be applied in case of seismic loads, result from the dead loads from the permanent loads and the variable loads, which are reduced by the factor $\psi_{E,i}$.

$$\sum G_{k,i} + \sum \psi_{E,i} \cdot Q_{k,i} \tag{16}$$

| Type of loading | Floor | $\varphi$ |
|---|---|---|
| Category A – C according EC 1 | housetop | 1.0 |
| | floors with an related usage | 0.8 |
| | independent usage of floors | 0.5 |
| Category D – F according EC 1 und archives | - | 1.0 |

Table 6. Values for the calculation of $\psi_{E,i}$

### 3.5.4 Computation of structures

Modelling has to ensure, that the distribution of masses and stiffness is described correctly, in case of nonlinear analysis the distribution of strength, too. The structural model should also take account of the connecting zones to the deformation of the structure, e.g. ending zones of beams or columns. In case of structures made of concrete, reinforced concrete, composite constructions or masonry, the stiffness of the load-bearing structural members should be determined by considering the formation of cracks. If no calculative analysis of the cracked members is available, approximately the half of the stiffness of the uncracked members can be assumed.

Bracing elements in masonry, which have an essential impact on the horizontal stiffness, should be considered. The ductility of the foundation has to be regarded, if it influences the model in a negative way and can optionally be regarded if there is a positive impact. The definitions of EC 8 recommend that the stiffness and the masses of the structure should be summarized to a substitute beam. Then the seismic loads are calculated for each floor and then redivided to the load-bearing structural members. This procedure presumes that the load distribution takes place by floor slabs with an adequate stiffness

Alternatively, the seismic loading can also be calculated directly at a spatial model, if the recommendations from EC 8 are used analogously. This model even allows the calculation of structures, which do not have a sufficient panel effect of the floor slabs. In Table 7 the calculation methods for the load case seismic loads are summarized.

| | Force method | Response spectrum | Pushover | Timehistory |
|---|---|---|---|---|
| Method | static | static | static | dynamic |
| Model | linear plane | linear plane, spatial | nonlinear plane, spatial | nonlinear plane, spatial |
| Torsion | simplified approach | plane: simplified approach, spatial: in model | plane: simplified approach, spatial: in model | plane: simplified approach, spatial: in model |
| Considering nonlinearity | global by coefficient of behaviour | global by coefficient of behaviour | in model | in model |
| Load | response spectrum | response spectrum | response spectrum | time response |
| Calculation | analysis with static resultant forces | modal analysis with quadratic superposition of state variables | Pushover-calculation, continuous increasing external forces | at least three time response calculations with static analysis |
| Uncertainties | modelling, dynamic of the structure, material behaviour | modelling, dynamic of the structure, material behaviour | modelling, dynamic of the structure | modelling |
| Regularity | very high | plane: high spatial: none | plane: high spatial: none | plane: high spatial: none |
| Traceability | very easy | easy | easy | difficult |
| Utilisation of bearing reserves | low | low | good | very good |
| Computational effort | low | middle | high | very high |

Table 7. Overview about the calculation methods for seismic loads (Zimmermann & Strauss, 2010b)

### 3.5.5 Resultant force method

The resultant force method or the simplified response spectrum method can be applied, if the structure complies with the requirements of regularity in vertical section from Sec. 3.5.2 and if the natural period $T_1$ in each main direction is lower than $4\,T_C$ and 2 s respectively. Then it can be assumed, that the higher mode shapes have no influence on the total seismic load and that they can be neglected. The total seismic force $F_b$ for each main direction is calculated by:

$$F_b = S_d(T_1) \cdot m \cdot \lambda \tag{17}$$

in which $S_d(T_1)$ = ordinate of the design spectrum, $m$ = total mass of the structure and $\lambda$ = correction coefficient for the participation of the mass. For more than two floors and $T_1 \leq 2\,T_C$ it is recommended $\lambda = 0.85$, otherwise $\lambda = 1.0$. The natural period $T_1$ can be determined by an approximation procedure, e.g. the energy method (Flesch, 1993). For buildings up to a height of 40 m, the following approach can be used:

$$T_1 = C_t \cdot H^{3/4} \tag{18}$$

in which $C_t = 0.085$ for flexural resistant steel frameworks, $C_t = 0.075$ for frameworks made of reinforced concrete and $C_t = 0.050$ for all other structures; $H$ = height of the structure.
Alternatively, in case of structures with shear walls made of concrete or masonry, the value for $C_t$ can be defined as follows:

$$C_t = 0.075 \Big/ \left( \sqrt{\sum \left[ A_i \cdot \left( 0.2 + \left( l_{w,i} / H \right)^2 \right) \right]} \right) \tag{19}$$

in which $A_i$ = effective cross section of the shear wall $i$ and $l_{w,i}$ = length of the shear wall $i$, under the condition of $l_{w,i}/H \leq 0.9$.
The distribution of the horizontal seismic loads is based on the mode shapes or can be assumed as a triangular distributed load over the height, if the horizontal displacement of the eigenmode is approximated to be linear over the height.

$$F_i = F_b \cdot \frac{s_i \cdot m_i}{\sum s_j \cdot m_j} \qquad F_i = F_b \cdot \frac{z_i \cdot m_i}{\sum z_j \cdot m_j} \tag{20}$$

Thereby $F_i$ = the applying horizontal load on floor $i$, $s_i$, $s_j$ = displacement of the masses $m_i$, $m_j$ and $z_i$, $z_j$ = height of the masses. If the horizontal seismic loads are calculated as loads of the floors, the assumption of load transfer by rigid floor panels has to be fulfilled. In case of the separate structural members, the additional loading resulting from accidentally torsion load has to be considered with a coefficient $\delta = 1 + 0.6(x/L_e)$, which is a multiplying factor for the seismic load. Thereby $x$ = distance from the structural member from centre of mass and $L_e$ = distance between the outside structural members perpendicular to the considered direction of seismic impact.

### 3.5.6 Multimodal response spectrum

If the criteria of regularity in respect of vertical section are not fulfilled and other modal shapes than the natural eigenmode are decisive, the multimodal response spectrum method has to be applied instead of the simplified response spectrum method, see Sec. 3.5.2. This method can be used for each type of structure. In this dynamic calculation method the whole structure is divided into individual single degree of freedoms and the reaction under applied dynamic load is identified for each single degree of freedom $i$ with a natural period $T_i$. This reaction can be determined from the design spectrum, the shear force $F_{b,i}$ is then:

$$F_{b,i} = S_d(T_i) \cdot m_{i,eff} \tag{21}$$

in which $S_d(T_i)$ = ordinate of the design spectrum for the natural period $T_i$ and $m_{i,eff}$ = effective modal mass of the $i$-th eigenmode.

All decisive mode shapes have to be considered, which have a significant impact on the structural response. An eigenmode is decisive, if the sum of the effective modal mass is at least 90 % of the total mass. In addition, no eigenmode may be neglected, which modal mass has more than 5 % of the total mass. If these requirements cannot be fulfilled, the number $k$ of the modal inputs which have to be taken into account should be at least k ≥ 3 $n^{0.5}$; thereby $n$ = number of floors.

The period of the last eigenmode $T_k$ which should be considered may not exceed 0.2 s. For each eigenmode $i$ the maximal loading values can be determined. As a result, the singular modal parts can be added to the reaction of the whole structure. If all decisive mode shapes can be assumed as independent from each other, the maximal value of the seismic loading follows with the SRSS-formula (Square Root of Sum of Squares):

$$E_E = \sqrt{\sum E_{E,i}} \tag{22}$$

in which $E_E$ = seismic load value and $E_{E,i}$ = seismic load value of the $i$-th eigenmode. Mode shapes are oscillating independently from each other, if the difference between the singular eigenfrequencies is large enough. This is fulfilled, if for two consecutive periods $i$ and $j$ $T_j \le$ 0.9 $T_i$. If this condition is not reached, other methods for combination have to be applied, e.g. the CQC-method (Complete Quadratic Combination). Thereby the load value is:

$$E_E = \sqrt{\sum \sum E_{E,i} \cdot \rho_{ij} \cdot E_{E,j}} \tag{23}$$

in which $\rho_{ij}$ = factor of interaction. The factor of interaction takes account of the modal damping in reference of the mode shapes $i$ and $j$ and the ratio of the circular eigenfrequencies $\omega_i$ and $\omega_j$. This method is documented e.g. in (Clough & Penzien, 1995; Flesch, 1993). Torsion loadings at spatial models can be incorporated by additional torsional moments $M_{a,i}$ around the vertical axis of each floor $i$:

$$M_{a,i} = e_{a,i} \cdot F_i \qquad with \ e_{a,i} = \pm 0.05 \cdot L_i \tag{24}$$

in which $F_i$ = horizontal force in floor $i$, $e_{a,i}$ = accidental excentricity of floor mass $i$ and $L_i$ = floor dimension perpendicular to the direction of seismic load.

### 3.5.7 Nonlinear static method (Pushover)

Alternatively to response spectrum methods, nonlinear methods, e.g. the nonlinear static pushover method can be applied (Chopra & Goel, 1995). The characterisation of the material behaviour has to be done with a bi-linear force-deformation-relation. For concrete and masonry, the linear-elastic stiffness of the bi-linear relationship should coincide with that from cracked cross sections. For ductile members the secant stiffness to the yield point should be taken for the bi-linear relation. After the yield point, the tangential stiffness can be appropriated. In case of brittle materials, the tangential stiffness of the force-deformation-relation should be considered. If there are no further specifications, the material characteristics should be based on average values. For new structures, the material parameters can be taken from the codes EC 2 – EC 6 or from other appropriate European standards.

At nonlinear static calculation, the horizontal loads are increased monotonous under constant dead loads and the gained load-deformation-curves of the singular load-bearing structural members are superposed. As a result the capacity curve of the structure is achieved. This calculation method can be used for both determining the bearing capacity of existing and of new structures (Chopra, 2002; Clough & Penzien, 1995). Instead of the behaviour factor $q$, which incorporates the energy dissipation and the nonlinear effects in a global way, the real nonlinear material behaviour is considered. Depending on the criteria of regularity, in the calculation either two plane models for each of the both horizontal main directions are set up or a spatial model is used. In case of low masonry structures ($\leq$ 3 floors), whose load bearing walls are loaded mainly by shear loads, each floor can be considered singularly. For distributing the horizontal loads two approaches should be applied, on the one hand a modal, and on the other hand a mass-proportional distribution of the horizontal loads, referring to eq (20).

The horizontal loads have to be applied in in the centres of mass, whereby accidental excentricities according eq (24) have to be considered. From the nonlinear static calculation, the curve of capacity of the structure has to be defined in a range of 0 – 150 % of the aimed displacement, whereby the control displacement of the capacity curve can be assumed in the centre of mass of the top of the structure. The aimed displacement is determined by the displacement of an equivalent single degree of freedom. The method for determining the aimed displacement is regulated in EC 8, Annex B.

### 3.5.8 Nonlinear dynamic calculation (Timehistory)

Seismic loads can also be determined by means of simulated or measured time responses of the ground acceleration. Solving eq. (25) yields to the variation in time of the responded oscillations of the system in the considered degrees of freedom (Chopra, 2002; Clough & Penzien, 1995).

$$[M]\{\ddot{x}\} + [C]\{\dot{x}\} + [K]\{x\} = \{f(t)\} \tag{25}$$

Thereby $\ddot{x}$ = acceleration, $\dot{x}$ = velocity and $x$ = displacement vector, $f(t)$ = load vector, $\mathbf{M}$ = mass matrix, $\mathbf{C}$ = damping matrix and $\mathbf{K}$ = stiffness matrix. The time responses of all decisive parameters have to be quantified separately, because the maximum values of the displacement parameters $x_j(t)$ do not appear at the same time. The structural response oscillation depends on the characteristics of the applied variation of time and therefore at least three time responses have to be considered. For solving the differential equation system in eq. (25), (a) the modal method for linear systems, or (b) the direct integration method for linear and nonlinear systems can be used.

By means of the **modal method** the linked differential equation system is decoupled by transformation of the variables. As a result for linear elastic systems the displacements can be described as a linear combination of the mode shapes. Solving the decoupled differential equation yields to the time response of the $i$-th modal response oscillation. An advantage in contrast to the response spectrum method is that the maximal response can be determined more accurately and in terms of the direct integration the calculation effort decreases. The main disadvantage is that only linear material behaviour can be incorporated.

The method of the **direct integration** solves eq. (25) directly with the aid of a numeric integration. As a result of this, variable constitutive equations and damping mechanisms (apart from Rayleigh-damping) can be regarded. The main disadvantage of this method is the huge calculation effort, and the need of adequate numerical models for the description of the nonlinear material behaviour under cyclic loading.

The scope of the two described time-history methods is primarily in the assessment of existing structures. Further information is given in (Bachmann, 2002b; Chopra, 2002; Clough & Penzien, 1995; Flesch, 1993).

## 4. Damage quantification

The quantification of damage is an important task for evaluating the condition of the structure and the degradation over time caused by loads and/or environmental impacts. Damage indexes can be used for structural assessment and further such indexes can be used for decision making of repair and of demolition respectively. Additionally different cost factors can be considered for the decision process and life cycle assessment of structures (Frangopol et al., 2009; Strauss et al. 2010; Strauss et al., 2008).

Damage indexes are mathematical models for a quantitative assessment and they are of substantial importance to estimate critical condition states of structures. The calculation of damage indexes and different studies can be found in the literature, e.g. (Fajfar, 1992; Cosenza et al., 1993; Moustafa, 2011).

Damage indexes can also be correlated with experimental test results. If a structure is subjected to repeated pseudo-dynamic load reversals the increasing degree of damage can be described by damage indexes (Tomazevic, 1998).

## 5. References

Bachmann, H. (2002a). *Erdbebengerechter Entwurf von Hochbauten - Grundsätze für Ingenieure, Architekten, Bauherren und Behörden*, Richtlinien des BWG, Bern.

Bachmann, H. (2002b). *Erdbebensicherung von Bauwerken*. Birkhäuserverlag, Basel-Boston-Berlin, ISBN: 3-7643-6941-8

Bakes, H.-P. (1983). *Zugfestigkeit von Mauerwerk und Verformungsverhalten unter Zugbeanspruchung*, TU Aachen

Bargmann, H. (1993). *Historische Bautabellen*, Werner Verlag

Benedetti, D.; Carydis, P. & Pezzoli, P. (1998). *Shaking table test on 24 simple masonry buildings*, Earthquake Engineering and Structural Dynamics, Vol. 27, pp. 67-90

Conrad, D. (1990). Kirchenbau im Mittelalter - Bauplanung und Bauausführung. 1. Auflage Leipzig

Cosenza, C.; Manfredi, G. & Ramasco, R. (1993). *The use of damage functionals in earthquake engineering: comparison between different methods*, Earthquake Engineering & Structural Dynamics, Vol. 22, pp. 855-868

Chopra, A. (2002). *Dynamics of structures: Theory and Application to Earthquake*, Prentic-Hall Inc. A Simon & Schuster Company, Englewood Cliffs, NJ, USA, 3rd Edition.

Chopra, A. & Goel, R.K. (1995). *Capacity-Demand-Diagram Methods for Estimating Seismic Deformations of Inelastic Structures: SDF Systems*, Technical Report, Pacific Earthquake Engineering Research Centre, UCA Berkeley, CA, USA.

Clough, R.W. & Penzien, J. (1995). *Dynamics of structures*, Computer & Structures Inc. Berkeley, CA, USA, 3rd Edition.

de Vekey, R. (1988). *General recommendations for methods of testing load-bearing unit masonry*, Materials and Structures, Vol. 21, No. 3, pp. 229-231

Eberhard, M.O.; Baldridge, S.; Marshall, J.; Mooney, W. & Rix, G.J. (2010). *The MW 7.0 Haiti earthquake of January 12, 2010*, USGS/EERI Advance Reconnaissance Team report: U.S. Geological Survey Open-File Report 2010–1048, available online: http://pubs.usgs.gov/of/2010/1048/

Egermann, R. & Mayer, K. (1987). *Die Entwicklung der Ziegelherstellung und ihr Einfluss auf die mechanischen Eigenschaften von Mauerziegeln*, In: *Erhalten historischer Bauwerke*, Sonderforschungsauftrag 315, Universität Karlsruhe, Ernst & Sohn, Berlin 1987, pp. 107-130

ESECMaSE – Forschungsprojekt: Enhanced Safety and Efficient Construction of Masonry Structures in Europe, *http://www.esecmase.org*

Fajfar, P. (1992). *Equivalent ductility factors, taking into account low cyclic fatigue*, Earthquake Engineering & Structural Dynamics, Vol. 21, pp. 837-848

Flesch, R. (1993). *Baudynamik: praxisgerecht*, Bauverlag, Wiesbaden – Berlin, ISBN: 3-7625-3010-6

Frangopol, D.M.; Strauss, A. & Bergmeister, K. (2009). *Lifetime cost optimization of structures by a combined condition-reliability approach*, Engineering Structures, Vol. 31, No. 7, pp. 1572-1580

Furtmüller, T. & Adam, C. (2009). *Numerische Simulation des seismischen Verhaltens von Mauerwerk in Gründerzeithäusern*, SEISMID: Report No. 06/2319-23, TU Universität Innsbruck

Glitzka, H. (1988). *Druckbeanspruchung parallel zur Lagerfuge*, In: *Mauerwerk Kalender 1988*, Ernst & Sohn

Grimm, H. D. (1989). *Historische Mörtel - Auswertung von Untersuchungsergebnissen hinsichtlich historischer und geographischer Herkunft*, Technische Universität Braunschweig.

Hilsdorf, H. K. (1965). *Untersuchungen über die Grundlagen der Mauerwerksfestigkeit*, Materialprüfamt für das Bauwesen der TU München, Report No. 40

Ingham, J. & Griffith, M. (2011). *Performance of unreinforced masonry buildings during the 2010 darfield (Christchurch, NZ) earthquake*, Australian Journal of Structural Engineering, Vol. 11, No. 3, pp. 207-224

Ingham, J.; Biggs, D.T. & Moon, L.M. (2011). *How did unreinforced masonry buildings perform in the February 2011 Christchurch earthquake?*, Structural Engineer, Vol. 89, No. 6, pp. 14-18.

Knox, M. & Ingham, J. (2011). *Experimental Testing to Determine Failure Patterns in URM Pier/Spandrel Sub-Structures*, In: *Proceedings of 9th Australasian Masonry Conference*, Queenstown, New Zealand, February 2011

Mann, W. & Müller, H. (1978). *Mauerwerk Kalender 1978*, Ernst & Sohn

Maier, J. (2002). *Handbuch historisches Mauerwerk*, Birkhäuser

Moustafa, A. (2011). *Damage-Based Design Earthquake Loads for Single-Degree-Of-Freedom Inelastic Structures,* Journal of Structural Engineering, Vol. 137, No. 3, pp. 456-467

Müller, F. P. & Keintzel, E.: (1984). *Erdbebensicherung von Hochbauten,* Ernst & Sohn, Berlin.

Paulay, T. & Priestley, M.J.N. (1996). *Seismic Design of Reinforced Concrete and Masonry Buildings,* John Wiley & Sons Inc.

Pech, A. (2010). *Forschungsprogramm zur Verifizierung der konstruktiven Kennwerte von altem Vollziegelmauerwerk nach EC 6,* Testreport MA39 – VFA 2009-1396.01

Pech, A. & Zach, F. (2009). *Mauerwerksdruckfestigkeit – Bestimmung bei Bestandsobjekten,* Mauerwerk Vol. 13, No. 9, pp. 135-139, Ernst & Sohn

Schäfer, J. & Hilsdorf, H. K. (1990). *Historische Mörtel in historischem Mauerwerk,* In: *Erhalten historischer bedeutsamer Bauwerke,* Ernst & Sohn, Berlin

Schubert, P. & Brameshuber, W. (2011). *Eigenschaften von Mauersteinen, Mauermortel, Mauerwerk und Putzen,* In: *Mauerwerk Kalender 2011,* Ernst & Sohn, Berlin

Strauss, A.; Frangopol, D.M. & Kim, S. (2008). *Use of monitoring extreme data for the performance prediction of structures: Bayesian updating,* Engineering Structures, Vol. 30, No. 12, pp. 3654-3666

Strauss, A.; Frangopol, D. & Bergmeister, K. (2010). *Assessment of Existing Structures Based on Identification,* Journal of Structural Engineering, Vol. 136, No. 1, pp. 86-97

SEISMID – Forschungsprojekt: Seismic System Identification, *http://www.seismid.com*

Takewaki, I.; Murakami, S.; Fujita, K.; Yoshitomi, S. & Tsuji, M. (2011). *The 2011 off the Pacific coast of Tohoku earthquake and response of high-rise buildings under long-period ground motions,* Soil Dynamics and Earthquake Engineering, In Press

Tomazevic, M.; Lutman, M. & Weiss, P. (1996a). *Seismic Upgrading of Old Brick-Masonry Urban Houses: Tying of Walls with Steel Ties,* Earthquake Spectra, Vol. 12, No. 3, pp. 599-622

Tomazevic, M.; Lutman, M. & Petkovic, L. (1996b). *Seismic Behaviour of Masonry Walls: Experimental Simulation,* Journal of Structural Engineering Vol. 122, pp. 1040-4047

Tomazevic, M. (1998). *Correlation between damage and seismic resistance of masonry walls and buildings,* Bernardini, Alberto (ed.): *Seismic damage to masonry buildings: Proceedings of the International workshop on measures of seismic damage to masonry buildings,* Monselice, Padova, Italy, June 1998, Rotterdam; Brookfield: A. A. Balkema, pp. 161-167, 1999

Tomazevic, M. (2007). *Damage as measure for earthquake-resistant design of masonry structures: Slovenian experience,* Canadian Journal of Civil Engineering Vol. 34, pp. 1403-1412

Wisser, S. & Knöfel, D. (1987). *Untersuchungen an historischen Putz- und Mauermörteln. T.1 - Analysengang,* In: *Bautenschutz + Bausanierung* Vol. 10, No. 3, pp. 124-126

Zimmermann, T. & Strauss, A. (2010a). *Leistungsfähigkeit von alten Ziegelmauerwerk in Bezug auf zyklische Belastung,* Bauingenieur Vol. 85, pp. S2-S9

Zimmermann, T. & Strauss, A. (2010b). *Gründerzeit Mauerwerk unter Erdbebenbelastung - Vergleich zwischen normativen Ansätzen und messtechnischen Ergebnissen,* Bautechnik Vol. 87, No. 9 pp. 532-540

Zimmermann, T.; Strauss, A. & Bergmeister, K. (2010a). *Numerical investigations of historic masonry walls under normal and shear load,* Construction and Building Materials Vol. 24, No. 8, pp. 1385-1391

Zimmermann, T.; Strauss, A.; Lutman, M. & Bergmeister, K. (2010b). *Stiffness Identification and Degradation of Masonry under Seismic Loads*, In: *Proceedings of 8th IMC*, Dresden, Germany, July 2011

Standards (all standards are published by Austrian Standards, Austria and available online via: *www.as-search.at*)

EN 771-1 (2011). *Specifications for masonry units – Part 1: Clay masonry units*

EN 772-1 (2011). *Methods of test for masonry units – Part 1: Determination of compressive strength*

EN 772-13 (2000). *Methods of test for masonry units – Part 13: Determination of net and gross dry density of masonry units*

EN 772-16 (2005). *Methods of test for masonry units – Part 16: Determination of dimensions*

EN 1015-11 (2007). *Methods of test for mortar for masonry – Part 1: Determination of flexural and compressive strength of hardened mortar*

EN 1052-1 (1999). *Methods of test for masonry – Part 1: Determination of compressive strength*

EN 1052-2 (1999). *Methods of test for masonry – Part 2: Determination of flexural strength*

EN 1052-3 (2007). *Methods of test for masonry – Part 3: Determination of initial shear strength*

EN 1052-5 (2005). *Methods of test for masonry – Part 5: Determination of bond strength*

EN 1990 (2006). *Eurocode 0 – Basis of structural design*

EN 1991-1-1 (2011). *Eurocode 1 – Actions on structures – Part 1-1: General actions – Densities, self-weight, imposed loads for buildings*

EN 1996-1-1 (2006). *Eurocode 6 – Design of masonry structures – Part 1-1: General rules for reinforced and unreinforced masonry structures*

EN 1998-1 (2011). *Eurocode 8 – Design of structures for earthquake resistance – Part 1: General rules, seismic actions and rules for buildings*

EN 1998-3 (2005). *Eurocode 8 – Design of structures for earthquake resistance – Part 3: Assessment and retrofitting of buildings*

# 7

# The Equivalent Non-Linear Single Degree of Freedom System of Asymmetric Multi-Storey Buildings in Seismic Static Pushover Analysis

Triantafyllos K. Makarios

*Hellenic Institute of Engineering Seismology & Earthquake Engineering,*
*Greece*

## 1. Introduction

In order to estimate seismic demands at low seismic performance levels (such as life safety or collapse prevention), the application of Non-Linear Response History Analysis (NLRHA) is recommended for reasons related to its accuracy. However, for the purposes of simplification, the application of Pushover Analysis is also often recommended. Here, in order to obtain the seismic demands of asymmetric multi-storey reinforced concrete (r/c) buildings, a new seismic non-linear static (pushover) procedure that uses inelastic response acceleration spectra, is presented. The latter makes use of the optimum equivalent Non-Linear Single Degree of Freedom (NLSDF) system, which is used to represent a randomly-selected, asymmetric multi-storey r/c building. As is proven below, for each asymmetric multi-storey building, a total of twelve suitable non-linear static analyses are required according to this procedure, while at least two hundred and twenty-four suitable non-linear dynamic analyses are needed in the case of NLRHAnalysis, respectively, while if accidental ecentricity is ignored (or external floor moments loads around the vertical axis are used) then a total number of fifty six NLRH Analyses are required. The seismic non-linear static procedure that is presented here is a natural extension of the documented seismic equivalent static linear (simplified spectral) method that is recommended by the established contemporary Seismic Codes, with reference to torsional provisions. From the numerical parametric documentation of this proposed seismic non-linear static procedure, we can reliably evaluate the extreme values of floor inelastic displacements, with reference to results provided by the Non-Linear Response History Analysis.

More specifically, it is well-known from past research that, as regards the pushover procedure (i.e. pushover analyses of buildings with inelastic response acceleration spectra), its suitability for use on asymmetric multi-storey r/c buildings is frequently questioned because of the following five points:

1.  For a known earthquake, an ideal equivalent NLSDF system, which represents the real asymmetric multi-storey r/c building, must first be defined, in order to calculate the seismic target-displacement. In other words, the above-mentioned r/c NLSDF system has an equivalent viscous damping ratio $\xi = 0.05$ and is combined with "inelastic acceleration spectra with an equivalent viscous damping ratio $\xi = 0.05$"; thus its demand seismic target-displacement is obtained for each earthquake level. The latter

displacement is transformed into a demand extreme seismic displacement of the monitoring point of the real Multi-Degree of Freedom (MDoF) system, namely the real multi-storey building. After that, the remaining demand seismic displacements (and seismic stress on the members) at the other points of the building are easily calculated using the pushover analysis image. Therefore, successfully defining the ideal equivalent NLSDF system is always the most important part of this procedure, and for this reason, it also constitutes a major concern. Many attempts have been made in the past to resolve this problem in the planar frames (Saiidi & Sozen, 1981; Fajfar & Fischinger, 1987a,b; Uang & Bertero, 1990; Qi & Moehle, 1991; Rodriquez, 1994; Fajfar & Gaspersic, 1996; Hart & Wong, 1999; Makarios, 2005;). On the other hand, in the case of asymmetric multi-storey buildings, a mathematically documented optimum equivalent Non-Linear Single Degree of Freedom System can be defined (Makarios, 2009). According to the latter proposal, three coupled degrees of freedom (two horizontal displacements and a rotation around a vertical axis that passes through the floor mass centre) are taken into account for each floor mass centre. The above-mentioned definition of the optimum equivalent NLSDF system was mathematically derived by observing suitable dynamic loadings on the floor mass centres of the multi-storey building, using simplified assumptions. The use of this optimum equivalent NLSDF system, in combination with the inelastic design spectra, provides an acceptable evaluation of the extreme (maximum/minimum) seismic floor displacements required for a known design earthquake, and for this reason constitutes the core issue of the present chapter.

2. It is known that in the case of Response History Analysis, three uncorrelated accelerograms (Penzien & Watabe, 1975;) are used simultaneously along the three axes $x$, $y$ & $z$. For this reason, in the case of Response Spectrum Analysis (RSA) or Simplified Spectral Method (SSM) or the equivalent static method, the three response acceleration spectra are "statistically independent" (sect.3.2.2.1(3)P of Eurocode EN-1998.01). This practically means that the response spectra (in the case of RSA), or the lateral floor static forces (in the case of SSM), must be inserted separately for each main direction of the building, and then a suitable superposition (i.e. rule of Square Root of Sum of Squares – SRSS) must be applied to the results of each independent analysis, always in the linear area. However, in the non-linear area, as in the example of static pushover analysis, where superposition is generally forbidden due to its non-linearity, a basic question arises; which is the most suitable way of taking the "spatial seismic action" of the three seismic components into account? In order to answer this question, the rule of Eq.(1) that has resulted from a parametric numerical analysis, provides an approximate evaluation of the spatial seismic action during the static pushover procedure (Makarios, 2011):

$$E_{,\text{ex}} = \pm\left(E_{,\text{I}}^{\kappa} + E_{,\text{II}}^{\kappa}\right)^{1/\kappa} \tag{1}$$

where $\kappa = 0.75$ is a mean value in the case of displacements/deformations, and $E_{,\text{I}}$ is the extreme seismic demand inelastic displacement/deformation due to static pushover analysis using the first seismic component only, along the horizontal real (or fictitious) principal *I*-axis of the building. Similarly, $E_{,\text{II}}$ is the extreme seismic demand inelastic displacement/deformation due to pushover analysis using the second seismic component only, the other horizontal real (or fictitious) principal *II*-axis of the building.

On the contrary, when $E_{,I}$ and $E_{,II}$ represent stress, then $\kappa = 2.00$, as is the case in the linear area. In Eq.(1) above, we consider that the vertical seismic component is often ignored (for example, when the vertical ground acceleration is less than $0.25g$, according to sect.4.3.3.5.2(1)/EN-1998.01; $g$ is the acceleration of gravity).

3.  What is the most suitable "monitoring point" and its suitable characteristic degree of freedom (control displacement) that is related to the degree of freedom of the equivalent NLSDF system, in the case of asymmetric buildings? As has been proven by the mathematical analysis provided below, concerning the definition of the optimum NLSDF system, the centre of mass of the last floor at the top of the building can play the role of the "monitoring point", whilst its "control displacement" must be parallel to the lateral floor static forces.

4.  What is the most suitable distribution of lateral floor static forces in elevation? There are various opinions on this matter, since the distribution can be "triangular", "uniform" or in accordance with the pure translational "fundamental mode-shape". Moreover, the distribution can be adapted to the fundamental mode-shape of the building in each step of the pushover analysis, an issue related to the action of higher order mode-shapes in tall buildings. According to sect.4.3.3.4.2.2(1)/EN-1998.01, at least two different distributions ("uniform" and "first mode-shape") in elevation should always be taken into account. On the other hand, fundamental mode-shape distribution is often applied to common asymmetric buildings, while in the case of very tall buildings, Non-Linear Response History Analysis is mainly reccommended.

5.  At which point in the plan should the lateral floor static forces be applied during the pushover analysis of asymmetric buildings? According to sect.4.3.3.4.2.2(2)P/EN-1998.01, the lateral static loads must be applied at the location of the mass centres, while simultaneously, accidental eccentricities should also be taken into account. Moreover, it is known that in the case of dynamic methods, such as "Response Spectrum Analysis" or "Response History Analysis", accidental eccentricities are also considered, with a suitable movement of the floor mass centres by $e_{ai} = \pm 0.05L_i$ or $\pm 0.10L_i$, where $L_i$ is the dimension of the building's floor that is perpendicular to the direction of the seismic component (sect.4.3.2(1)P/EN-1998.01). However, in the case of static seismic methods, such as the "equivalent static method", "simplified spectrum method", "lateral force method" and "static pushover analysis", the accidental eccentricities are examined with a suitable movement (by $e_{ai} = \pm 0.05L_i$ or $\pm 0.10L_i$) of the point at which the lateral floor static forces are applied. In the latter case, and for one seismic component, the final two design eccentricities ($\max e_{II,i}$ & $\min e_{II,i}$ for $i$ floor) consist of the following two parts: In order to take into account the dynamic amplification of the torsional effects of the building, two dynamic eccentricities $e_{f,i}$ & $e_{r,i}$ are examined (first part). According to several Seismic Codes (NBCC-95, EAK/2003), dynamic eccentricities are defined as $e_{f,i} = 1.50 \cdot e_u$ & $e_{r,i} = 0.50 \cdot e_u$, where $e_u$ is the static eccentricity of the building along the examined horizontal principal axis of the building, perpendicular to the direction of the seismic component. Dynamic eccentricities are measured from the real/fictitious stiffness centre of the building to its mass centre (CM); more accurate closed mathematical relationships of the former have been developed by Anastassiadis *et al* (1998). As we can observe, a calculation of (a) the location of the real/fictitious stiffness

centre of the asymmetric building in the plan, (b) the orientation of its horizontal principal directions $I$ & $II$ and (c) the magnitude of its torsional-stiffness radii, must be carried out before the seismic static pushover analysis begins. For this triple purpose, the fictitious elastic Reference Cartesian System $P_o(I, II, III)$ of the asymmetric multi-storey building is the most rational, documented method for use (Makarios & Anastassiadis 1998a,b; Makarios, et al 2006; Makarios, 2008;). On the contrary, it is worth noting that, various other floor centres, such as centres of rigidity/twist/shear, are not suitable for seismic design purposes, because they are dependent on external floor lateral static loads (Cheung & Tso, 1986; Hejal & Chopra, 1987;). At a next stage, the static eccentricities $e_{o,I}$ and $e_{o,II}$ of an asymmetric multi-storey building are defined as the distance between the floor mass centre (CM) and the real/fictitious stiffness centre $P_o$ of the building, along the two horizontal principal $I$ & $II$ –axes.

In relation to my previous point, and in order to account for uncertainties as regards the real location of floor mass centres, as well as the spatial variation of the seismic motion, accidental eccentricities are also examined (second part). However, it is well-known that according to documented Seismic Codes with reference to torsional provisions, the critical horizontal directions for static lateral loading (in linear equivalent static seismic methods) are the real/fictitious principal $I$ & $II$-axes of a building. Therefore, the documented static pushover procedure must define the same point; the lateral floor static loads that will be used in the static pushover analysis must be oriented along the real/fictitious principal $I$ & $II$-axes of the building. In this case, and since the final design eccentricities ($\max e$ & $\min e$ along the two principal $I$ & $II$-axes) have also been used, we can observe that this non-linear static pushover analysis is, undoubtedly, a natural extension of the linear equivalent static seismic method.

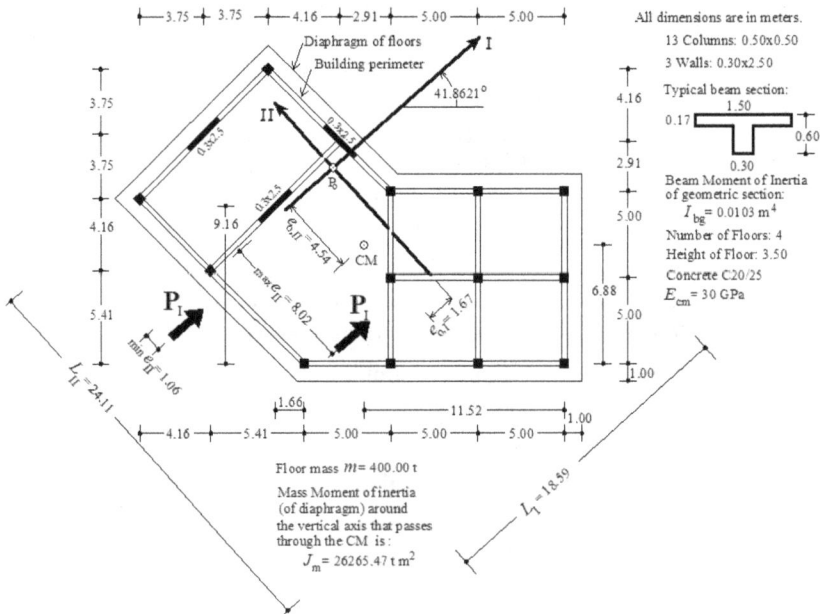

Fig. 1. Typical plan of an asymmetric four-storey r/c building.

For illustration purposes, we can see the numerical example of an asymmetric four-storey r/c building with a typical plan, whose real/fictitious principal axes are initially unknown (Fig.1). The location in the plan and the orientation of the $P_0(I,II,III)$ system have been calculated according to the above-mentioned relative references, using a 50% reduction of the stiffness properties of all structural elements, due to cracks (sect.4.3.1(7)/EN-1998.01). The position of all floor mass centres (CM) coincides with the geometric centre of the floor rigid diaphragms. Next, the static eccentricities are calculated as $e_{o,I} = 1.67$ & $e_{o,II} = 4.54$ metres, along the fictitious principal $I$ & $II$-axes, respectively. The two torsional-stiffness radii $r_I$ & $r_{II}$ are calculated as $r_I = 9.35$ & $r_{II} = 7.90$ metres on level $0.8H$ of the building (namely, on the 3rd floor; $H$ is the total height of the building) according to the theory of *Dynamics of Structures* (Makarios, 2008;), and along $I$ & $II$-axes, respectively. Note that, in the case of a single-storey building, the torsional-stiffness radius $r_I$ is for example calculated by the square root of the ratio of the torsional stiffness $k_{III}$ of the building around its Stiffness Centre to the lateral stiffness $k_{II}$ along principal $II$-axis (i.e. $r_I = \sqrt{k_{III}/k_{II}}$ , $r_{II} = \sqrt{k_{III}/k_I}$ ). Also, the radius of gyration of a typical floor rigid diaphragm is calculated as $l_s = \sqrt{J_m/m} = \sqrt{26265.47/400} = 8.10$ metres (see data in Fig.1). In order to calculate the final design eccentricities for loading $P_I$ along the principal $I$-axis of the building, we have:

$$\max e_{II} = e_{f,i} + e_{ai} = 1.50 \cdot e_{o,II} + 0.05 \cdot L_{II} = 1.50 \cdot 4.54 + 0.05 \cdot 24.11 = 8.02 ,$$

$$\min e_{II} = e_{r,i} + e_{ai} = 0.50 \cdot e_{o,II} - 0.05 \cdot L_{II} = 0.50 \cdot 4.54 - 0.05 \cdot 24.11 = 1.06$$

The envelope of the two individual static loading states, using the above final design eccentricities, produces the results $E_{,I}$ of the first seismic component (along principal $I$-axis). The results $E_{,II}$ of the second seismic component (along principal $II$-axis) are obtained in a similar way.

Please note the following important point: According to sect.4.2.3.2(6)/EN-1998.01, the building does not satisfy the criterion of regularity in plan, because $e_{o,II} = 4.54$ is not lower than $0.3 \cdot r_{II} = 0.3 \cdot 7.90 = 2.37$ . Furthermore, $r_{II} = 7.90$ is lower than $l_s = 8.10$ and, according to sect.4.3.3.1(8)d/EN-1998.01, the condition $r_{II}^2 > l_s^2 + e_{o,II}^2$ is not true. Consequently, the present building has to be simulated as a fully spatial model, according to both the elastic analysis (sect.4.3.3.1(7)/EN-1998.01) and the non-linear static (pushover) analysis (sect.4.3.3.4.2.1(2)P/EN 1998.01).

To conclude, the real/fictitious principal elastic Cartesian system $P_0(I,II,III)$ of the asymmetric multi-storey building is initially calculated. In order to calculate the real/fictitious principal elastic Cartesian system $P_0(I,II,III)$ of the asymmetric multi-storey r/c building, all structural elements (columns, walls, beams, cores) of the building model must have effective flexural and shear stiffness, constant along their whole length, corresponding to 50% of their geometric section values. This effective stiffness is more rational, since it leads to a realistic "total lateral stiffness" of the building in the post-elastic, non-linear area. However, if we use much lower values of effective stiffness (i.e. equal to $0.20EI$ or less), then an artificial increase of the eigen-periods occurs. In such a case, one possible result is that the building model is inadequately seismically loaded, because the state of co-ordination between the fundamental eigen-period of the building and the fundamental period of the actual seismic ground strong motion is removed. Existing views about the use of very small effective flexural stiffness values stem from the following assumption, that "...*plastic hinges appeared simultaneously at both ends of each structural element of the multi-storey building.*" However, this is rather incorrect, because many structural elements do not yield, even in the "near collapse state" of the building, as we can observe from the seismic NLRHA. On the other hand, cracks appeared on the plastic hinges of the r/c elements with yielding steel bars, while at the next moment, when the bending of the elements has the opposite sign, some cracks on the plastic hinges cannot fully close, because the yielded steel bars obstruct their closure. Additionally, in the frame of the static pushover analysis of irregular in the plan, asymmetric multi-storey buildings, a Cartesian system $CM(xyz)$ that has the same orientation with the known fictitious elastic system $P_o(I,II,III)$ of the building must be adopted as a global reference system. Thus, the lateral floor static loads, according to the static pushover analyses, must be oriented along the two horizontal principal $I$ & $II$–axes of the building. Otherwise, it is well-known from the linear area that it is not possible to calculate the envelope of the floor displacements/stress of an asymmetric building. In other words, the loading along the principal axes of a building creates the most unfavourable state for static methods. Moreover, as a general conclusion, final design eccentricities ($\max e_I$, $\min e_I$, $\max e_{II}$ & $\min e_{II}$) must always be taken into account in pushover analysis, and thus the present non-linear static pushover procedure is a natural extension of the equivalent linear static method, with mathematical consistency.

With reference to the simulation for the non-linear analysis, we consider that each structural column/beam consists of two equivalent sub-cantilevers (Fig.2a,b). Each sub-cantilever is represented by an elastic beam element with a non-linear spring at its base. This inelastic spring represents the inelastic deformations that are lumped at the end of the element. In order to identify the characteristics of a plastic hinge, it is assumed that each member (beam/column) deforms in antisymmetric bending. The point of contraflexure is approximately located at the middle of the element and for this reason, two sub-cantilevers appear for each member. Therefore, each sub-cantilever has a "shear length" $L_{s,i}$, where its chord slope rotation $\theta$ characterizes the Moment-Chord Slope Rotation ($M-\theta$) diagram of the non-linear spring in the end section. In practice, the two critical end-sections of the clear length of the beams and columns, as well as the critical section at the base of the walls/cores of the building model, must be provided with suitable non-linear springs. In order to calculate the $M-\theta$ diagram of a plastic hinge, two methods are frequently implemented: firstly, the most rational method involves the use of suitable semi-empirical, non-linear diagrams of Moment-Chord Slope Rotation ($M-\theta$) of the "shear length" of each sub-cantilever, taken from a large database of experimental results, despite their considerable scattering (Panagiotakos & Fardis 2001). Such relations are proposed by Eurocode EN-1998.03 (Annex A: sections A.3.2.2 to A.3.2.4). Secondly, the calculation of $M-\theta$ diagrams can alternatively be achieved via Moment-Curvature

The Equivalent Non-Linear Single Degree of Freedom System of Asymmetric Multi-Storey
Buildings in Seismic Static Pushover Analysis

185

($M$-$\varphi$) diagrams, which can be calculated using the "fiber elements" (i.e. XTRACT, 2007;).

Next, using a suitable experimental lengh $L_p$ for each plastic hinge, $M-\theta$ diagrams are calculated (Fig.2c) by the integral of curvatures in length $L_p$. More specifically, the yielding rotation can be calculated by $\theta_y = \varphi_y \cdot L_s/3$ and the plastic rotation by $\theta_p \approx \left(\varphi_u - \varphi_y\right) \cdot L_p$, where $\varphi_y$ is the yielding curvature of the end section and $\varphi_u$ is the ultimate one by "fiber element" method, whilst the ultimate rotation $\theta_u$ is always calculated from the sum of both the yielding rotation $\theta_y$ and the plastic rotation $\theta_p$. However, using the second method, only flexural deformations are taken into account, while other sources of inelastic properties are ignored. For example, shear & axial deformations, slippage and extraction of the main longitudinal reinforcement and opened cracks with yielding steel bars, should also be taken into account. In order to avoid all these issues, the initial method of semi-empirical Moment-Chord Slope Rotation ($M-\theta$) diagrams, either a combination of the two methods, is often selected. Moreover, the maximum available ultimate rotation of a plastic hinge in Fig.(2a) very closely approximates (in practice) the ultimate chord slope rotation $\theta$ of the equivalent sub-cantilever in Fig.(2b). On the other hand, other simulations of plastic hinges, such as the one based on the "dual component model", where each member is replaced by an elastic element (central length of the element) and an elasto-plastic element (plastic hinge) in parallel, are not suitable for analyzing reinforced concrete members, because then the "stiffness degradation" for repeated dynamic seismic loading cycles is ignored. Also, in many cases, and especially for existing r/c structures, a "shear failure" on structural members can occur because the mechanism of the shear forces in the plastic hinges is destroyed, and therefore doubts arise concerning the validity of the well-known modified "Morsch's truss-model" after the yielding. Therefore, the available plastic rotation $\theta_p$ of a plastic hinge is reduced (by up to 50% in several cases); (see Eq.(A.12) of EN-1998.03). Also, in the case where plastic hinges appeared simultaneously at both ends of an structural element, the flexural effective stiffness of the section is calculated by $EI_{eff} = M_y \cdot L_s/3\theta_y$, (arithmetic mean value of the two antisymmetric bending states of the element).

Last but not least, in order to calculate the non-linear Moment-Chord Slope Rotation ($M-\theta$) diagrams, in contrast to the initial design state, the mean values of the strength of the materials must be used (namely, $f_{cm} = f_{ck} + 8$ in MPa for concrete and $f_{sm} = 1.10 f_{sk}$ for steel, where $f_{ck}$ and $f_{sk}$ are the characteristic values). If some structural elements have a low available ductility ($\mu < 2$), such as low-walls with a shear span ratio $a_s < 1.50$ at their base, columns/beams with inadequate strength in shear stress, brittle members such as short-columns with $a_s < 2.50$ etc, then these must be checked in relation to stress only. With reference to the "shear length $L_s$" of a sub-cantilever, we can note that it is calculated by $L_s = M/V$, where $M$ is the flexural yielding moment and $V$ is the respective shear force on a plastic hinge. In the case of tall-walls, the "shear length" $L_s$ can be calculated by the distance of the zero-moment point to the base in elevation, from a temporary lateral static floor force.

The "shear span ratio" is calculated by $a_s = M/Vh = L_s/h$, where $h$ is the dimension of the section that is perpendicular to the axis of the flexural moment $M$. Moreover, the yielding moment $M$ of the diagram $M - \theta$ cannot be greater than the value $\max M = V_R L_s$, where $V_R$ is calculated by Eq.(A.12) of Eurocode EN/1998.03, with the axial force $N$ of the column at zero (i.e. due to the action of the vertical seismic building vibration of vertical and horizontal seismic components) and the plastic ductility of the element is equal to $\mu_\Delta^{pl} = \theta_p/\theta_y = 5$ (conservative value), because then the "brittle shear failure" appears prematurely, due to the doubts of the validity of the modified "Morsch's truss-model" after the yielding.

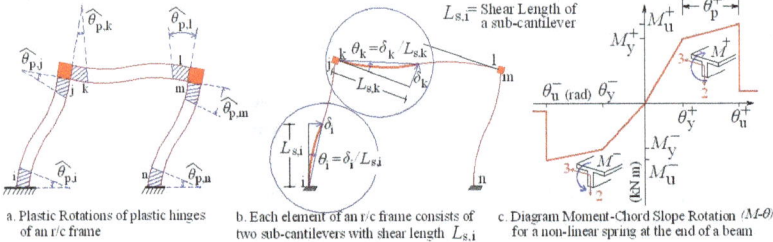

a. Plastic Rotations of plastic hinges of an r/c frame    b. Each element of an r/c frame consists of two sub-cantilevers with shear length $L_{s,i}$    c. Diagram Moment–Chord Slope Rotation $(M-\theta)$ for a non-linear spring at the end of a beam

Fig. 2. a. Plastic rotations of plastic hinges; b. Two equivalent sub-cantilevers of each structural column/beam; c. Diagram $M - \theta$ of inelastic springs at the ends of the elements (where $\theta$ is the chord slope rotation).

On the other hand, in the case of seismic Non-Linear Response History Analysis (NLRHA), we have to use suitable pairs (or triads) of uncorrelated accelerograms as seismic action. However, the seismic demands of asymmetric multi-storey buildings are often unreliable for the following reasons:

i.    It is possible for the accelerograms to be deemed unsuitable (they often do not have the necessary frequency content or are inadequate as regards the number of strong cycles of the dynamic loading or the strong motion duration). In order to cope with this problem, according to contemporary seismic codes, at least seven triads (or pairs, in the case where the vertical seismic component is ignored) of accelerograms must be taken into consideration and then, the average of the response quantities from all these analyses should be used as the "final design values" of the seismic action effect. Note that, for each pair (or triad) of accelerograms, the latter must act simultaneously (sect.3.2.3.1.1(2)P/EN-1998.01) and be "statistically independent". This means that the "correlation factor" among these accelerograms must be zero (uncorrelated accelerograms). However, each accelerogram of each pair must be represented by the same response acceleration spectrum (sect.3.2.2.1(3)P/EN-1998.01). The acceleration response spectra (with 0.05 equivalent viscous damping ratio) of artificial (or recorded) accelerograms should match the elastic response spectra for 0.05 equivalent viscous damping ratio. This should occur over a wide range of periods (or at least between 0.2T and 2T, where T is the fundamental period of the structure), as is defined by sect.3.2.3.1.2(4c)/EN-1998.01. Also, no value of the "mean elastic spectrum", which is calculated based on all the used accelerograms, should be less than 90% (or greater than 110%) of the corresponding value of the tagret elastic response spectrum. The peak ground acceleration of each accelerogram must always be equal to $a_g \cdot S$, where $a_g$ is the design ground acceleration of the local area and

$S$ is the soil factor according to EN-1998.01. Simultaneously, for strong earthquakes, the minimum duration of the strong motion of each accelerogram must be sufficient (about 15s in common cases), whilst, from personal observations, I consider that there should be a minimum number of approximately ten "large" and thirty-five "significant" loading loops, due to the dynamic seismic cyclic loading (i.e. a "large" cycle has an extreme ground acceleration of over $0.75PGA$ and a "significant" cycle has an extreme ground acceleration of between $0.30PGA$ and $0.75PGA$). However, the subject of the exact definition of the Design Basis Earthquake and the Maximum Capable Earthquake of a seismic hazard area is open to question, while recently, a new framework for the simulation and definition of the seismic action of Design Earthquake Basis for the inelastic single degree of freedom system, using the Park & Ang damage index on the structures, has been developed (Moustafa, 2011,).

ii. It is possible for the numerical models from the building simulation to be inadequate, because many false assumptions about the non-linear dynamic properties of plastic hinges can be inserted into the building model. For example, each non-linear spring must possess a suitable dynamic model $M - \theta$ for cyclic dynamic seismic loading (Dowell et al, 1998; Takeda et al, 1970; Otani, 1981; Akiyama, 1985;), where after each loading cycle, a suitable "stiffness degradation" must be taken into account (Fig.3).

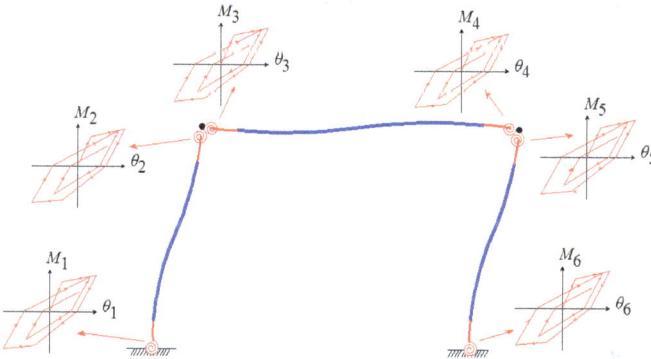

Fig. 3. Non-linear model of a planar frame, where each spring possesses a suitable diagram for cyclic dynamic seismic loading.

iii. It is possible for the numerical integration method to be inadequate, as regards accuracy & stability.

iv. The critical dynamic loading orientation of the pair of horizontal seismic components is unknown or does not exist and leads to the examination of various other orientations (at least one more orientation with a rotation of 45 degrees relative to the initial principal horizontal orientation must be examined). Alternative, a special methodology about the examination of various orientations usimg pairs of accelerograms, gives the envelope of the most unfavourable results (in the linear area only) was presented by Athanatopoulou (2005). Note that, for each pair of accelerograms, all possible sign combinations of seismic components must be taken into account, resulting in four combinations for each pair. Moreover, seven (7) pairs must be taken into account. In addition, due to the accidental eccentricities, four different positions of the floor mass centres must be considered. Therefore, (2 orientations)x(4 combinations of signs per pair)x(7 pairs)x(4 positions of the mass centres)=224 NLRH Analyses. Alternatively, in order to reduce the number of said analyses, the action of accidental eccentricities can be taken into account using equivalent external floor moments $M_{m,i}$ around the vertical axis, which are calculated by the floor lateral static forces, $M_{f_i} = \pm F_{f_i} \times e_{ai,m}$ , where $F_{f_i}$ is the lateral floor static force and $e_{ai,m}$ is a mean accidental eccentricity ($e_{ai,m} = \sqrt{e_{ai,I}^2 + e_{ai,II}^2}$ ) for the case where the base shear is same for both principal directions of the building. Otherwise, the external moments can derive by $M_{m,i} = \pm\sqrt{M_{I,i}^2 + M_{II,i}^2} = \sqrt{\left(F_{f_{I,i}} \times e_{ai,II}\right)^2 + \left(F_{f_{II,i}} \times e_{ai,I}\right)^2}$ . In this case, we arrive at a total number of fifty six (56) NLRH Analyses.

Consequently, the most important issues that we come across during the static pushover analysis are the following: (1) Why is the use of a single degree of freedom system that represents the real structure required? (2) How can the spatial action of the two horizontal seismic components during the static pushover analysis be taken into account? (3) What is the most suitable monitoring point in the case of asymmetric buildings? (4) Which distribution of lateral floor static forces is most suitable? and (5) At which point, in the plan of an asymmetric building, should the lateral floor static forces be applied? The design dynamic eccentricities and accidental eccentricities must also be taken into account, which means that the real or fictitious elastic axis of the building with the real or fictitious horizontal main elastic axes, where the lateral static forces must be oriented during the static pushover analysis, must be calculated. On the other hand, the Non-Linear Response History Procedure has a high computational cost and, simultaneously, many of its results present a high sensitivity. For these reasons, in practice, the mathematically documented static non-linear (pushover) analysis, presented below in this chapter, is a reliable alternative approximate method, which can envelope the accurate seismic demand floor inelastic displacements by the above Non-Linear Response History Procedure, in a rational way. However, it is known that the remaining seismic demands, such as the inter-storey drift ratios, real distribution of the plastic hinges on the building, yielding/failure mechanism of the building, etc, are only predicted correctly by using Non-Linear Response History Analysis.

## 2. The equivalent non-linear SDF system of asymmetric multi-storey buildings in non-linear static pushover analysis

Let us consider an asymmetric multi-storey building consisting of a rigid deck (diaphragm) on each floor, where the total mass $m_{,i}$ of each floor is concentrated at the floor geometric centre $CM_{,i}$ of the diaphragm, and the multi-storey building has a vertical mass axis, where all floor concentrated masses are located. Each centre $CM_{,i}$ therefore has the translational mass $m_{,i}$ for any horizontal direction, while it has the floor mass inertia moment $J_{m,i} = m_{,i} \, l_{s,i}^2$ due to its diaphragm (where $l_{s,i}$ is the floor inertia radius). Every diaphragm is supported on massless, axially loaded columns and structural walls, in reference to the global Cartesian mass system $CM(x,y,z)$ that is parallel to the known fictitious principal Cartesian reference system $P_o\left(I,II,III\right)$, as presented in the previous paragraph. Next, the original point CM has three degrees of freedom for each floor; two horizontal displacements $u_{x,i}$ and $u_{y,i}$ relative to the ground, along axes $x$ and $y$ respectively, and a rotation $\theta_{z,i}$ around the vertical $z$-axis (Fig.4). Moreover, for the needs of this mathematical analysis, we consider that this N-storey building is loaded with the 3N-dimensional loading vector **P** of external lateral floor static loading, where the floor forces are oriented in parallel to one fictitious principal axis (i.e. along the $II$-axis) of the building. The loading vector $\mathbf{P}$ relating to the 3N-dimensional vector $\mathbf{u}$ of the degrees of freedom of the floor masses of the asymmetric multi-storey building is formed according to the pure translational (along the principal $II$-axis) mode-shape's distribution of lateral floor static forces in elevation (Eqs.2a,b). For the needs of this analysis, the lateral floor static forces of the loading vector $\mathbf{P}$ in the plan, are located at $\max e_1$ distance from the fictitious stiffness centre $P_o$ to the mass centre of the building (Fig.4a). Next, a primary static pushover analysis of the asymmetric N-storey building is performed, using this increased static loading vector $\mathbf{P}$, until the collapse of the building model.

$$\mathbf{u} = \begin{Bmatrix} \mathbf{u}_{,1} \\ \mathbf{u}_{,2} \\ \cdots \\ \mathbf{u}_{,i} \\ \cdots \\ \mathbf{u}_{,N} \end{Bmatrix}, \quad \mathbf{P} = \begin{Bmatrix} \mathbf{P}_{,1} \\ \mathbf{P}_{,2} \\ \cdots \\ \mathbf{P}_{,i} \\ \cdots \\ \mathbf{P}_{,N} \end{Bmatrix} \qquad (2a,b)$$

where $\mathbf{u}_{,i} = \begin{Bmatrix} u_{x,i} \\ u_{y,i} \\ \theta_{z,i} \end{Bmatrix}$ and $\mathbf{P}_{,i} = \begin{Bmatrix} P_{x,i} \\ P_{y,i} \\ M_{z,i} \end{Bmatrix} = \begin{Bmatrix} 0 \\ P_{,i} \\ P_{,i} \cdot (\max e_{1,i} - e_{0,1}) \end{Bmatrix}$ for each floor $i$ of the building.

Next, let us consider an intermediate step of this pushover analysis near the middle of the inelastic area. On each floor $i$, the vector of inelastic displacements of the mass centre of level $i$ of the building is $\mathbf{u}_{,i,o} = \{u_{x,i,o} \quad u_{y,i,o} \quad \theta_{z,i,o}\}^T$, while the global vector of inelastic displacements of all floor mass centres constitutes the 3N-dimensional vector $\mathbf{u}_o$, Eq.(3). It follows that this vector constitutes the initial state (index "o") of displacements in the following mathematical analysis:

$$\mathbf{u}_o = \begin{Bmatrix} u_{x,1,o} \\ u_{y,1,o} \\ \theta_{z,1,o} \\ u_{x,2,o} \\ u_{y,2,o} \\ \theta_{z,2,o} \\ \cdots \\ \cdots \\ u_{x,N,o} \\ u_{y,N,o} \\ \theta_{z,N,o} \end{Bmatrix} = \begin{Bmatrix} u_{x,1,o}/u_{y,N,o} \\ u_{y,1,o}/u_{y,N,o} \\ \theta_{z,1,o}/u_{y,N,o} \\ u_{x,2,o}/u_{y,N,o} \\ u_{y,2,o}/u_{y,N,o} \\ \theta_{z,2,o}/u_{y,N,o} \\ \cdots \\ \cdots \\ u_{x,N,o}/u_{y,N,o} \\ 1.00 \\ \theta_{z,N,o}/u_{y,N,o} \end{Bmatrix} u_{y,N,o} = \begin{Bmatrix} \psi_{x,1,y,N} \\ \psi_{y,1,y,N} \\ \psi_{z,1,y,N} \\ \psi_{x,2,y,N} \\ \psi_{y,2,y,N} \\ \psi_{z,2,y,N} \\ \cdots \\ \cdots \\ \psi_{x,N,y,N} \\ 1.00 \\ \psi_{z,N,y,N} \end{Bmatrix} u_{y,N,o} = \boldsymbol{\Psi}_{y,N,o} \cdot u_{y,N,o} \qquad (3)$$

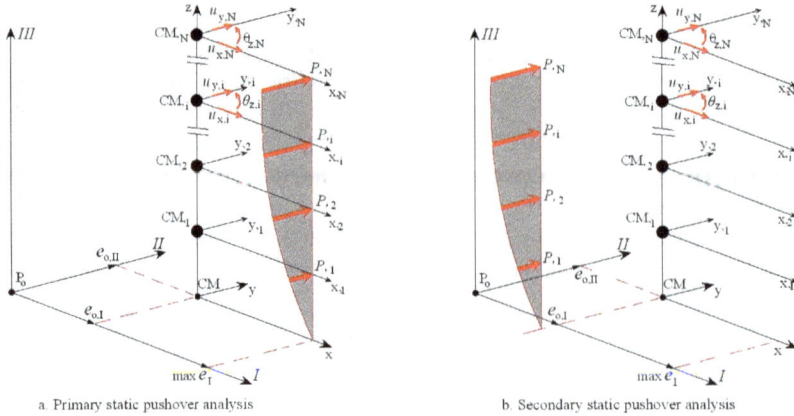

a. Primary static pushover analysis          b. Secondary static pushover analysis

Fig. 4. Spatial Asymmetric Multi-storey System. Primary and Secondary pushover analysis.

Next, consider the known pure translational mode shape distribution $\mathbf{Y}$ of the lateral floor forces in elevation (from the linear modal analysis), where, Eq.(4);

$$\mathbf{Y} = \left\{ Y_1 \quad Y_2 \quad \ldots \quad Y_i \quad \ldots \quad Y_N \right\}^{\mathrm{T}} \tag{4}$$

with $Y_N = 1.00$.

In order to define the optimum equivalent Non-Linear SDF system, the following two assumptions are set (Makarios, 2009):

*1st Assumption*: Vector $\boldsymbol{\psi}_{y,N,o}$ of the inelastic displacement distribution of the floor mass centres of the asymmetric multi-storey building is a 'notional mode-shape'.

*2nd Assumption*: The lateral force $\mathbf{P}$ causes translational and torsional accelerations on the concentrated mass of the asymmetric multi-storey building. Therefore, if we consider that the vector loading $\mathbf{P}$ of the external loads is a function of time $t$, i.e. $\mathbf{P}(t)$ is a dynamic loading, this function is written as follows in Eq.(5):

$$\mathbf{P} = \left\{ \begin{array}{c} P_{x,1} \\ P_{y,1} \\ M_{z,1} \\ P_{x,2} \\ P_{y,2} \\ M_{z,2} \\ \ldots \\ \ldots \\ P_{x,N} \\ P_{y,N} \\ M_{z,N} \end{array} \right\} = \left\{ \begin{array}{c} 0.00 \\ P_{,1} \\ \left(\mathrm{max}e_{\mathrm{I},1} - e_{\mathrm{o,I}}\right) \cdot P_{,1} \\ 0.00 \\ P_{,2} \\ \left(\mathrm{max}e_{\mathrm{I},2} - e_{\mathrm{o,I}}\right) \cdot P_{,2} \\ \ldots \\ \ldots \\ 0.00 \\ P_{N} \\ \left(\mathrm{max}e_{\mathrm{I},N} - e_{\mathrm{o,I}}\right) \cdot P_{,N} \end{array} \right\} \mathbf{f}(t) = \left\{ \begin{array}{c} 0.00 \\ Y_1 \\ \left(\mathrm{max}e_{\mathrm{I},1} - e_{\mathrm{o,I}}\right) \cdot Y_1 \\ 0.00 \\ Y_2 \\ \left(\mathrm{max}e_{\mathrm{I},2} - e_{\mathrm{o,I}}\right) \cdot Y_2 \\ \ldots \\ \ldots \\ 0.00 \\ 1.00 \\ \left(\mathrm{max}e_{\mathrm{I},N} - e_{\mathrm{o,I}}\right) \end{array} \right\} P_N \cdot \mathbf{f}(t) = \mathbf{i}\, P_N \cdot \mathbf{f}(t), \tag{5}$$

where $f(t)$ is a known, suitable, increasing, linear, monotonic time function (i.e. $f(t) = a \cdot t + b$, where a & b are coefficients) and $P_{,i}$ is the value of the floor lateral force.

Therefore, the equation of the motion of the masses of the multi-storey building, which are loaded with the dynamic loading $P(t)$, is described by the mathematical system of linear second order differential N-equations, which can be written in matrix form, Eq.(6):

$$\mathbf{M} \ddot{\mathbf{u}}_0(t) + \mathbf{C} \dot{\mathbf{u}}_0(t) + \mathbf{K} \mathbf{u}_0(t) = \mathbf{P}(t) \tag{6}$$

where $\mathbf{u}_0(t)$ is the 3N-dimensional vector of displacements from Eq.(3), $\mathbf{K}$ and $\mathbf{M}$ are the 3Nx3N-dimensional square symmetric matrices (Eqs.7a,b) of lateral elastic stiffness and masses respectively, while $\mathbf{C}$ is the damping 3Nx3N matrix.

$$\mathbf{K} = \begin{bmatrix} k_{,1} & k_{,12} & \dots & k_{,1i} & \dots & k_{,1N} \\ & k_{,2} & \dots & k_{,2i} & \dots & k_{,2N} \\ & & \dots & \dots & \dots & \dots \\ & & & k_{,i} & \dots & k_{,iN} \\ & & & & \dots & \dots \\ & & & & & k_{,N} \end{bmatrix}, \mathbf{M} = \begin{bmatrix} \mathbf{M}_{,1} & & & & \\ & \mathbf{M}_{,2} & & & \\ & & \dots & & \\ & & & \mathbf{M}_{,i} & \\ & & & & \dots \\ & & & & & \mathbf{M}_{,N} \end{bmatrix} \tag{7a,b}$$

where, $\mathbf{k}_{,i} = \begin{bmatrix} k_{xx,i} & k_{xy,i} & k_{xz,i} \\ k_{yx,i} & k_{yy,i} & k_{yz,i} \\ k_{zx,i} & k_{zy,i} & k_{zz,i} \end{bmatrix}$, $\mathbf{M}_{,i} = \begin{bmatrix} m_{,i} & 0 & 0 \\ 0 & m_{,i} & 0 \\ 0 & 0 & J_{m,i} \end{bmatrix}$ for each floor $i$

and where $J_{m,i} = m_{,i} \, l_{s,i}^2$ is the mass inertia moment of the diaphragm (floor) around the vertical axis passing through the mass centre, and $l_{s,i}$ is the floor inertia radius. The floor lateral stiffness sub-matrix $\mathbf{k}_{,i}$, as well as the global lateral stiffness matrix $\mathbf{K}$ of the building, are calculated with reference to the global Cartesian mass system CM(x,y,z) using a suitable numerical technique pertaining to the "stiffness condensation" of the total stiffness matrix. Alternatively, we enforce a displacement (that is equal to one) of a floor mass centre along one degree of freedom and simultaneously all the remaining degrees of freedom of the building are fixed. The generated reactions on the degrees of freedom provide the coefficients of a column of the global lateral stiffness matrix $\mathbf{K}$, and this procedure is repeated for all other mass degrees of freedom of the building.

Using Eqs.(2), (3), (5) & (7), Eq.(6) is written as follows:

$$\mathbf{M} \, \psi_{y,N,o} \cdot \ddot{\mathbf{u}}_{y,N,o}(t) + \mathbf{C} \, \psi_{y,N,o} \cdot \dot{\mathbf{u}}_{y,N,o}(t) + \mathbf{K} \, \psi_{y,N,o} \cdot \mathbf{u}_{y,N,o}(t) = \mathbf{i} \, P_N \cdot f(t) \tag{8}$$

Next, we pre-multiply Eq.(8) by the $\psi_{y,N,o}^T$ vector:

$$\psi_{y,N,o}^T \mathbf{M} \, \psi_{y,N,o} \cdot \ddot{\mathbf{u}}_{y,N,o}(t) + \psi_{y,N,o}^T \mathbf{C} \, \psi_{y,N,o} \cdot \dot{\mathbf{u}}_{y,N,o}(t) + \psi_{y,N,o}^T \mathbf{K} \, \psi_{y,N,o} \cdot \mathbf{u}_{y,N,o}(t) =$$

$$= \psi_{y,N,o}^T \, \mathbf{i} \cdot P_N \cdot f(t) \tag{9}$$

where $\psi_{y,N,o}^T\, \mathbf{i}$ is a factor.

In this case, it is clear that the base shear $V_{oy}(t)$ of the building, for every time $t$, is given as follows:

$$V_{oy}(t) = P_N \cdot f(t) \cdot \sum_{i=1}^{N} Y_i \tag{10}$$

Therefore, the lateral force $P_N \cdot f(t)$ at the top of the frame is given by:

$$P_N \cdot f(t) = \frac{V_{oy}(t)}{\sum\limits_{i=1}^{N} Y_i} \tag{11}$$

Eq.(9) is divided by the number $\psi_{y,N,o}^T\, \mathbf{i}$ and by inserting Eq.(3), we arrive at:

$$\frac{\psi_{y,N,o}^T\, \mathbf{M}\, \psi_{y,N,o} \cdot \sum\limits_{i=1}^{N} Y_i}{\psi_{y,N,o}^T\, \mathbf{i}} \cdot \ddot{u}_{y,N,o}(t) + \frac{\psi_{y,N,o}^T\, \mathbf{C}\, \psi_{y,N,o} \cdot \sum\limits_{i=1}^{N} Y_i}{\psi_{y,N,o}^T\, \mathbf{i}} \cdot \dot{u}_{y,N,o}(t) +$$

$$\tag{12}$$

$$+ \frac{\psi_{y,N,o}^T\, \mathbf{K}\, \psi_{y,N,o} \cdot \sum\limits_{i=1}^{N} Y_i}{\psi_{y,N,o}^T\, \mathbf{i}} \cdot u_{y,N,o}(t) = V_{oy}(t)$$

Eq.(12) presents an equation of the motion of a single-degree-of-freedom (SDF) system. This equation must be transformed, so that the "optimum equivalent Non-linear SDF system" represents the initial asymmetric multi-storey building. As we can see in Eq.(12), the degree of freedom is the displacement $u_{y,N}$ of the mass centre CM, while loaded with the base shear $V_{oy}$. Thus, the "effective lateral stiffness" $k^*$ of this SDF system must be obtained through an additional, special, secondary, non-linear (pushover) analysis of the initial, asymmetric multi-storey building, where the lateral static force is applied to the mass centre CM along the y-axis (Fig.4b). Hence, the lateral stiffness $k^*$ of the first branch and the lateral stiffness $\alpha \cdot k^*$ of the second branch of the bilinear diagram $V_{oy} - u_{y,N}$ are the result of this special, secondary pushover analysis (Fig.5a). Therefore, we can consider that the slopes $k^*$ and $\alpha \cdot k^*$ are already known from the secondary pushover analysis. Thus, for reasons of convergence between the known analytic SDF system of Eq.(12) and the (unknown at present) required optimum equivalent NLSDF system of the asymmetric multi-storey building, Eq.(12) is multiplied with the coefficient $L = k^* / k_o$, hence:

$$m^* \cdot \ddot{u}_{y,N,o}(t) + c^* \cdot \dot{u}_{y,N,o}(t) + k^* \cdot u_{y,N,o}(t) = L \cdot V_{oy}(t) \tag{13}$$

where,

$$m^* = \frac{k^*}{k_o} \cdot \frac{\boldsymbol{\psi}_{y,N,o}^T \mathbf{M} \, \boldsymbol{\psi}_{y,N,o} \cdot \sum_{i=1}^{N} Y_i}{\boldsymbol{\psi}_{y,N,o}^T \, \mathbf{i}} \tag{14}$$

$$k_o = \frac{\boldsymbol{\psi}_{y,N,o}^T \mathbf{K} \, \boldsymbol{\psi}_{y,N,o} \cdot \sum_{i=1}^{N} Y_i}{\boldsymbol{\psi}_{y,N,o}^T \, \mathbf{i}} \tag{15}$$

$\boldsymbol{\psi}_{y,N,o}^T \mathbf{M} \, \boldsymbol{\psi}_{y,N,o}$ = Coefficient, $\boldsymbol{\psi}_{y,N,o}^T \, \mathbf{i}$ = Coefficient,

$\boldsymbol{\psi}_{y,N,o}^T \mathbf{K} \, \boldsymbol{\psi}_{y,N,o}$ = Coefficient, $\sum_{i=1}^{N} Y_i$ = Coefficient,

Therefore, the optimum equivalent NLlinear SDF system is represented by Eq.(13) and is presented in Fig.(6a). In addition, we can assume that the NLSDF system possesses the equivalent viscous damping $c^*$, hence:

$$c^* = \frac{k^*}{k_o} \cdot \frac{\boldsymbol{\psi}_{y,N,o}^T \mathbf{C} \, \boldsymbol{\psi}_{y,N,o} \cdot \sum_{i=1}^{N} Y_i}{\boldsymbol{\psi}_{y,N,o}^T \, \mathbf{i}} = 2 \cdot m^* \cdot \omega^* \cdot \xi \tag{16}$$

where, $\omega^*$ is the circular frequency (in rad/s) of the equivalent vibrating SDF system in the linear range:

$$\omega^* = \sqrt{k^*/m^*} \tag{17}$$

and $\xi$ is the equivalent viscous damping ratio that corresponds to the critical damping of the SDF system (about 5% for reinforced concrete).

In order to transform the diagram $V_{oy} - u_{y,N}$ of Fig.(5a,b) into the capacity curve $P^* - \delta^*$ of Fig.(5b) of the equivalent NLSDF system, factor $\Gamma$ is directly provided by:

$$\Gamma = m_{tot}/m^* \tag{18}$$

where,

$$m_{tot} = \mathbf{l}_y^T \mathbf{M} \, \mathbf{l}_y = \left\{ \mathbf{l}_{,1} \quad \mathbf{l}_{,2} \quad \cdots \quad \mathbf{l}_{,i} \quad \cdots \quad \mathbf{l}_{,N} \right\}^T \begin{bmatrix} \mathbf{M}_{,1} & & & & \\ & \mathbf{M}_{,2} & & & \\ & & \ddots & & \\ & & & \mathbf{M}_{,i} & \\ & & & & \ddots \\ & & & & & \mathbf{M}_{,N} \end{bmatrix} \left\{ \begin{array}{c} \mathbf{l}_{,1} \\ \mathbf{l}_{,2} \\ \cdots \\ \mathbf{l}_{,i} \\ \cdots \\ \mathbf{l}_{,N} \end{array} \right\}$$

with $\quad \mathbf{1}_{,i}^{T}=\{0 \quad 1 \quad 0\}^{T}$

Moreover, $\mathbf{1}_y$ is the global 'influence vector', that represents the displacements of the masses resulting from the static application of a unit horizontal ground displacement of the building along the seismic loading (i.e. $y$-direction). Hence, the optimum equivalent NLSDF system is characterized by Eq.(13) and has a bilinear capacity curve ($P^*$-$\delta^*$) according to Fig.(5b).

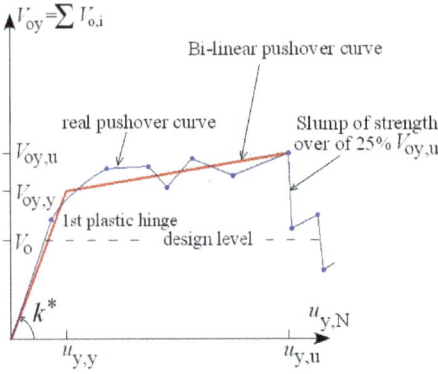

a. Pushover Curve of MDoF system by the Secondary Analysis        b. Capacity Curve of equivalent NLSDF system

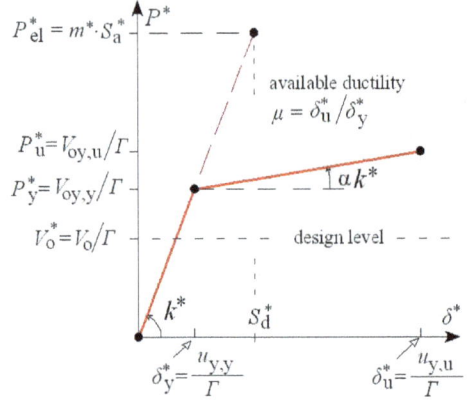

Fig. 5. Transformation of the Pushover Curve of the Multi Degree of Freedom system into the Capacity Curve of the optimum equivalent NLSDF system.

The maximum elastic base shear $P_{el}^*$ of the infinitely elastic SDF system, the yielding base shear $P_y^*$ of the respective equivalent non-linear SDF system, its ultimate base shear $P_u^*$, its yielding displacement $\delta_y^*$ and its ultimate displacement $\delta_u^*$, are given (Fig.5b):

$$P_{el}^* = m^* \cdot S_a^* , \; P_y^* = \frac{V_{oy,y}}{\Gamma} , \; P_u^* = \frac{V_{oy,u}}{\Gamma} \qquad (19a,b,c)$$

$$\delta_y^* = \frac{u_{y,y}}{\Gamma} , \; \delta_u^* = \frac{u_{y,u}}{\Gamma} \qquad (20a,b)$$

The "effective period" $T^*$ of the optimum equivalent NLSDF system is given by Eq.(21), Fig.(6a):

$$T^* = 2\pi \sqrt{\frac{m^*}{k^*}} = 2\pi \sqrt{\frac{m^* \cdot \delta_y^*}{P_y^*}} \qquad (21)$$

Next, in order to evaluate the demand seismic displacement of the asymmetric multi-storey r/c building, we must first calculate the seismic target-displacement $\delta^*_{t,inel}$ of the optimum equivalent NLSDF system using the known Inelastic Response Spectra, since this is the only credible option; otherwise, a Non-Linear Response History procedure must be performed. Therefore, the inelastic seismic target-displacement $\delta^*_{t,inel}$ is calculated directly using Eq.(22), Fig.(6b). Secondly, we must transform the inelastic seismic target-displacement $\delta^{\hat{}}_{t,inel}$ into the target-displacement of the monitoring point (namely, the centre mass at the top level of the real initial asymmetric multi-storey building, along the examined direction), multiplying it with factor $\Gamma$, Eq.(28). Finally, the maximum required seismic displacements of the other positions of the multi-storey building can be obtained from that step in the known primary pushover analysis, where the seismic target-displacement of Eq.(28) of the monitoring point appears.

Seismic target-displacement $\quad \delta^*_{t,inel}$ of NLSDF system

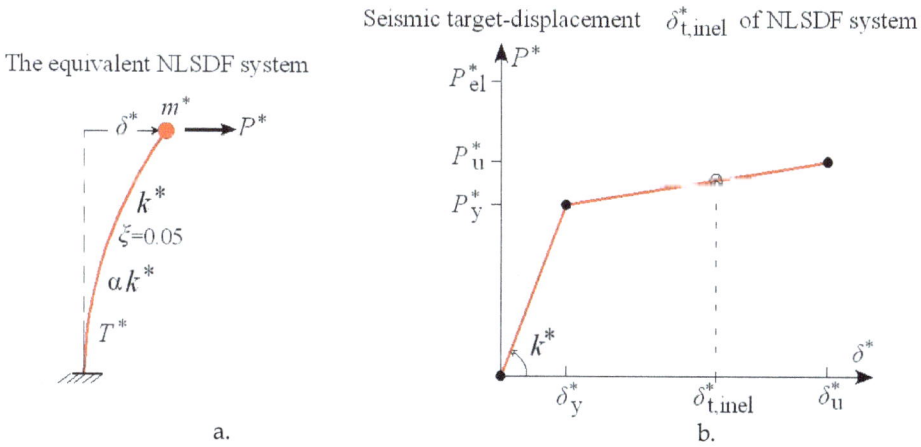

a.

b.

Fig. 6. a. The optimum equivalent NLSDF system, b. The seismic target-displacement of the equivalent NLSDF system.

More specifically, with reference to the optimum equivalent NLSDF system, its elastic spectral acceleration $S^*_a$ is calculated by the elastic response acceleration spectrum, for the known period $T^*$. Consequently, the maximum required inelastic seismic displacement (target-displacement) $\delta_t$ of the new equivalent NLSDF system arises from the inelastic spectrum of a known earthquake, according to the following equations:

If $\quad \dfrac{P^*_y}{m^*} < S^*_a \quad$, then the response is non-linear and post-elastic and thus the target-displacement is inelastic, because this means that the yielding base shear $P^*_y$ of the NLSDF system is less than its elastic seismic base shear; therefore this system yields:

$$\delta^*_{t,inel} = L \cdot \frac{\mu_d}{R_y} \cdot \frac{S^*_a}{(\omega^*)^2} = L \cdot \frac{\mu_d}{R_y} \cdot \frac{S^*_a}{4\pi^2} \cdot \left(T^*\right)^2 = L \cdot \frac{\mu_d}{R_y} \cdot S^*_d \tag{22}$$

where,

$S_a^* = S_a(T^*)$ and $S_d^* = S_d(T^*)$ are the "elastic spectral acceleration" and the "elastic spectral displacement" from the elastic response acceleration and displacement spectrum respectively, for equivalent viscous damping ratio $\xi = 0.05$ for r/c structures,

$R_y = P_{el}^*/P_y^*$ is the reduction factor of the system (Fig.5b)

$L = k^*/k_o$ the "convergence factor" between the initial Multi Degree of Freedom system and the NLSDF

$\mu_d$ is the demand ductility of this non-linear SDF system (Eq.23):

$$\mu_d = 1 + \frac{\left(R_y^c - 1\right)}{c} \tag{23}$$

$c$ is a coefficient due to the second branch slope from Eqs.(24, 25, 26), using a suitable slope ratio $\alpha = 0\%, 2\%, 10\%$ of the equivalent NLSDF system (Fig.5b), according to Krawinkler & Nassar (1992):

$$c = \frac{T^*}{1 + T^*} + \frac{0.42}{T^*} \text{ for the second branch slope } \alpha = 0\% \tag{24}$$

$$c = \frac{T^*}{1 + T^*} + \frac{0.37}{T^*} \text{ for the second branch slope } \alpha = 2\% \tag{25}$$

$$c = \frac{\left(T^*\right)^{0.8}}{1 + \left(T^*\right)^{0.8}} + \frac{0.29}{T^*} \text{ for the second branch slope } \alpha = 10\% \tag{26}$$

Note that the demand ductility $\mu_d$ by Eq.(23) simultaneously satisfies both the rule of "equal energies" and the rule of "equal displacements" according to Veletsos & Newmark (1960) and Newmark & Hall (1982).

If $\dfrac{P_y^*}{m^*} \geq S_a^*$ , then the response is linear elastic and thus the target-displacement $\delta_{t,el}^*$ is

elastic, because this means that the yielding base shear $P_y^*$ of the NLSDF system is greater than the elastic seismic base shear and therefore this system remains in the linear elastic area,

$$\delta_{t,el}^* = L \cdot \frac{S_a^*}{(\omega^*)^2} = L \cdot \frac{S_a^* \cdot \left(T^*\right)^2}{4\pi^2} \tag{27}$$

It is worth noting that, in order to achieve adequate seismic stability of the building, the demand ductility $\mu_d$ must be less than the available ductility $\mu = \delta_u^*/\delta_y^*$ . Next, in the first

The Equivalent Non-Linear Single Degree of Freedom System of Asymmetric Multi-Storey
Buildings in Seismic Static Pushover Analysis

197

approach (index "1"), the maximum required seismic displacement $u_{y,N,1}$ of the mass at the top level of the asymmetric multi-storey building, along y-direction, is directly given by:

$$u_{y,N,1} = \Gamma \cdot \delta_t^*$$ (28)

Finally, as mentioned above, the other maximum required seismic displacements $u_{x,i}$ and $\theta_{z,i}$ of the mass of the multi-storey building are obtained from that step in the known primary pushover analysis, where the displacement $u_{y,N,1}$ appears. If an optimum approach is required, then we can repeat the calculation of Eqs.(3-28), using the new displacements $u_{x,i}, u_{y,i}, \theta_{z,i}$ instead of $u_{x,i,o}$ $u_{y,i,o}$ $\theta_{z,i,o}$; the third approach is not usually needed. Finally, for each lateral loading, we follow the same methodology with a suitable index alternation (i.e. if the lateral floor forces are parallel to x-axis, then $\mathbf{u}_o = \psi_{x,N,o} \cdot u_{x,N,o}$, $\mathbf{P}_{,i} = \{P_{x,i} \quad 0 \quad M_{z,i}\}^T$, $\mathbf{1,}_i^T = \{1 \quad 0 \quad 0\}^T$, etc).

To sum up, the above-mentioned optimum equivalent Non-Linear Single Degree of Freedom System allows us to provide the definition of the behavior factor of the multi-storey building. Indeed, it is well known that the global behavior factor $q$ of a system is mathematically defined by the single degree of freedom system only. Therefore, in order to define the behavior factor of an asymmetric multi-storey building, an equivalent SDF system (such as the above optimum equivalent NLSDF system) must be estimated at first, and then its behavior factor, which refers to the asymmetric multi-storey building, is easy to determine. According to recent research (Makarios 2010), the "available behavior factor" $q_{av}$ of the optimum equivalent NLSDF system indirectly represents the "global available behavior factor" of the asymmetric multi-storey building, while $q_{av}$ is given by Eq.(29). Note that, in order for a building to cope with seismic action, the total "available behavior factor" $q_{av}$ has to be higher than the "design behavior factor" $q$, that was used for the initial design of the building.

$$q_{av} = q_o \cdot q_m \cdot q_p$$ (29)

where $q_o$ is the "overstrength partial behavior factor" that is part of the $q_{av}$ and is related to the "extent of static-indefiniteness / superstaticity" of the asymmetric multi-storey building, as well as the constructional provisions of the Reinforced Concrete Code used, and can be calculated by Eq.(30), where $V_o^*$ is the design base shear of the NLSDF system ($V_o^* = m^* \cdot S_a$):

$$q_o = P_y^*/V_o^*$$ (30)

Furthermore, $q_m$ is the "ductility partial behavior factor" that is another part of $q_{av}$ that is related to the ductility of the equivalent elasto-plastic SDF system, and is calculated by Eq.(31), which is an inversion of Eq.(23):

$$q_m = \left[c(\mu - 1) + 1\right]^{1/c}$$ (31)

where $\mu = \delta_u^* / \delta_y^*$ is now the "available ductility" of the NLSDF system.

Next, $q_p$ is the *"post-elastic slope partial behavior factor"* that is the last part of the $q_{av}$ and is related to the slope of the post-elastic (second) branch of the bilinear capacity curve of the equivalent NLSDF system and is calculated by Eq.(32):

$$q_p = P_u^* / P_y^* \tag{32}$$

Note that, if the second (post-elastic) branch has a negative slope (i.e. intense action of second order phenomena), then the *"post-elastic slope partial behavior factor"* $q_p$ is less than one.

Next, with reference to the spatial seismic action (signs of seismic components, simultaneous action of the two horizontal seismic components) in the presented non-linear static (pushover) procedure, we can observe the following issues:

Being aware, from the above mathematical analysis, that two static pushover analyses are needed for one location of lateral floor static forces, we can conclude that, if we use the final four design eccentricities, then twelve (12) individual static pushover analyses are required, as following:

-      For the first seismic horizontal component along principal $I$-axis, six load cases are needed, Fig.(7). Note that from each primary analysis (where one "final design eccentricity" out of the four is used), the demand seismic displacements of the asymmetric multi-storey building are obtained, whilst from the secondary analysis (where the lateral static forces are located on the mass centres) the effective lateral stiffness $k^*$ of its NLSDF is defined. Then, the twelve individual pushover analyses are analytically provided:

1st case: Lateral static loads along positive principal (+)$I$-axis, with final design eccentricity $\max e_{II}$ (primary pushover analysis, $E_1$).

2nd case: Lateral static loads along negative principal (–)$I$-axis, with final design eccentricity $\max e_{II}$ (primary pushover analysis, $E_2$).

3rd case: Lateral static loads along positive principal (+)$I$-axis, with final design eccentricity $\min e_{II}$ (primary pushover analysis, $E_3$).

4th case: Lateral static loads along negative principal (–)$I$-axis, with final design eccentricity $\min e_{II}$ (primary pushover analysis, $E_4$).

5th case: Lateral static loads that are located on the floor mass centres along positive principal (+)$I$-axis (secondary pushover analysis, useful for the 1st and 3rd cases).

6th case: Lateral static loads that are located on the floor mass centres along negative principal (–)$I$-axis (secondary pushover analysis, useful for the 2nd and 4th cases).

The envelope of the results from the previous primary pushover analyses (1st, 2nd, 3rd & 4th) is symbolized as $E_{,I}$ ; it represents the demands of the first seismic component along principal $I$-axis:

$$E_{,I} = \max\left( |E_1|, \; |E_2|, \; |E_3|, \; |E_4| \right) \tag{33}$$

- For the second seismic horizontal component along principal $II$-axis, six more load cases are needed, Fig.(8):

  7th case: Lateral static loads along positive principal (+)$II$-axis, with final design eccentricity $\max e_I$ (primary pushover analysis, $E_7$).

  8th case: Lateral static loads along negative principal (-)$II$-axis, with final design eccentricity $\max e_I$ (primary pushover analysis, $E_8$).

  9th case: Lateral static loads along positive principal (+)$II$-axis, with final design eccentricity $\min e_I$ (primary pushover analysis, $E_9$).

  10th case: Lateral static loads along negative principal (-)$II$-axis, with final design eccentricity $\min e_I$ (primary pushover analysis, $E_{10}$).

  11th case: Lateral static loads that are located on the floor mass centres along positive principal (+)$II$-axis (secondary pushover analysis, useful for the 7th & 9th cases).

  12th case: Lateral static loads that are located on the floor mass centres along negative principal (-)$II$-axis (secondary pushover analysis, useful for the 8th & 10th cases).

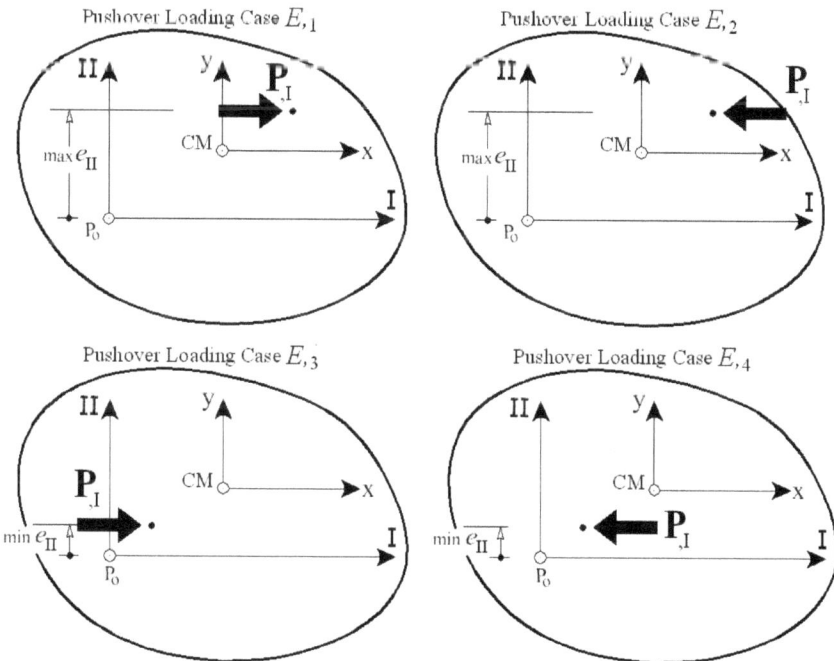

Fig. 7. The envelope of these four pushover analyses represents the results of the first seismic component $E_{,I}$.

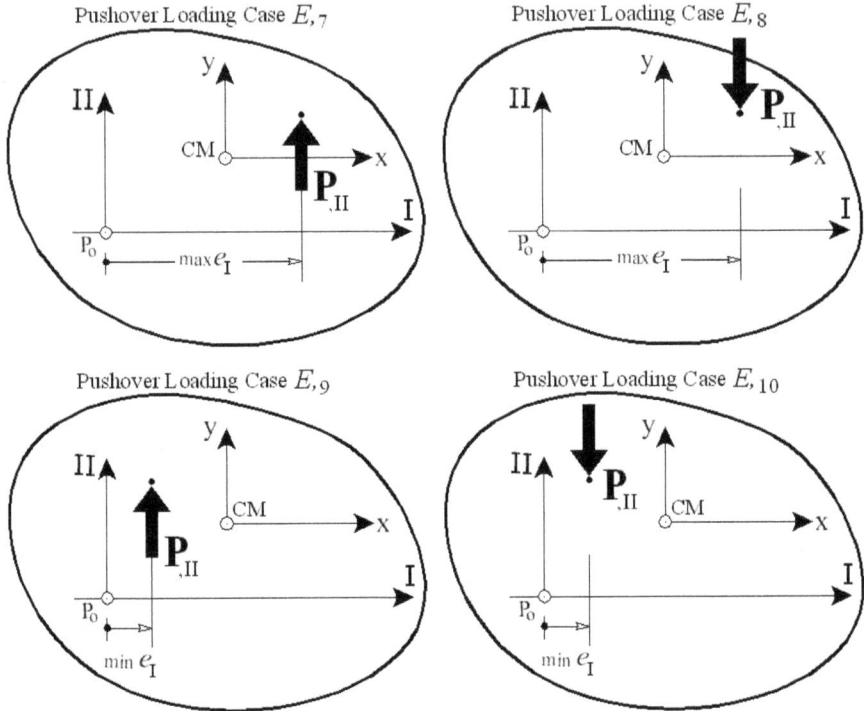

Fig. 8. The envelope of these four pushover analyses represents the results of the second seismic component $E_{,\mathrm{II}}$ .

The envelope of the results from the pushover analyses ($7^{th}$, $8^{th}$, $9^{th}$ & $10^{th}$) is symbolized as $E_{,II}$ and represents the demands of the second seismic component along principal $II$-axis:

$$E_{,II} = \max\left(\left|E_7\right|, \left|E_8\right|, \left|E_9\right|, \left|E_{10}\right|\right) \qquad (34)$$

Note that, the use of the two signs of seismic components is necessary, because superpositions are generally forbidden in the non-linear area (sect.4.3.3.4.1(7)P/EN-1998.01). With reference to the simultaneous action of the two horizontal seismic components, Eq.(1) must be applied in order to obtain the final result of the extreme seismic demands by the spatial seismic action (where $E_{,I}$ & $E_{,II}$ are calculated by Eqs.(33-34). Note that suitable numerical examples, which demostrate the applicability of the proposed method, have been provided in other papers (Makarios, 2009 & 2011), where the correctness of these final results has been verified by the Non-Linear Response History Analysis.

In addition, contemporary Sesmic Codes, such as Eurocode EN-1998, do not give details and quidance regarding the static seismic pushover analyis of irregular in plan, asymmetric, multistorey r/c buildings. In reality, Annex B of EN-1998.01 refers to the calculation of the seismic target-displacement of planar frames only, whilst for spatial irregular in plan multi-storey buildings there are no relevant provisions. The present work intends to precisely cover this gap. The present non-linear static (pushover) procedure is a natural extension of the established equivalent linear (simplified spectral) static method, recommended by the Seismic Codes (EAK/2003, NBCC/1995), which are adequately documented as regards torsional provisions. The twelve static pushover analyses of the irregular asymmetric building involve an important computational cost, but on the other hand, this is much lower compared to the two hundred and twenty-four (224) non-linear dynamic analyses that the Non-Linear Response History Procedure requires. Thus, the presented non-linear static (pushover) procedure is a very attractive option, despite its twelve non-linear static analyses. In addition, within the framework of the proposed procedure, there is also natural supervision, which constitutes a very important issue.

## 3. Conclusions

Consequently, the most important issues that we come across during the static pushover analysis, in order to avoid the exact non-linear response history analysis, have been noted. More specifically, we have explained (1) why it is necessary to use a single degree of freedom system that represents the real structure; (2) how the spatial action of the two horizontal seismic components during the static pushover analysis can be taken into account; (3) what is the most suitable monitoring point in the case of asymmetric buildings; (4) which distribution of the lateral floor static forces is suitable; and (5) at which point in the plan of an asymmetric building, lateral floor static forces should be applied. Design dynamic eccentricities plus accidental eccentricities must be taken into account. This point leads to the calculation of the real or fictitious elastic axis of the building and the real or fictitious horizontal main elastic axes, where the lateral static forces must be oriented during the static pushover analysis (Fig.1). Extensive guidance has been given about the simulation of plastic hinges using Eurocode EN/1998.03. Moreover, various important issues have been discussed concerning seismic non-linear response history analysis. Next, a documented proposal of the optimum equivalent NLSDF system that represents asymmetric multi-storey buildings is presented. The definition of the NLSDF system is mathematically derived, by studying suitable dynamic loadings on the masses of each r/c system and using simplified assumptions. The coupling of the translational and the torsional degrees-of-freedom of the asymmetric building has also been taken into consideration, without dividing the asymmetric building into various, individual (two planar) subsystems. This NLSDF system is used in combination with the inelastic design spectra in order to calculate the seismic demands. The natural meaning of the characteristics of the asymmetric building, such as the fundamental eigenperiod, is not distorted by the optimum equivalent NLSDF system, since both the fundamental period of the multi-storey building and the period of the NLSDF system are very close, due to the fact where the total methodology is derived in a mathematical way. Extended numerical comparisons have verified the correctness of the use of the equivalent NLSDF system (Makarios, 2009 & 2011;). However, on the other hand, it is a known fact that, the sequence of plastic hinges on a building and the yielding/failure mechanism of multi-storey buildings are not correctly calculated by the various pushover analyses, which can only provide approximate results; nevertheless, this question is beyond the scope of the present chapter.

## 4. References

Akiyama, H. 1985. *Earthquake-resistant limit-state design for buildings,* University of Tokyo, Press, Tokyo.

Anastassiadis K, Athanatopoulou A, Makarios T. 1998. Equivalent Static Eccentricities In the Simplified Methods of Seismic Analysis of Buildings. *Earthquake Spectra the Profes. Jour. of the Earth. Engin. Research Inst.;* 1998; vol. 14, Number 1, 1-34.

Athanatopoulou AM. 2005. Critical orientation of three correlated seismic components. *Journal Engineering Structures,* 27, 2, January, pp.301-312.

Cheung V WT, Tso WK. 1986. Eccentricity in Irregular Multistorey Buildings. *Can. Journ. Civ. Eng.* 13, 46-52.

Dowell R, Seible F, Wilson E L. 1998. Pivot Hysterisis Model for Reinforced Concrete Members. *ACI Structural Journal,* Technical Paper, Title no. 95-S55, September-October, pp 607-617.

EAK/2003: *Hellenic Seismic Code*/2003; FEK184B /20.12.1999, FEK 781/18.07.2003, FEK 1154/12.08.2003 (in Greek).

EN-1998.01. 2004. Eurocode 8: *Design of structures for earthquake resistance – Part 1: General rules, seismic actions and rules for buildings.* European Commitetee for Standardization, Brussels.

EN-1998.03. 2005. Eurocode 8: *Design of structures for earthquake resistance – Part 3: Assessment and retrofitting of building.* European Commitetee for Standardization, Brussels.

Fajfar P, Fischinger M. 1987a. Non-linear Seismic Analysis of RC Buildings: Implications of a Case Study. European Earthquake Engineering;1:31-43.

Fajfar P, Fischinger M. 1987b. N2 - a method for non-linear seismic analysis of regular buildings. *Proc. 9th World conf. Earthquake Eengin.,* Tokyo, Kyoto; vol. 5:111-116.

Fajfar P, Gaspersic P. 1996. The N2 Method for the Seismic Damage Analysis of RC Buildings. *Earthq. Engin.& Struct. Dynamics.* vol.25: 31-46.

Hart G, Wong K. 1999. *Structural Dynamics for Structural Engineers.* Willey, New York, USA.

Hejal R, Chopra AK. 1987. Earthquake Response of Torsionally-Coupled Building. *Report No U.S.B./e.e.r.c.-87/20*, Berkeley.

Krawinkler H, Nassar AA. 1992. Seismic design based on ductility and cumulative damage demands and capacities. *Nonlinear Seismic Analysis and Design of Reinforced Concrete Buildings*, eds P.Fajfar and H. Krawinkler. New York: Elsevier Applied Science, 23-39.

Makarios T, Anastassiadis K. 1998a. Real and Fictitious Elastic Axis of multi-storey Buildings: Theory. *The Structural Design of Tall Buildings Journal*, 7,1, 33-55.

Makarios T, Anastassiadis K. 1998b. Real and Fictitious Elastic Axis of multi-storey Buildings: Application. *The Structural Design of Tall Buildings Journal*, 7, 1,57-71.

Makarios T. 2005. Optimum definition of equivalent non-linear SDF system in pushover procedure of multistory r/c frames. *Engineering Structures Journal* v.27, 5, April, pp.814-825.

Makarios T, Athanatopoulou A, Xenidis H. 2006. Numerical verification of properties of the fictitious elastic axis in asymmetric multistorey buildings. *The Structural Design of Tall and Special Buildings Journal*, 15, 3, 249-276.

Makarios T. 2008. Practical calculation of the torsional stiffness radius of multistorey tall buildings. *Journal of the Structural Design of Tall & Special Buildings*. 2008; 17, 1, 39-65.

Makarios T. 2009. Equivalent non-linear single degree of freedom system of spatial asymmetric multi-storey buildings in pushover procedure. Theory & applications. *Journal of the Structural Design of Tall & Special Buildings*, 18,7, pp.729-763.

Makarios T. 2011. Seismic non-linear static new method of spatial asymmetric multi-storey r/c buildings. *Journal of the Structural Design of Tall & Special Buildings*, DOI: 10.1002/tal.640, (in press).

Moustafa A. 2011. Damage-based design earthquake loads for single-degree-of-freedom inelastic structures. *Journal of Structural Engineering*, ASCE, 137(3), 456-467.

National Building Code of Canada (NBCC/95): Associate *Committee on the National Building Code*. National Research Council of Canada, 1995.

Newmark NM, Hall WJ. 1982. Earthquake Spectra and Design. *Earthquake Engineering Research Institute*, Berkeley, CA.

Panagiotakos T, Fardis M. 2001. Deformations of reinforced concrete members at yielding and ultimate. *ACI Structural Journal*, March-April; v.98, No.2. pp 135-148.

Penzien J, Watabe M. 1975. Characteristics of 3-Dimensional Earthquake Ground Motions. *Earthquake Engin. & Struct. Dynamics*. vol. 3, pp365-373.

Qi X, Moehle JP. 1991.Displacement design approach for reinforced concrete structures subjected to earthquakes. *Report No. UCB/EERC-91/02, Univercity of California*, Berkeley.

Otani, S. 1981. Hysteretic models of reinforced concrete for earthquake response analysis. *J. Fac. Eng.*, University of Tokyo, 36(2), 407-441.

Rodriguez M.1994. A measure of the capacity of earthquake ground motions to damage structures. *Earthquake Engin. Struct. Dynam.*; 23:627-643.

Saiidi M, Sozen M A. 1981. Simple Nonlinear Seismic Analysis of r/c Structures. *Journal of the Structural Division*, ASCE 1981; 107:937-952.

Takeda T, Sozen M.A, Nielsen N.N . 1970. Reinforced Concrete Response to Simulated Earthquakes. *Journal Structures Engin*. Div, ASCE, v.96, No 12, pp 2557-2573

Uang C, Bertero VV.1990. Evaluation of Seismic Energy in Structures. *Earthq. Eng. & Struct. Dynam.* vol.19:77-90.

Veletsos AS, Newmark NM. 1960. Effects of inelastic behavior on the response of simple system to earthquake motions. *Proccedings of the 2nd World Conference on Earthquake Engineering.* Japan; 2: pp895-912.

XTRACT. v.3.0.8. 2007. *Cross-sectional X sTRuctural Analysis of ComponenTs.* Imbsen Software System. 9912 Business Park Drive, Suite 130, Sacramento CA 95827.

# 8

# Seismic Performance of Masonry Building

Xiaosong Ren, Pang Li, Chuang Liu and Bin Zhou
*Institute of Structural Engineer and Disaster Reduction, Tongji University,*
*China*

## 1. Introduction

The frequent occurrence of huge earthquake in the recent years results catastrophic losses to the people's life and property, which is mainly caused by the devastation of many buildings. The Tangshan Earthquake (Ms 7.8,1976) and Wenchuan Earthquake (Ms 8.0,2008) should be mentioned here for the destructive influence on the development of China in the recent ages(Housner & Xie, 2002; Liu & Zhang,2008; Wang,2008). After the trial version of seismic design code as TJ 11-74, the Chinese seismic design code was published first in 1978 (as TJ 11-78), revised in 1989 (as GBJ 11-89) , 2000 (as GB 50011-2001) and 2010 (as GB 50011-2010). The minor earthquake, moderate earthquake and major earthquake are concerned in the seismic fortification. The moderate earthquake is defined to be of local fortification intensity. The minor earthquake means the frequent earthquake, whose intensity is about 1.5 degree lower, while the major earthquake means the rare earthquake, whose intensity is about 1 degree higher. The exceeding probability during 50 years as $P_f$ , the maximum acceleration of ground $a_{max}$ and the maximum coefficient of horizontal earthquake action as $\alpha_{max}$ for local fortification intensity 7 are summarized in Table 1. It is seen that the relative major earthquake action, which is defined as the ratio of major earthquake action and minor earthquake action is 6.3 for local fortification intensity 7.

| Condition | $P_f$ | $a_{max}$ (cm/s$^2$) | $\alpha_{max}$ |
|---|---|---|---|
| Minor earthquake | 63.2% | 35 | 0.08 |
| Moderate earthquake | 10% | 100 | 0.23 |
| Major earthquake | 2-3% | 220 | 0.50 |

Table 1. Main parameters for local fortification intensity 7 according to the Chinese code.

In order to realize the seismic objective in China, which is defined as no failure under minor earthquake, repairable damage under moderate earthquake and no collapse under major earthquake, the seismic design procedure should be finished in two steps. The strength and lateral deformation in the elastic range must be checked under minor earthquake action, while for some specified structures, the elasto-plastic deformation analysis should also be done to verify the collapse-resistant capacity of structures under major earthquake action. Because of the relative low construction cost, masonry building is the widely used structural type in China. For this reason, the seismic damage of masonry buildings was specially investigated by the first author just after the 5.12 Wenchuan Earthquake of 2008. Based on the seismic damage collected in the disaster area, the seismic performance of masonry building is discussed. In order to ensure the collapse-resistant capacity, the ductility of structures should be involved. It is necessary to set more margin of shear strength in the design. The method of parcelling masonry structure with reinforced concrete members is suggested to retrofit the existing masonry buildings.

## 2. Seismic damage caused by 5.12 Wenchuan earthquake

The severe damage of buildings caused by 5.12 Wenchuan Earthquake of 2008 is really a good lesson for engineering community to understand more about the seismic performance of structures. The first author took part in the work of site urgent structural assessment in a small mountainous county, Qingchuan County. Among the concerned 133 buildings, there are 6 buildings of reinforced concrete structure and 127 masonry buildings. Till now, there are 44 big after-shocks of magnitudes larger than Ms 5.0, while 9 big after-shocks took place in Qingchuan County, including the largest one (Ms 6.4,16:21, May 25th, 2008). Qingchuan County belongs to the extremely heavy disaster area. The actual seismic intensity of Qingchuan County is 9, which exceeds the level of major earthquake (about intensity 8.5) of the previous fortification intensity 7 in this region.

No steel structure was found in Qingchuan County. The amount of the reinforced concrete structure is about 10% and the others are masonry buildings. The masonry buildings were constructed mainly after 1980 and seismic proof was generally considered by the previous design code. About 50% of masonry buildings collapsed or nearly collapsed while the ratio of reinforced concrete structure is about 20%. The seismic damage of masonry buildings investigated in this area is mainly stated (Lu and Ren, 2008).

### 2.1 Through diagonal cracks or through X-shape cracks on the wall

Through diagonal cracks or through X-shape cracks are the common earthquake induced damages on the walls of the masonry buildings, as shown in Fig.1.This kind of earthquake damage belongs to shear failure, which is caused by the principal tensile stress exceeding shear strength of masonry. The through X-shape cracks are very popular on the longitudinal walls, especially between the door or window openings of nearly every floor. The diagonal cracks usually appear mostly on the bearing transverse walls. These cracks may lead to the obvious decrease of structural capacity and even collapse of the buildings.

Fig. 1. Through diagonal or X-shape cracks on the wall

## 2.2 Horizontal crack on the wall

Another main seismic damage is the horizontal crack on the wall. Horizontal cracks usually appear at the wall near the elevation of floor or roof, which enlarges the damage and results in collapse of pre-cast hollow slab. Meanwhile, horizontal cracks also appear on the end of some bearing brick columns, which lead to the decrease and even loss of the structural capacity. This kind of cracks means horizontal shear failure of walls. It is deduced that the large vertical ground motion may lead to this kind of earthquake damage. Typical phenomenon of horizontal cracks on the wall is shown in Fig. 2.

Fig. 2. Horizontal cracks on the wall or the bearing brick column

## 2.3 Damage of the stair part

Comparing with the other parts, the damage of stair part is relative severe because of the relative large stiffness of the slope structural members. Fig. 3 shows the severe damage of the stair part in the building with irregular plan. From the layout of this building, it is seen that the stair part is the convex part of the T-shaped plan.

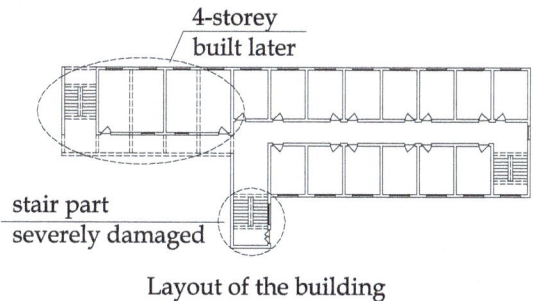

Fig. 3. Partial collapse of stair part

Layout of the building

## 2.4 Damages of nonstructural components

Severe damages on nonstructural components, such as horizontal crack, diagonal crack, even partial collapse can be easily found due to no reliable connections with the main structures. Fig.4 shows the partial collapse of the parapet, which even leads to the damage of the roof slab. Fig. 5 shows the falling of the corridor fence, severe damage of the protruding member in the roof. These are typical phenomena of the seismic induced damage of the nonstructural components.

Fig. 4. Failure of the parapet wall

Fig. 5. Falling of the corridor fence and horizontal crack of the protruding member

## 2.5 Damage caused or aggravated by structural irregularity

Two examples are given here to show the harmful influence of structural irregularity on the building damage. Fig. 6 shows the severe damage of the L-shaped building with unequal height. Fig.7 shows the severe damage in the part of staggered elevation.

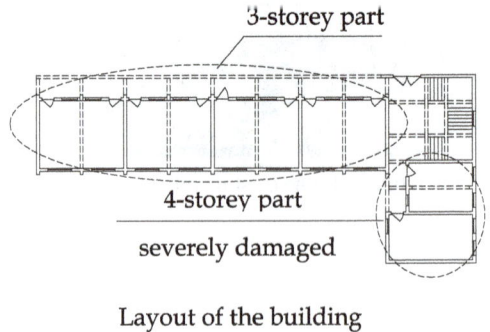

3-storey part

4-storey part

severely damaged

Layout of the building

Fig. 6. Severe damage of masonry buildings with plan irregularity

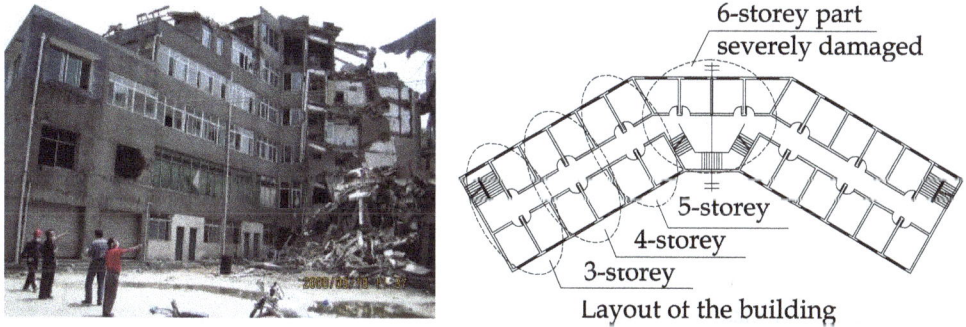

Fig. 7. Severe damage of masonry buildings with elevation irregularity

## 3. Discussion on the current seismic design method

### 3.1 Conceptual design

In order to get better seismic performance, the conceptual design is always very important to achieve besides the seismic analysis, especially in the early stage of architectural scheme. The current seismic code GB50011-2010 stipulates regulations for seismic conceptual design in detail. The consensus is reached by the engineering community that strictly following the seismic conceptual requirement should perform better seismic performance and at least minimize the possibility of the structural collapse (Wang, 2008; Zhang et al., 2008; Ren et al., 2008). The main key points for seismic conceptual design are emphasized here.

### 3.1.1 Structural regularity

Reasonable architecture arrangement may play an important role in the seismic conceptual design, with emphasis on the simplicity and symmetry in plan and elevation for uniform stiffness distribution of structures. Detailed description for regularity is given in the code.

More attention to enhance the structural ductility should be paid to the irregular structures if the seismic joint is not feasible to set, although separating the irregular structure into regular parts by seismic joint is a simple and good way in usual condition. Try to avoid the structural system of one bay transverse bearing wall with outside corridor supported by the cantilever beam.

### 3.1.2 Structural integrity

In order to get the largest possible number of redundancies subjected to earthquake action, structures should be fully integrated by structural members. Good structural integrity will guarantee the good seismic performance of structures.

The measure for structural integrity of masonry buildings should be the reinforced concrete members, such as tie beam, column and cast-in-site slab. The enhancement of the small size masonry wall segments for seismic protection is very important, as these segments are proved to be the weak parts subjected to earthquake in practice. The valid connection between the nonstructural components and the main structure should also be set properly.

## 3.2 Shear strength check

In the current seismic design code, only the shear strength check under minor earthquake is stipulated for seismic design analysis of masonry buildings.

The seismic shear capacity is contributed not only by the masonry wall segment but also by reinforced concrete member. It is checked by

$$V \leq R = \frac{1}{\gamma_{RE}}\left[ \eta_c f_{VE} A + \xi f_t A_c + 0.08 f_y A_s \right] \tag{1}$$

In Equation (1), $V$ is the shear force on the wall, $R$ is the structural resistance, $\gamma_{RE}$ is the seismic adjusting factor which is taken as 1 for bearing wall and 0.75 for self-bearing wall, $\eta_c$ is the confined factor of wall which is usually taken as 1, $f_{VE}$ is the design value for seismic shear strength of wall, $A$ is the net cross area of the wall, $\xi$ is the participation factor of reinforced concrete tie column in the middle which is taken as 0.4 or 0.5 by the number of tie column, $f_t$ and $A_c$ is the design tensile strength of concrete and the cross area of the tie column in middle, $f_y$ and $A_s$ is the design tensile strength and the total area of reinforcements of tie column in middle.

It should be mentioned here that the design value of the shear strength $f_{VE}$ is got by the primary design value of shear strength $f_V$ and the normal stress influence factor $\xi_N$, i.e.

$$f_{VE} = \xi_N f_V \tag{2}$$

The normal stress influence factor is determined by the pressure of the cross section corresponding to the gravity load. It is in the range of 1.0 to 4.8. The beneficial influence of normal stress on the shear strength is caused by the friction in the wall.

For convenience, the strength check parameter, which is defined as the structural resistance divided by the shear force, is set for shear strength check for the wall segment. Satisfactory result means the strength check parameter is no less than 1 as

$$SI = \frac{R}{V} \geq 1 \tag{3}$$

## 3.3 Introspection on the design analysis

The engineering practice showed that strength and deformation are two import factors to evaluate the structural performance. The deformation check is valuable to proceed. The elastic deformation analysis is quite helpful to find some seismic weak parts, such as the torsional irregularity, discontinuous displacement. And the elasto-plastic deformation check can directly verify the structural performance under major earthquake action.

The seismic design of masonry structure is dominated by the shear strength check under minor earthquake. As neither elastic nor elasto-plastic deformation check is involved, it is somewhat questionable to guarantee the collapse resistant capacity under major earthquake (Ren, Weng & Lu,2008).

1.   The investigated seismic damage shows the two way relationship between shear strength and axial strength of the wall. It is different from the theoretical assumption in the current design code that only the shear strength is affected by the axial strength. Once the shear failure happened, the mortar will break and have crack on it, which means that the axial strength will decrease. For the difference of masonry block or mortar and the influence of

construction quality, it is difficult to get the unified model for elasto-plastic deformation analysis. As no convincing progress in the elasto-plastic analysis of masonry structure, the current strength check method should be improved.

2. In practice, the structural ductility is proved to be the key for the collapse-resistant capacity of masonry buildings under the major earthquake. As the ductility of masonry structure is about 1-3, which is less than the normal range as 3-5 and 5-10 for the ductility of reinforced concrete structure and steel structure, the masonry structure is more like the brittle structure. In usually condition, the major earthquake action is about 4.5 to 6.3 times the minor earthquake action. If the strength check parameter is close to 1, which means not many margins for strength, severe damage and even collapse will happen on the structure. It is conflicted with the demand of no collapse under the major earthquake and necessary for more margins of shear strength in the seismic design.

3. The seismic design code emphasizes on the design details such as tie columns and beams for better structural regularity and integrity. Although the shear strength check method in current code can take these factors into consideration, no quantitative index can be got to evaluate the collapse resistant capacity under major earthquake action. To process the shear strength with the demand of deformation capacity may be the feasible way to evaluate the collapse resistant capacity of masonry structure.

4. The earthquake action on the structure is in any arbitrary direction. The earthquake damage shows that once the failure or partial failure of the wall happens in one direction, the wall will easily be broken in another direction due to the out of plan stability problem. Hence it is important to keep uniform seismic capacity in two directions. In usual conditions, the door and window will bring very different seismic capacity in two directions. These wall segments near the door and windows are usually the weak parts of the structure. The ratio of seismic capacity in two directions should be limited to a certain value.

## 4. Analysis of a severely damaged building

A typical severely damaged 3-storey masonry school building is found in Qingchuan County (Ren & Tao, 2011). Typical damaged longitudinal and transverse walls in first storey are shown in Fig. 8. The detailed position of damaged walls is marked in the layout (Fig. 9). As the through cracks are on the load-bearing walls, the structural capacity of this building decreases remarkably. Specially mentioned here, the damaged wall segment in longitudinal direction is in a quite dangerous state as it may collapse or partially collapse in the strong after-shock.

Fig. 8. Through cracks on the longitudinal and transverse load-bearing wall

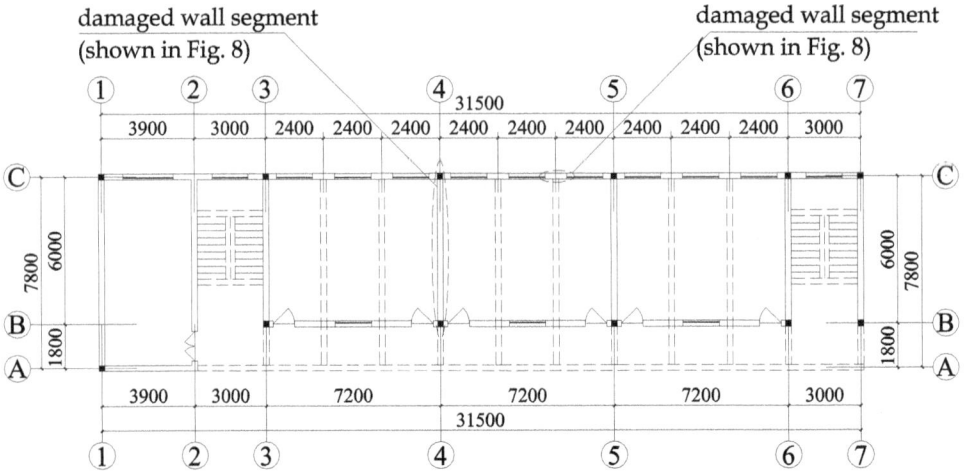

Fig. 9. Structural layout and the position of the severely damaged wall segments

## 4.1 Shear strength check under minor earthquake

An authorized design and analysis software PMCAD/PKPM is used for shear strength check of masonry structure (http://www.pkpm.com.cn/). In site, the mortar, the brick and the concrete are deduced to be M5, MU7.5 and C20. The wall thickness is 240mm. The storey height is 3.2m. The live and dead load on the floor is 2.0 and 4.0kN per square meter respectively.

Qualified result under minor earthquake is shown in Fig.10, as the strength check parameter is larger than 1. The damaged wall segments as shown in Fig. 9 are verified to be the most dangerous segment in transverse and longitudinal direction because of the smallest value of shear strength check parameter, which are 1.26 and 2.19 respectively.

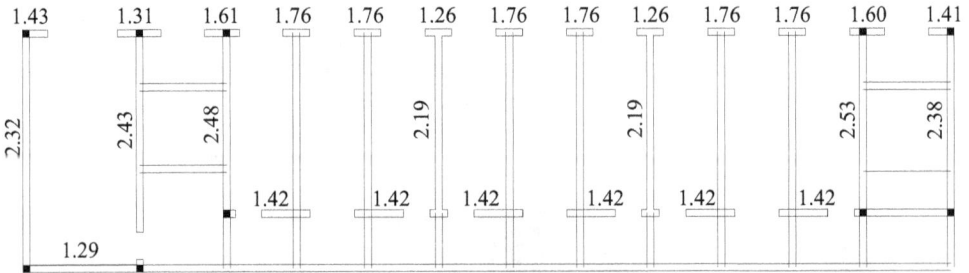

Fig. 10. Shear strength check parameter of the bottom storey

## 4.2 Further analysis

Although satisfactory check for intensity 7 is found according to the current design code, why so severe damage happened on the wall segments? The poor ductility of the masonry structure is the main reason. Here give a simple explanation for it.

Assuming the masonry structures has an idealized curve for shear force V and storey drift Δ as shown in Fig. 11, line 1-2-3 means the elasto-plastic behaviour while line 1-2-4 represents the elastic model. Here point 1, 2 indicate the design state and the yield state of the structure. Point 4 can be determined by point 1 multiplying the relative real earthquake, which is the ratio of the real earthquake action vs. the design earthquake action (or the minor earthquake action). The failure point 3 can be determined by the equal area of two shadowed region in Fig. 11. The deformation ratio of point 3 vs. point 2 is defined as the ductility.

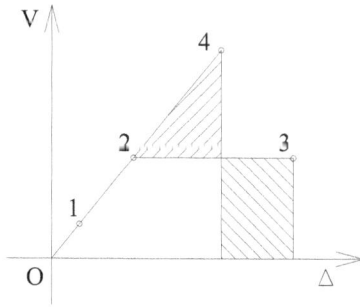

Point 1:design state
   (subjected to minor earthquake)
Point 2:yield state
Point 3:ultimate state for elasto-plastic model
   (subjected to actual earthquake)
   $\Delta_3 / \Delta_2$ is the ductility of structure
Point 4:ultimate state for elastic model
   (subjected to actual earthquake)

Fig. 11. Idealized curve of storey shear force vs. storey drift

When subjected to major earthquake, point 3 can be determined by the equal area of two shadowed regions in Fig. 11. The ratio of ultimate strength vs. design strength ($V_3 / V_1$ or $V_2 / V_1$) is usually about 2.5 to 3, an average value 2.7 is used here. For different fortification intensity, the minimum ductility can be got. From Table 2, it is found that the objective of no collapse under major earthquake may not be easily realized by the shear strength check under minor earthquake, as the ductility of masonry buildings is usually about 1-3.

| Local fortification intensity | 7 | 8 | 9 |
|---|---|---|---|
| Relative major earthquake action | 6.3 | 5.6 | 4.4 |
| Minimum ductility required | 3.22 | 2.65 | 1.83 |

Table 2. Minimum ductility required for different local fortification intensity

In usual condition, the ductility for this kind of masonry buildings with not many reinforced concrete members is no large than 2, the severe damage will happen in the condition of an earthquake action at the level of 5 to 6 times the minor earthquake action, which is smaller than the major earthquake action. This means the possibility of losing structural capacity under major earthquake action, which is about 6.3 times the design earthquake action (minor earthquake action) for fortification intensity 7. So it is not difficult to understand the severe damage on the load bearing walls, as shown in Fig. 8.

## 5. Suggestion for more margin

### 5.1 Improved shear strength check under minor earthquake
As the difficulty to make the elasto-plastic deformation check of masonry buildings, the feasible way for better seismic performance is to set more margin of shear strength under minor earthquake action. Enough shear resistant capacity may be got under the major earthquake action to prevent the loss on axial strength due to the horizontal crack.

For the improvement on Equation (3), the shear strength check under minor earthquake is suggested to satisfy

$$SI = \frac{R}{V} \geq \psi \qquad (4)$$

Where, $\psi$ is the modified limitation with consideration of the structural ductility. Equation (4) means higher requirement of structural capacity comparing with Equation (3). By the simplified model, the suggested parameter $\psi$ can be got by the equal area of two shadowed regions in Fig. 11. It is seen that larger modified limitation should be set for the structure with smaller ductility.

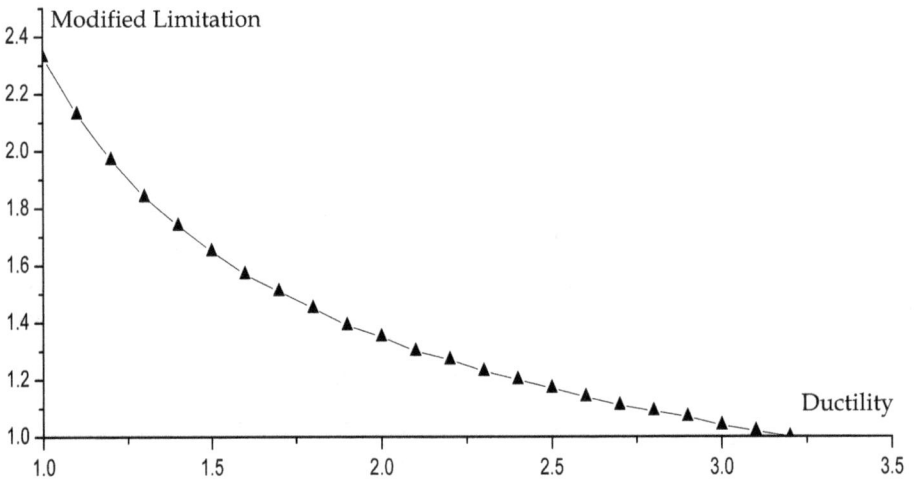

Fig. 12. Modified limitation of shear strength check for different ductility

In the meanwhile, to keep uniform seismic capacity in two directions is also very import for better seismic performance. The strength check parameter in two directions should be close to each other. Referring to regulations concerning the stiffness regularity in the current seismic design code, the ratio of the strength check parameter in two directions should be no less than 0.8. Considering the openings on the wall in the actual condition, some reinforced concrete member s should be used to replace the small masonry wall segment. The structure is transformed to the composite structure of masonry and reinforced concrete, which is quite different from the ordinary masonry structure. Due to the large number and section size of the reinforced concrete member, the structure has more strength along with more ductility. The collapse resistant capacity of masonry buildings should be greatly improved.

## 5.2 Illustrative analysis

The damaged school building is illustrated here as an example here to demonstrate the authors' suggestion. As less tie columns are set, it is suggested to strengthen the structure in the longitudinal direction. As shown in Fig. 13, the small wall segments in axis B and C should be substituted by reinforced concrete member with section 900mmX240mm, while the reinforced concrete tie-columns should be set in axis A at the outside corridor part.

Similar structural analysis by PKPM software is done for strength check under minor earthquake. The main results are shown in Fig, 14 and Table 3. By comparison with Fig.10, it can be seen that the shear strength check ratio along transverse direction is raised from 2.19-2.53 to 2.68-2.80 and the parameter along longitudinal direction is raised from 1.26-1.76 to 1.74-3.07. The strengthened scheme can greatly enhance the seismic capacity.

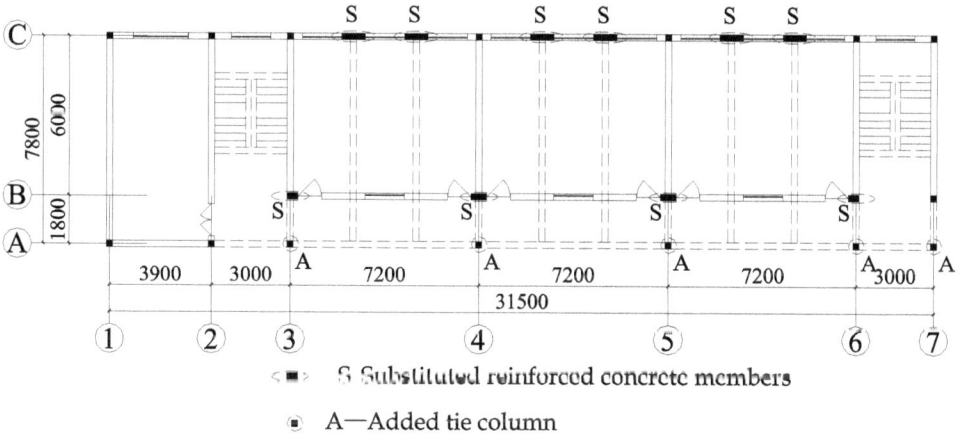

S—Substituted reinforced concrete members

A—Added tie column

Fig. 13. Strengthened structural layout

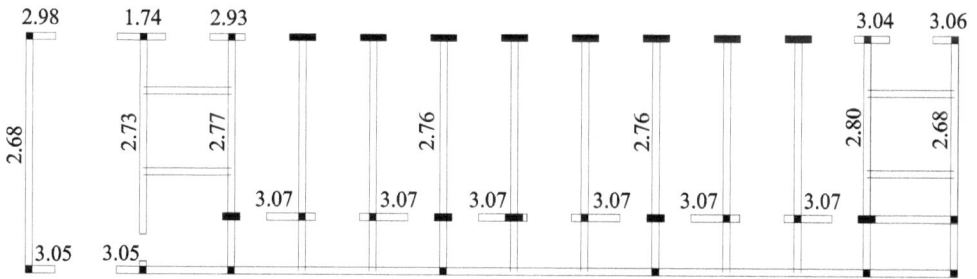

Fig. 14. The shear strength check parameter of the bottom storey of strengthened scheme

| Item | Original | Strengthened |
|---|---|---|
| Average strength check parameter in transverse direction | 2.37 | 2.74 |
| Average strength check parameter in longitudinal direction | 1.65 | 3.05 |
| Ratio of average strength check parameter in two directions | 0.70<0.8 | 0.89>0.80 |

Table 3. Average shear strength check parameter for the original and strengthened structure

## 5.3 Ductility evaluation

A simplified finite element plane model with two reinforced tie columns in the edge is used for ductility evaluation of the original and strengthened structure. It is used to simulate the longitudinal wall. The width of the wall is determined as 2.2m. The width of the reinforced concrete member is determined as 60mm and 300mm for. The area of the steel bar in the tie column is also determined as the average area of steel bar for the wall segment.

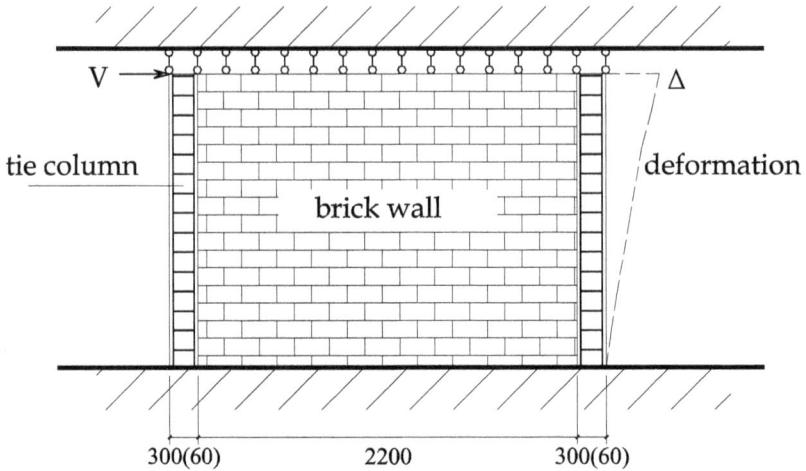

Fig. 15. Finite element model for ductility evaluation

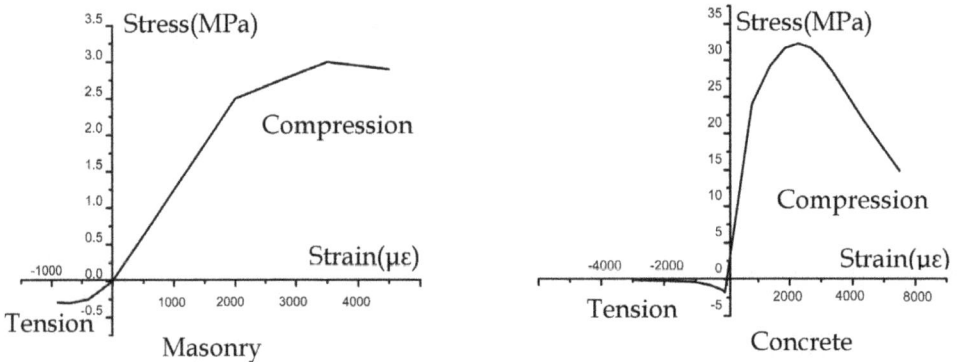

Fig. 16. Constitutive model for masonry and concrete

From the skeleton curve shown in Fig. 17, it is seen that the yield load and ultimate load of the strengthened structure is about twice the value of the original structure. Moreover, the ductility of the structure should be deduced from the hysteretic curve by the principle of no obvious decrease of the primary strength. The ductility is raised from 1.65 to 2.50. From Fig.12, the corresponding modified limitation is determined as 1.55 and 1.17 respectively. The original and the strengthened structure can not and can meet the authors' suggestion.

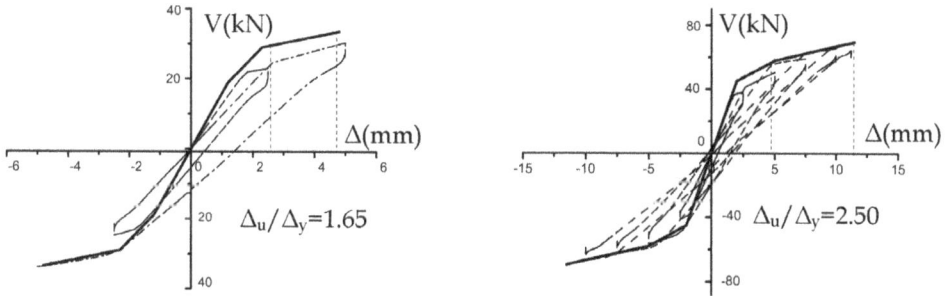

Fig. 17. Hysteretic curve and deduced skeleton curve for the different condition

| Item | Original condition | Strengthened condition |
|---|---|---|
| Elastic stiffness(N/mm) | $1.58\times10^5$ | $2.25\times10^5$ |
| Yield displacement(mm) | 1.20 | 2.00 |
| Crack load(N) | $1.90\times10^5$ | $4.5\times10^5$ |
| Ultimate load(N) | $3.10\times10^5$ | $6.95\times10^5$ |

Table 4. Main results for finite element analysis

## 6. Seismic retrofit by parcelled reinforced members

In 2001, a six storey masonry building in Shanghai was chosen as the first engineering case for the comprehensive transformation of residence. Duplex apartments with slope roof were added in the seventh floor. The ground floor residents moved to the corresponding apartment in the seventh floor. And the ground floor was used to be the space for public community. Elevators were also set. It is an active attempt to aim at the improvement on the residential function along with raising the level of seismic protection (Ren & Liu,2010).

### 6.1 Engineering background and strength check
The original structure was built in 1986. The building was found in a good condition by site test. The grade of masonry and mortar could meet the design demands of MU10 and M10, while the strength of concrete is deduced to be of grade C25.

The structure is a supported-on-transverse-wall system while the stair part is a supported-on-longitudinal-wall subsystem. Reinforced concrete ring beams are set in every floor, but no tie column is set. And prefabricated slab is used for floor. It is found that the potential capacity in vertical direction is exerted for the static strength demand of adding storey while the capacity of seismic resistance is insufficient, especially in longitudinal direction. For the difficulty to retrofit the structure by direct method, the strengthening strategy of load transferring is applied here for the improvement of poor seismic capacity. In Fig.18, the dashed lines represent the demolished walls while the black lines represent the parcelled reinforced concrete members.

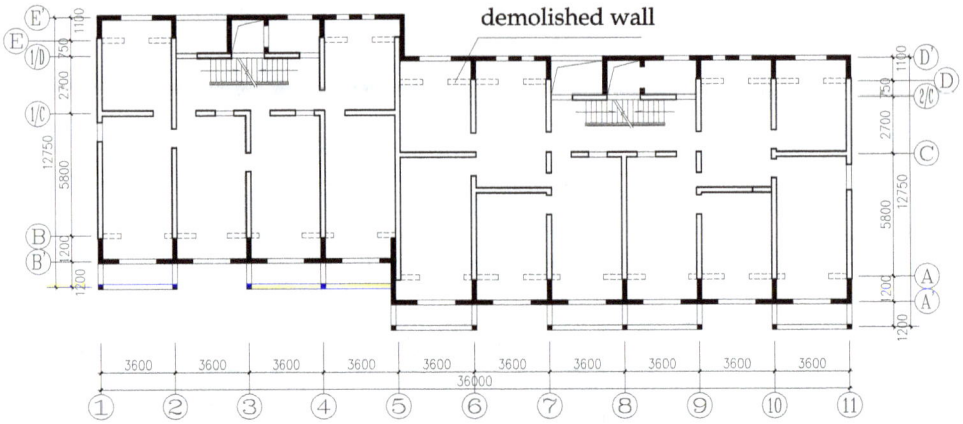

Fig. 18. Structural layout after comprehensive transformation

An eight storey model is established by PKPM software. The overhead floor is treated as the first storey in the model, and the added floor including the duplex part is treated as the eighth storey in the model with the load of one and half storey on it. Detailed results of shear strength check are got. The weakest structural member is located in the 2nd floor, which is shown in Fig. 19. The seismic performance is effectively enhanced.

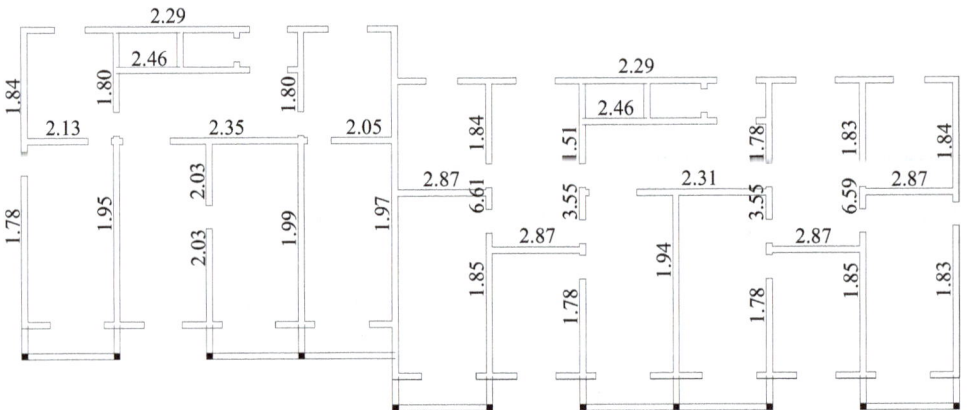

Fig. 19. The shear strength check parameter of the second storey

Moreover, the average parameter for the shear strength check under minor earthquake can be got. From Table 5, the satisfactory shear strength check can be found according to the authors' suggestion.

| Model storey | Transverse direction | Longitudinal driection | Ratio of two directions |
|:---:|:---:|:---:|:---:|
| 1 | 2.70 | 3.82 | 0.71 |
| 2 | 2.72 | 2.67 | 0.98 |
| 3 | 2.70 | 2.65 | 0.98 |
| 4 | 2.70 | 2.73 | 0.99 |
| 5 | 2.92 | 2.89 | 0.99 |
| 6 | 3.30 | 3.32 | 0.99 |
| 7 | 4.07 | 4.18 | 0.97 |
| 8 | 7.80 | 8.41 | 0.93 |

Table 5. The average shear strength check parameter under minor earthquake

Using the SATWE/PKPM program for further analysis under minor earthquake, the lateral storey stiffness can be got.(http://www.pkpm.com.cn). For comparison, another eight storey masonry model without the outside reinforced concrete walls is also established. The longitudinal analytical results are summarized in Table 6. Making comparison between the storey stiffness with and without reinforced concrete walls, the proportion of RC (reinforced concrete) part stiffness is got. It is seen that the earthquake action on the masonry part is greatly reduced. The first 3 natural modes are longitudinal, horizontal and torsional modes, and the corresponding periods are 0.46s, 0.35s, 0.30s. For the little influence of the mode of torsion, the analysis of longitudinal and transverse directions could be made separately. The largest storey drift is 1/2691, which is in fourth model storey or the third storey of actual structure.

| Model storey | Masonry stiffness (kN/m) | Storey stiffness (kN/m) | Proportion of RC part stiffness | Storey drift |
|:---:|:---:|:---:|:---:|:---:|
| 1 | $1.56 \times 10^7$ | $1.39 \times 10^8$ | 88.7% | <1/9999 |
| 2 | $1.19 \times 10^6$ | $5.09 \times 10^6$ | 76.6% | 1/4474 |
| 3 | $8.61 \times 10^5$ | $3.45 \times 10^6$ | 75.0% | 1/2946 |
| 4 | $7.54 \times 10^5$ | $3.00 \times 10^6$ | 74.8% | 1/2691 |
| 5 | $6.93 \times 10^5$ | $2.76 \times 10^6$ | 74.9% | 1/2780 |
| 6 | $6.47 \times 10^5$ | $2.61 \times 10^6$ | 75.3% | 1/3174 |
| 7 | $6.06 \times 10^5$ | $2.52 \times 10^6$ | 75.9% | 1/4126 |
| 8 | $3.64 \times 10^5$ | $2.04 \times 10^6$ | 82.2% | 1/10500 |

Table 6. Analytical results along the longitudinal direction

### 6.2 Lumped storey model for elasto-plastic analysis

For the status of more margins along transverse direction, the elasto-plastic analysis in longitudinal direction is done to verify the collapse resistant capacity under major earthquake.

The masonry part and RC (reinforced concrete) part of every storey are represented by two lumped joints connected by a rigid rod. The lumped storey model is shown in Fig. 20. As stated before, the stress-strain relationship for masonry part is a multi-line curve while it is a curve for reinforced concrete part. Here the proposed skeleton curves for the masonry and reinforced concrete part are demonstrated in Fig. 20.

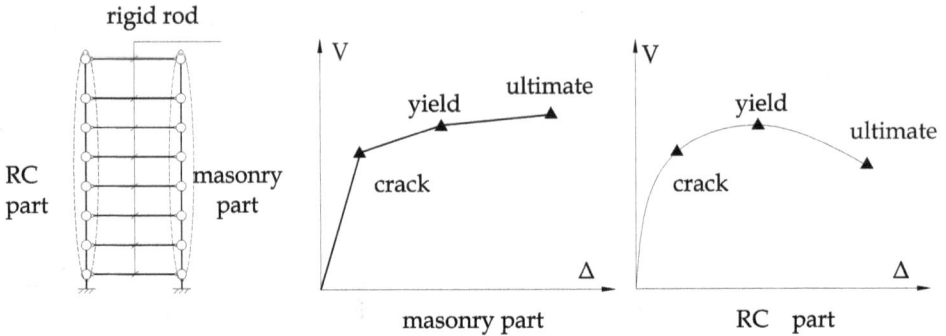

Fig. 20. Lumped storey model and proposed skeleton curve for two parts

From the past work carried out mainly by shaking table test and static push test (Zhu et al. 1980,1983; Xia, et al.,1989; Tomazevic, et al. 1996,1997; Benedett, et al. 1998,2009;Weng et al. 2002; Hori, et al. 2006;), three key points in the curve are denoted as primary crack, yield and ultimate. Although reaching the state of ultimate usually does not mean the complete failure of the structure, the descending stage from the ultimate state to the complete failure state is remarkably different from each other and usually neglected in the analysis. The values of storey drift of three key points are 1/1200, 1/500, 1/250, while the three key points' shear forces are 50%, 90%, 100% of the ultimate shear force. Detailed parameters for the skeleton cure will be gotten by the elastic storey stiffness resulted in the SATWE/PKPM analysis.

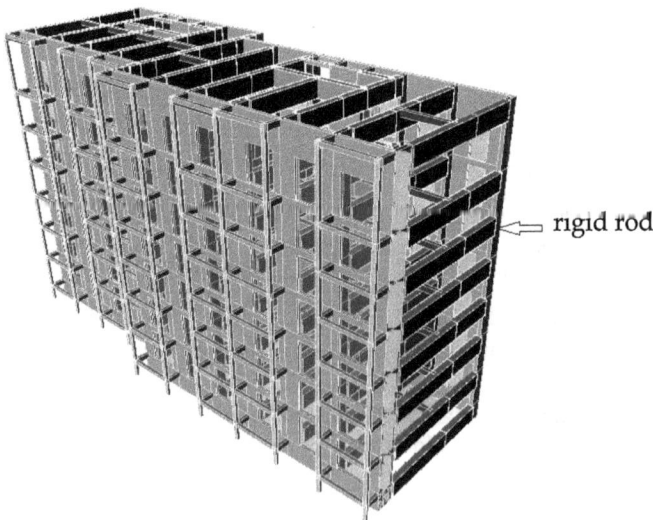

Fig. 21. Equivalent structure model for the analysis of skeleton curve of the RC part

An equivalent the PMCAD/PKPM model is established for the deduced skeleton curve of reinforced concrete part by the EPDA/PKPM software (http://ww.pkpm.com.cn). The masonry members are treated as the rigid rods with lumped mass on the model, which is marked as dark beam in Fig.21. The dynamic characteristic of this equivalent model is proved to be nearly the same as the actual structural model. In order to fit the actual case, the skeleton curve will be described as symmetric curve artificially. The three key points as crack point, yield point and ultimate point can be determined by the characteristics as obvious decrease in stiffness, stiffness degrading and the maximum deformation. The results are summarized in Table 7.

| Model storey | Crack deformation (mm) | Crack load (kN) | Yield deformation (mm) | Yield load (kN) | Ultimate deformaton (mm) | Ultimate load (kN) |
|---|---|---|---|---|---|---|
| 1 | 0.14 | $1.4 \times 10^4$ | 0.5 | $2.2 \times 10^4$ | 0.7 | $2.4 \times 10^4$ |
| 2 | 2.0 | $1.5 \times 10^4$ | 11.0 | $2.2 \times 10^4$ | 18.0 | $2.4 \times 10^4$ |
| 3 | 2.8 | $1.5 \times 10^4$ | 7.5 | $2.0 \times 10^4$ | 11.0 | $1.6 \times 10^4$ |
| 4 | 3.2 | $1.4 \times 10^4$ | 7.5 | $1.8 \times 10^4$ | 10.0 | $1.4 \times 10^4$ |
| 5 | 3.0 | $1.2 \times 10^4$ | 8.0 | $1.5 \times 10^4$ | 10.0 | $1.3 \times 10^4$ |
| 6 | 3.2 | $1.5 \times 10^4$ | 7.0 | $1.3 \times 10^4$ | 11.0 | $1.0 \times 10^4$ |
| 7 | 2.2 | $8.0 \times 10^3$ | 7.0 | $9.0 \times 10^3$ | 10.0 | $7.0 \times 10^3$ |
| 8 | 2.0 | $4.4 \times 10^3$ | 5.5 | $5.5 \times 10^3$ | 10.0 | $4.2 \times 10^3$ |

Table 7. Key parameters for skeleton curve of reinforced concrete part

The lumped storey model has similar structural characteristics of the space model by PKPM software. The first vibration mode is of the first grade transversal vibration mode with natural period 0.44s, while the result of space structural model is 0.45s. And the story drift by the method of earthquake spectrum for Shanghai region (maximum horizontal influence coefficient 0.08) is quite close to the results by PKPM software, which is shown in column "response spectrum" of Table 8.

### 6.3 Elasto-plastic analysis under major earthquake

Two artificial records and a natural record are used as the ground acceleration input for the lumped storey model. The peak acceleration is 220 cm/s². The response spectrum for the input seismic record and the design spectrum for Shanghai code are also shown in Fig.22. The symbol A-1,A-2 and N-1 in the following paragraphs represent the condition of two artificial seismic records and the natural seismic record.

The storey drift in different conditions is summarized in Table 8. The data in column "A-1", "A-2" and "N-1" is the elasto-plastic storey drift under three input ground acceleration. The data in column "average" is the average storey drift under three records. The data in column "response spectrum" is the elastic storey drift under minor earthquake action. Using the ratio of average storey drift divided by the elastic storey drift under major earthquake action, which the data in column "response spectrum" multiply 6.3, the amplifying factor for storey drift is listed in the last column of Table 8.

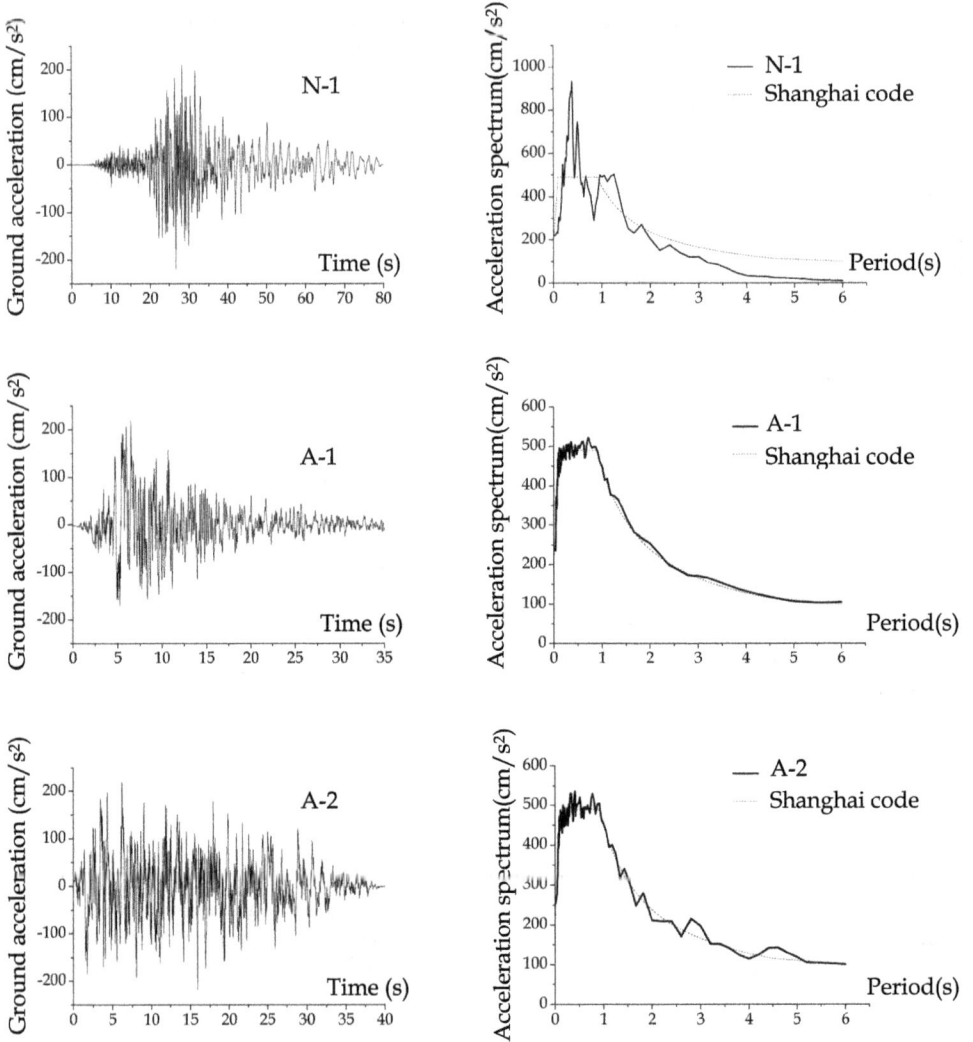

Fig. 22. Input ground acceleration and the response spectrum

| Model storey | A-1 | A-2 | N-1 | Average | Response spectrum | Amplifying factor for storey drift |
|---|---|---|---|---|---|---|
| 1 | 1/6292 | 1/5450 | 1/4846 | 1/5467 | 1/51395 | 1.12 |
| 2 | 1/572 | 1/418 | 1/322 | 1/414 | 1/4071 | 1.20 |
| 3 | 1/511 | 1/433 | 1/311 | 1/401 | 1/3141 | 1.11 |
| 4 | 1/453 | 1/399 | 1/304 | 1/375 | 1/2812 | 1.23 |
| 5 | 1/427 | 1/372 | 1/284 | 1/351 | 1/2919 | 1.34 |
| 6 | 1/491 | 1/449 | 1/407 | 1/446 | 1/2979 | 1.06 |
| 7 | 1/660 | 1/633 | 1/478 | 1/578 | 1/4000 | 1.10 |
| 8 | 1/579 | 1/676 | 1/509 | 1/580 | 1/3784 | 1.03 |

Table 8. Elasto-plastic storey drift, elastic storey drift and the amplifying factor

From Table 8, it is seen that the strengthened structure can satisfy the demand of no collapse under major earthquake as the maximum storey drift is less than 1/250, which is the drift of ultimate point of masonry part. It is shown that the fourth and fifth storey have more plastic deformation than other storeys for the relative large values of amplifying factor for storey drift. The whole degree of structural plasticity is not very large. Moreover, the structural response under natural record is large than the reponse in aritificial records due to the resonant period near the structual period, which can be found on its curve of response spectrum in Fig. 22.

The distribution of storey shear can also be obtained. Table 9 is the result of the proportion of storey shear on the reinforced concrete walls. The column "A-1", "A-2", and "N-1" is the proportion under three input ground acceleration, while the last column represents the average data of storey shear proportion. It is seen that the values are about 84%~90% and larger than the the corresponding values in elastic range. This means that parcelled reinforced concrete walls contribute much more strength in plastic stage than in elastic stage. And the small changes in the condition of A-1, A-2 and N-1 demonstrate the re-distribution of storey shear in the plastic range.

| Model storey | A-1 | A-2 | N-1 | Average |
|---|---|---|---|---|
| 1 | 86.2% | 89.7% | 86.9% | 87.6% |
| 2 | 86.0% | 86.0% | 86.1% | 86.0% |
| 3 | 86.2% | 86.2% | 86.1% | 86.2% |
| 4 | 85.7% | 85.1% | 85.4% | 85.4% |
| 5 | 85.3% | 85.3% | 85.2% | 85.3% |
| 6 | 84.1% | 84.3% | 84.3% | 84.2% |
| 7 | 85.7% | 85.7% | 85.7% | 85.7% |
| 8 | 85.4% | 85.1% | 86.2% | 85.6% |

Table 9. Proportion of storey shear distribution for reinforced concrete part

Although the earthquake action on the masonry part is limited after the adding of reinforced concrete walls, the masonry part is still the key for satisfying the demand of no collapse under major earthquake.

## 7. Conclusion

The discussion of the seismic performance of masonry building is presented here. The collapse-resistant capacity under major earthquake action is somewhat questionable according to the current design code. It is of great importance for better structure performance subjected to major earthquake. For the purpose of "no failure under minor earthquake, repairable damage under moderate earthquake and no collapse under major earthquake", the structural ductility should be improved. More shear strength margin along with structural regularity and integrity are suggested in order to get better seismic performance. The suggestion can be easily realized in the current design analysis procedure. Besides a typical damaged masonry building illustrated, a success engineering case of retrofitting existing masonry building by parcelled reinforced concrete members is also presented to proceed seismic design by the authors' suggestion. The seismic retrofit by parcelled reinforced concrete member is quite effective for the increment of the strength and the enhancement of the ductility. As a lot of masonry buildings exist in China and will not be demolished in a short time, parcelling reinforced members is the feasible and practical way for better seismic performance of existing masonry buildings.

## 8. Acknowledgment

Financial support from Shanghai Natural Science Foundation (No. 09ZR1433400) is gratefully acknowledged.

## 9. References

Benedett, D. Carydis, P. and Pezzoli, P. (1998), Shaking Table Test on 24 Simple Masonry Buildings, *Earthquake Engineering and Structural Dynamics*, Vol.27,No.1,pp. 67-90

Benedett, D. Carydis, P. and Pezzoli, P. (2009), Cyclic Behavior of Combined and Confined Masonry Walls, *Engineering Structures*, Vol.38,No.1,pp. 240-259

Bruneau, M. (1994) State of the art Report on Seismic Performance of Unreinforced Masonry Building. *Journal of Structural Engineering*, ASCE,Vol. 120,No.1,pp.230-251.

Hori, N. Inoue, N. Purushotam, D. and Kobayashi, T. (2006), Experimental and Analytical Studies on Earthquake Resistant Behavior of Confined Concrete Block Masonry Structures, *Earthquake Engineering and Structural Dynamics*, Vol.35,No.1, pp.1699-1719

Housner, G. W. and Xie, L. L. ed. (2002), *The Great Tangshan Earthquake of 1976*, Technical Report EERL 2002.001. California Institute of Technology, Pasadena, California.

http://www.pkpm.com.cn/

Liu, X. H., Zhang H.X.(1981) A Study of Aseismic Characteristics of Masonry Buildings with Reinforced Concrete Tie-columns , *Journal of Building Structures*, Vol. 2,No.6,pp.47-55. (in Chinese)

Lu, X.L. and Ren, X.S.(2008), Site Urgent Structural Assessment of Buildings in Earthquake-hit Area of Sichuan and Primary Analysis on Earthquake Damages, *Proceedings of 14th World Conference on Earthquake Engineering*, No. S31-034, Bejing, China,Oct. 2008.

National Standards of the People's Republic of China (2001). *Code for Design of Masonry Structures (GB 50003-2001) (English version)*, Architecture & Building Press. Beijing, China

National Standards of the People's Republic of China (2001). *Code for Design of Concrete Structures (GB 50010-2001) (English version)*, China Architecture & Building Press. Beijing, China

National Standards of the People's Republic of China (2001). *Code for Seismic Design of Buildings (GB 50011-2001) (English version)*, China Architecture & Building Press. Beijing, China

Ren X. S., Liu C. (2010). Seismic Analysis of Multi-story Masonry Structure Parceled with Reinforced Concrete Wall [J], *Journal of Building Structures*,2010,31 (Supplementary Issue No.2): 334-339. (in Chinese)

Ren X. S., Weng D.G., Lu X.L.(2008). Earthquake Damage of Masonry Buildings in Sichuan Province and Discussion on Seismic Design of Pimary and Middle School Buildings, *Earthquake Engineering and Retrofitting*, Vol.30,No.4,pp.71-76. (in Chinese)

Ren X. S., Tao Y.F.(2011),Discussion on The Seismic Design Analysis Method of masonry Building,*Proceedings of International Conference on Structure and Building Material*, pp.3952-3957, Guangzhou,China, Jan. 2011

Tomazevic, M. Lutman, M. and Petkovic, L. (1996), Seismic Behavior of Masonry Walls: Experimental Simulation, *Journal of Structural Engineering, ASCE*, Vol.122,No.9, pp.1040-1047.

Tomazevic, M. and Klemenec, I. (1997), Seismic Behavior of Masonry Walls: Experimental Simulation, *Earthquake Engineering and Structural Dynamics*, Vol.26,No.1, pp.1059-1071

Wang Y.Y. (2008). Lessons Learned from the "5.12" Wenchuan Earthquake: Evaluation of Earthquake Performance Objectives and the Importance of Seismic Conceptual Design Principles, *Earthquake Engineering and Engineering Vibration*, Vol.7,No.3,pp. 255-262.

Weng D.G., Lu X.L., Ren X.S. et al. (2002), Experimental Study on Seismic Resistant Capacity of Masonry Walls, *Proceedings of 4th Multi-lateral Workshop on Development of Earthquake Technologies and their Integration for the Asia-Pacific Region(EQTAP)*,Tokyo, Japan,Sept. ,2002

Xia J. Q., Huang Q. S.(1989), Model Test of Brick Masonry Buildings Strengthened with Reinforced Concrete Tie Columns, *Earthquake Engineering and Engineering Vibration*, Vol. 9, No.2,pp. 83-96. (in Chinese)

Zhu B.L., Wu M.S., Jiang Z.X. (1980) , The Test Study on the Behavior of Masonry under the
    Reversed Load, *Journal of Tongji University*, No.2,pp.1-14. (in Chinese)
Zhu B.L., Jiang Z.X., Wu M.S. (1983). Study on the Seismic Behavior of Masonry Buildings
    with Reinforced Concrete Tie-columns, *Journal of Tongji University*, No.1,pp.21-43.
    (in Chinese)

# 9

# Seismic Response of Reinforced Concrete Columns

Halil Sezen and Muhammad S. Lodhi
*Department of Civil and Environmental Engineering and Geodetic Science,*
*The Ohio State University, Columbus, Ohio*
*USA*

## 1. Introduction

There are a large number of reinforced concrete buildings in seismically active areas of the world that are not built in accordance with modern seismic design provisions such as those published by American Concrete Institution ([ACI] Committee 318, 2008). In the United States and other parts of the developed world, these buildings were constructed between 1930s to mid 1970s according to the building code requirements of that time. Even today, in low to moderate seismic regions and in some developing countries that are in process of developing and implementing their seismic codes, reinforced concrete structures are being designed and built without essential seismic details deemed vital to withstand large lateral loads. These buildings often have low lateral displacement capacities and undergo rapid degradation of shear strength and axial load carrying capacity during strong ground motions and hence are extremely vulnerable to excessive structural damage or collapse during future earthquakes.

In the past, the earthquakes have caused wide spread damage to the reinforced concrete structures with inadequate seismic design and construction practices. For example, during Kashmir (Pakistan) earthquake of 2005 and Haiti earthquake of 2010, extensive structural damage to residential, commercial and government buildings was observed (Earthquake Engineering Research Institute [EERI], 2005; Mid-America Earthquake [MAE] Center, 2005; U.S. Geological Survey [USGS]/EERI, 2010). The damage was attributed largely to lack of earthquake-resistant design, poor standard of construction and inferior quality of building materials. In majority of the collapsed or damaged structures, structural types, member dimensions and detailing practices (insufficient lap length, improper lap location and lack of confinement in columns etc) were found inadequate to resist forces imposed by these earthquakes. The 2011 off The Pacific Coast of Tohoku (Japan) earthquake is a modern day example of large scale devastation to a highly industrial nation in which building and infrastructure is well designed and constructed. Although, majority of causalities and large scale destruction of infrastructure was caused by ensuing tsunami, limited damage to the buildings due to ground shaking was reported (Pacific Earthquake Engineering Research Center [PEER]/EERI/Geotechnical Extreme-Event Reconnaissance [GEER]/Tsunami Field Investigation Team, 2011; Takewaki et al. 2011). However, extensive and severe structural damage was observed in older residential and commercial buildings that were constructed prior to 1978 code revision of Japan, whereas modern structures built to withstand seismic demands did not sustain any substantial and widespread damage (Aydan & Tano, 2011).

The existing reinforced concrete buildings with deficient seismic design can, however, be retrofitted to enhance their performance during future earthquakes. The need to assess their vulnerability to earthquake damage, and suggest desired level of retrofit, requires evaluation of their expected behavior in terms of strength and deformation capacity. In addition to the retrofitting requirements, there may be many situations where structures are required to be analyzed accurately to evaluate their structural responses. For example, in performance- and displacement-based design philosophy (Priestley et al, 2007, Structural Engineers Association of California [SEAOC], 2002), important existing buildings and planned future structures may need to be evaluated to determine their maximum load carrying capacity, ultimate deformation capacity, progression of the damage, and collapse mechanism. However, a realistic seismic damage analysis, in pre- or post-earthquake scenario, requires development of analytical models to accurately predict non-linear structural behavior during the seismic event (Mergos & Kappos, 2010).

The expected behavior of a structure can be evaluated by determining load-deformation responses of the concrete elements, such as beams, columns and shear walls, considering all potential failure mechanisms associated with axial, flexure and shear behavior. The pattern of damage observed during past earthquakes suggests that columns are the most critical elements that sustain damage and lead the potential building failure. Hence, understanding of their response to applied seismic loads is vital for overall assessment of the structural performance. This paper presents a procedure for response estimation of reinforced concrete columns subjected to lateral loads, with focus on modeling columns commonly found in the older existing buildings. Traditionally, these columns have insufficient and widely spaced transverse reinforcement and non-seismic details such as 90-degree end hooks and splicing of the longitudinal bars in the regions experiencing largest inelastic deformations near column ends. Due to such deficiencies, columns may not have sufficient shear strength to develop plastic hinges at the ends. Also, wide spacing of column ties does not provide good confinement to the core concrete resulting in non-ductile behavior and sudden brittle failure. Although, this study focuses on modeling columns with poor seismic details, the proposed analytical procedure is equally applicable to predict structural response of the columns designed to meet the requirements of modern seismic codes.

## 2. Research background and overview

When a typical fixed ended reinforced concrete column is subjected to the lateral loads at its ends, it undergoes total lateral deformation that is mainly comprised of three components due to flexure, reinforcement slip and shear mechanisms (Setzler & Sezen, 2008; Sezen & Moehle, 2004) as shown in Figure 1.

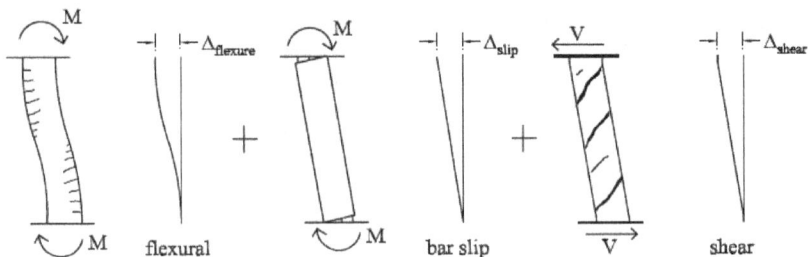

Fig. 1. Components of the total lateral deformation in a fixed ended column

For a column whose behavior is dominated by flexure, issues regarding its performance evaluation have been studied well and current design procedures for strength estimation are generally considered well established. Among available approaches, most of which are either based on lumped plasticity models or distributed nonlinearity models, fiber models are considered advanced analytical procedures that can conveniently be employed for evaluating structural response. It must however be noted that fiber models are appropriate tools for analyzing flexural performance only and behavior of the columns dominated by shear related mechanisms can not be simulated.

For evaluating shear response of structural elements, such as beams and columns, many analytical models and theories have been presented in the past. Some of the most commonly used approaches are strut and tie models (Mörsh, 1902; Ritter, 1899; Schlaich et al., 1987) and empirical formulations/rational theories based on experimental observations such as Arakawa equation (Arakawa, 1970), Modified Compression Field Theory (MCFT) (Vecchio & Collins, 1986) and Disturbed Stress Field Model (DSFM) (Vecchio, 2000). These approaches are fundamentally different in their theoretical modeling and conceptual development. Their applicability to structural members, computational demand and accuracy also vary in wide range from one approach to the other. Hence, accurate modeling of the shear behavior in beams and columns still remains elusive. MCFT is a powerful tool to model the response of reinforced concrete elements subjected to in-plane shear and normal stresses. However, in order to evaluate flexure-shear response of the reinforced concrete columns by MCFT, the member needs to be discretized into large number of biaxially stressed elements and analyzed using nonlinear finite element procedure (Vecchio, 1989). Vecchio and Collins (Vecchio & Collins, 1988) extended concept of MCFT to fiber model approach for response estimation of reinforced concrete beams loaded in combined axial, shear and flexural forces. In this approach, concrete fibers are treated as biaxially stressed elements in the cross section and analyzed for in-plane stress field based on MCFT. Later, this approach was improved for accurate determination of shear stress distribution on the cross section and advanced formulations were implemented successfully into a non-linear section analysis computer program called Response-2000 (Bentz, 2000). The application of the MCFT in finite element approach or sectional analysis approach yields reliable flexure-shear response, but results in fastidious computations which are not simple for practical applications.

Total lateral deformation of a concrete column is mainly comprised of the flexure and shear components. These mechanisms interact with each other and corresponding deformations do not occur independently. For example, in the web of a reinforced concrete column, axial strain due to flexural mechanism will increase principal tensile strain and width of the shear crack resulting in lower shear capacity of the element. On the other hand, it has been established by experimental evidence that principal compressive stress in the concrete is function of principal compressive strain as well as of principal tensile strain (Vecchio & Collins, 1986). Compressive strength and stiffness of concrete decrease as tensile strains increase. The concrete in the web of laterally loaded element is subjected to shear stresses in addition to the normal stresses due to axial load and flexure. As the shear stresses increase, principal tensile strains increase which will decrease compressive strength of the concrete resulting in lower flexural strength of the element.

Therefore, any numerical procedure that aims to model overall lateral load-displacement relationship must take the interaction of flexural and shear mechanisms into account. Recently, Mostafaei and Kabeyasawa (Mostafaei & Kabeyasawa, 2007) presented Axial-Shear-Flexural Interaction (ASFI) approach for the displacement-based analysis of reinforced concrete elements such as beams, columns and shear walls by considering interaction between axial, shear and flexural mechanisms. This macro-model based approach consists of two models evaluating axial-flexural and axial-shear responses simultaneously to obtain total response of elements subjected to axial, flexural and shear loads. In this approach, axial-flexural behavior is simulated by employing conventional section analysis or fiber model whereas axial-shear response is determined through MCFT by considering one integration point in the in-plane stress conditions. The axial-flexural and axial-shear mechanisms are coupled in average stress-strain field considering axial deformation interaction and softening of concrete compression strength while satisfying compatibility and equilibrium conditions. Although, ASFI approach reduces computational demand considerably as compared to other models implementing MCFT into finite element analysis approach or sectional analysis approach, computational process is still intense and complicated due to coupling of the axial-flexure and axial-shear mechanisms and requires a deliberate iterative scheme at each loading step. However, few concepts from ASFI approach are utilized in the model proposed in this study.

Few studies in the recent past have also addressed the issues of stiffness degradation and strength deterioration in the reinforced concrete elements dominated by shear or shear-flexure behaviors. These studies represent advanced formulations for fiber-based element (Ceresa et al., 2007, 2009; Chao & Loh, 2007; Mullapudi & Ayoub, 2008, 2010; Xu & Zhang, 2011; Zhang & Xu, 2010) and Macro-element model (Mergos & Kappos, 2008, 2010) and consider interaction between inelastic shear and nonlinear flexural behaviors with different conceptual backgrounds, solution strategies and implementation complexities. A state-of-art review is presented on fiber elements with focus on concentrated plastic-hinge type model that can be implemented in displacement-based finite element programs (Ceresa et al., 2007).

Currently available studies for response estimation of non-ductile reinforced concrete columns show that the approaches that can predict structural behavior with good accuracy employ complicated and computation-intensive procedures that may not be amenable and are difficult to implement. As a result, many approaches try to simplify the process by making simplifying assumptions but in most cases this is done at the cost of accuracy. A suitable procedure is proposed here to address critical modeling issues while predicting the response accurately and keeping overall computational process simple with easy implementation. The authors believe that the proposed model can effectively be employed to predict the strength and total lateral displacement capacity, considering the deformation components due to flexure, shear and reinforcement slip. Implementation of the proposed procedure results in satisfactory response envelope curves which can be used for development of cyclic response models.

## 3. The proposed analytical procedure for response estimation of columns

Flexural and shear deformations in the proposed model are calculated independently while considering the interaction between these mechanisms and then combined together depending upon dominant failure mode. The flexural deformations are determined through fiber section model considering shear effects by employing compressive constitutive law for cracked concrete. Shear deformations are calculated by combination of MCFT (Vecchio & Collins, 1986) and shear response envelope by Sezen (Sezen, 2008), while considering effect of axial strains due to flexure on shear mechanism. Lateral deformation component due to reinforcement slip in beam-column joint regions is determined separately and added to the flexural and shear deformation components to obtain total response. The interaction between flexure and shear mechanisms allows for accurate response estimation while decoupled flexural analysis minimizes complexity of calculations and makes the analysis process relatively simple and easy. In addition, buckling of compression bars under large compressive strains is also incorporated in the analysis by employing separate stress-strain relationships for reinforcing steel in compression. The effects of concrete tension strength and softening of cracked concrete in compression are also considered. The details of the components deformation models and total deformation model are presented in the following sub-sections.

## 3.1 Flexural analysis and deformations

Flexural deformations in the proposed procedure are determined from fiber section analysis in one-dimensional stress field. For the reinforced concrete elements subjected to bending moment and axial load, such as beams or columns, fiber model approach is usually handy and accurate approach if actual stress distribution across the depth of the cross-section is considered. The reliability of the analysis is also directly related to the ability of the constitutive material models to accurately simulate material behavior and level of simplifying assumptions during the analysis. In this approach, a reinforced concrete cross-section is discretized into finite number of concrete and steel fibers. Each of the fibers is idealized as a uniaxial element with its unique stress-strain relationship. Bernoulli's principle, that plane section before bending remains plane after bending, is the main hypothesis in the analysis and implies that the longitudinal strain in concrete and steel at any point in the cross section is proportional to its distance from neutral axis resulting in linear strain distribution. Based upon the resulting strain profile, stress distribution for concrete and reinforcing steel can be determined in accordance with their respective stress-strain relationships. By satisfying equilibrium equations at the cross section, the moment capacity of the section is determined. The process is repeated number of times by incrementing longitudinal stain until either the concrete or steel fails as per defined failure criterion.

In conventional flexural section analysis, the concrete behavior is simulated by its response usually derived from standard cylinder test where it is subjected to uniaxial compression. The strain conditions for the concrete in the web of a laterally loaded reinforced concrete beam or column are significantly different from those in a cylinder test. The concrete in a cylinder test is subjected to only small tensile strains primarily due to Poisson's effect, whereas, the concrete in the web experience shear stresses in addition to the normal stresses due to axial load and flexure. Due to applied shear stresses, concrete in the web cracks diagonally in the direction normal to principal tensile strain. As mentioned earlier, experimental evidence has shown that that principal compressive stress in the concrete is not only the function of principal compressive strain but is also affected by the coexisting principal tensile strain in a way that compressive strength and stiffness of the concrete decrease as tensile strains increase (Vecchio & Collins, 1986). This implies that the concrete subjected to combined normal compressive and shear stresses is weaker in compression than the concrete subjected to normal compressive stresses only. Hence, the concrete in the web of a laterally loaded column must exhibit weaker and softer response as compared to the concrete subjected to uniaxial compression in cylinder test. The behavior of the cracked concrete in the manner explained above is called compression-softening and is illustrated in Figure 2. In this figure, $f_c'$ is compressive strength of the concrete, $f_{c2}$ is principal compressive stress in the concrete and $\varepsilon_{c2}$ is principal compressive strain in the concrete.

The effect of shear stress on degrading compressive strength of the concrete can be taken into account by considering compressive stress-strain relationships of diagonally cracked concrete in flexural section analysis instead of employing conventional constitutive relationship for uniaxially compressed concrete. This can be done by softening the response of concrete in uniaxial compression by a factor which decreases as shear deformations increase. This factor, known as compression softening factor $\beta$, is function of principal tensile strain in the concrete and is defined as following (Vecchio & Collins, 1986).

$$\beta = \frac{1}{0.8 - 0.34 \dfrac{\varepsilon_{c1}}{\varepsilon_{co}}} \leq 1.0 \tag{1}$$

where $\varepsilon_{co}$ is concrete strain corresponding maximum concrete cylinder strength and $\varepsilon_{c1}$ is principal tensile strain in the concrete which can be determined through in-plane shear analysis of the flexural element. The procedure for determining principal tensile strain in the concrete and compression softening factor is explained in the subsequent section.

In addition to considering cracked concrete behavior in the fiber model, enhancement in the strength and ductility of the concrete due to confinement and contribution of the concrete tensile properties to section moment capacity must also be considered in the analysis. For determining realistic moment capacity and analyzing the buckling of the longitudinal bars under excessive compressive strains, confined core concrete and unconfined cover concrete are modeled separately with their respective stress-strain relationships.

Fiber model analysis results in a moment-curvature relationship for given geometric and material properties, reinforcement details and applied axial load for the cross-section being analyzed. From here, lateral load $V$ corresponding to respective moment capacity, resulting average shear stress from flexural analysis $\tau_f$ and maximum lateral force sustainable by the column can be calculated with the help of following equations.

$$V = \frac{M}{a} \qquad \tau_f = \frac{V}{bd} \qquad V_p = \frac{M_p}{a} \tag{2}$$

where, $M$ is the flexural section moment capacity at any load level, $b$ is width of the section, $d$ is the effective depth of the section, $a$ is the shear span equal to cantilever column length and one half of the length a fixed ended column, and $M_p$ is maximum moment capacity from flexural section analysis.

Flexure deformations are calculated with the help of plastic hinge model in which elastic and inelastic curvatures are idealized separately. In this model, a linear curvature distribution is assumed in the elastic range over the length of the column, and the inelastic curvatures are lumped at the column end over the plastic hinge length. The conceptual illustration of the plastic hinge model for a cantilever column is presented in Figure 3. Hence, lateral displacement due to flexure $\Delta_f$ can be calculated by integrating curvature over the length of the column as per Equation 3.

$$\Delta_f = \int_0^a \phi(x)x\,dx \tag{3}$$

where $\phi(x)$ is section curvature at distance $x$ measured along column axis, and $\psi_y$ is curvature at yield point. The plastic hinge length $L_p$ is taken as one-half of the section depth $h$.

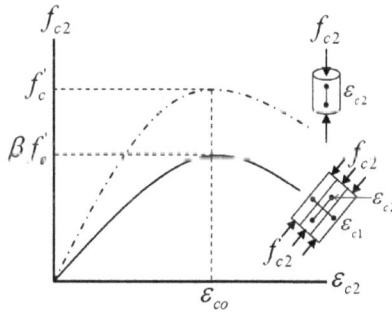

Fig. 2. Behavior of the cracked concrete in compression

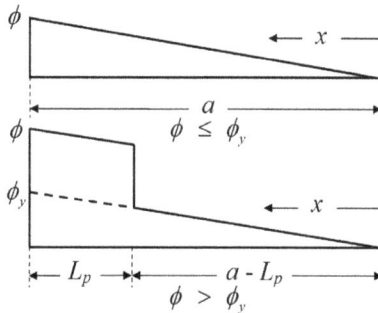

Fig. 3. Plastic hinge model for calculating flexural displacements

### 3.2 Reinforcement slip deformations

When a reinforced concrete column is subjected to bending moment, strain accumulates in the embedded length of the tensile reinforcing bars. This causes the bars to extend or slip relative to the anchoring concrete at column fixed end(s). The extension is commonly known as reinforcement slip and leads to rigid-body rotation of the column, as shown in Figur 1. This results in an additional lateral displacement component that can be as large as 25 to 40

% of the total lateral displacement (Sezen, 2002). Flexural deformations determined from conventional fiber section analysis (moment-curvature analysis) do not account for lateral deformations caused by reinforcement slip at column ends. Therefore, these deformations must be calculated separately and added to the other deformation components due to flexure and shear to calculate the total lateral displacement.

In this study, lateral displacement due to reinforcement slip is calculated using the model illustrated in Figure 4 (Sezen & Setzler, 2008).

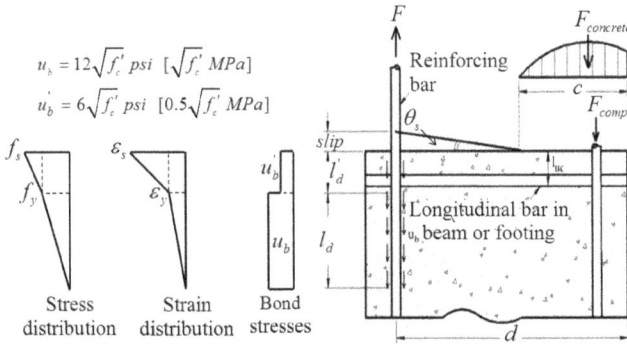

Fig. 4. Reinforcement slip model (Sezen & Setzler, 2008)

The model approximates the bond stress as bi-uniform function with different values for elastic and inelastic steel behaviors. The bond stress in the elastic and inelastic range is taken as $u_b = 12\sqrt{f_c'}\ psi\ (1\sqrt{f_c'}\ MPa)$ and $u_b' = 6\sqrt{f_c'}\ psi\ (0.5\sqrt{f_c'}\ MPa)$, respectively, where $f_c'$ is concrete compressive strength. Slip $s$ at the loaded end of the reinforcing bar can be calculated by integrating bi-linear strain distribution $\varepsilon_s(x)$ over the development length as follows.

$$s = \int_0^{l_d + l_d'} \varepsilon_s(x)dx \qquad (4)$$

where $l_d = \dfrac{f_s d_b}{4u_b}$ and $l_d' = \dfrac{(f_s - f_y)d_b}{4u_b'}$ are development lengths for the elastic and inelastic portion of the bar, respectively. Hence, integrating Equation 4, extension or slip of the reinforcing bars is

$$s = \begin{cases} \dfrac{\varepsilon_s l_d}{2} = \dfrac{\varepsilon_s f_s d_b}{8u_b} & \text{for } \varepsilon_s \le \varepsilon_y \\[4mm] \dfrac{\varepsilon_y l_d}{2} + \dfrac{(\varepsilon_s + \varepsilon_y)l_d'}{2} = \dfrac{\varepsilon_y f_y d_b}{8u_b} + \dfrac{(\varepsilon_s + \varepsilon_y)(f_s - f_y)d_b}{8u_b'} & \text{for } \varepsilon_s > \varepsilon_y \end{cases} \qquad (5)$$

where $\varepsilon_s$ is the strain at loaded end of the bar, $\varepsilon_y$ is steel yield strain, $f_s$ is stress at loaded end of the bar, $f_y$ is steel yield stress, and $d_b$ is diameter of the longitudinal bar. The reinforcement slip is assumed to occur in tension bars only and cause the rotation about the neutral axis as shown in Figure 4. Rotation caused due to reinforcement slip can be calculated as,

$$\theta_s = \frac{s}{d-c} \qquad (6)$$

where $d$ and $c$ are the distances from the extreme compression fiber to the centroid of the tension steel and neutral axis, respectively. The lateral displacement due to slip at free end of a cantilever column can be calculated as the product of slip rotation $\theta_s$ and length of the column $L$ as,

$$\Delta_s = \theta_s L \qquad (7)$$

## 3.3 Shear deformations

Shear deformations in reinforced concrete members have traditionally been ignored in design and research due to lack of their complete understanding and being difficult to measure, independent of other deformation components, in an experimental set up or a real structure. For a well designed reinforced concrete column, shear deformations are small as compared to the flexural deformations and are often less than 10 percent of total deformations. Contrary, for a reinforced concrete column not designed according to stricter seismic design provisions, shear behavior could be the governing failure criterion. Shear deformations in such shear critical reinforced concrete column could contribute large percentage towards total deformations and hence can not be ignored if an accurate analysis of deformation capacity is required.

Shear deformations in the proposed model are calculated using a combination of MCFT (Vecchio & Collins, 1986) and post-peak shear response envelope (Patwardhan, 2005; Sezen, 2008). In this model, pre-peak non-linear shear force-shear deformation response is obtained from in-plane analysis of the shear element based on MCFT while considering the interaction of the axial strain (Mostafaei & Kabeyasawa, 2007). Axial strain obtained from flexural section analysis is incorporated into the total axial strain of the shear element to include the effect of flexural behavior on shear response. After the peak strength has reached, shear strength is first assumed to remain constant at its peak value until the onset of the shear strength degradation and then declines linearly with increasing shear deformations to the point of axial load failure (Figure 5). At the point of axial load failure, lateral strength is assumed zero. The peak strength $V_{peak}$ in the proposed shear response model refers to the point where response estimation by MCFT terminates either due to shear failure or reaching the load step corresponding to the peak flexural strength prior to experiencing shear failure. Hence, the peak strength $V_{peak}$ is the minimum of the shear strength of the column $V_n$ and shear force corresponding to the maximum moment that can be carried by the section $V_p$.

Shear displacements at the onset of shear degradation $\Delta_{v,u}$ can be calculated as follows (Patwardhan, 2005; Sezen, 2008).

$$\Delta_{v,u} = \left(4 - 12\frac{v_n}{f_c'}\right)\Delta_{v,n} \qquad (8)$$

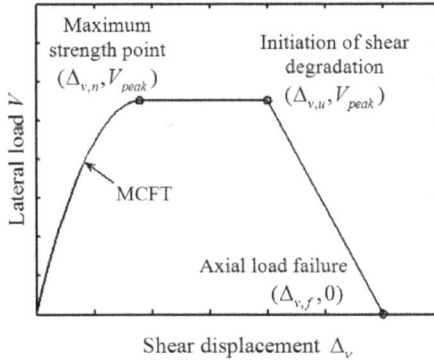

Fig. 5. Shear response model

where $v_n = \dfrac{V_{peak}}{bd}$ is the shear stress at peak strength, $f'_c$ is the concrete compressive strength, and $\Delta_{v,n}$ is the shear displacement corresponding to the peak strength as determined from MCFT analysis. The shear displacement at axial load failure $\Delta_{v,f}$ is calculated as,

$$\Delta_{v,f} = \Delta_{ALF} - \Delta_{f,f} - \Delta_{s,f} \geq \Delta_{v,u} \tag{9}$$

where $\Delta_{ALF}$ is the total displacement at axial load failure, $\Delta_{f,f}$ and $\Delta_{s,f}$ are the flexural and slip displacement at the point of axial load failure, respectively. The total displacement at axial load failure is determined by the expression based on a shear friction model and an idealized shear failure plane (Elwood & Moehle, 2005).

$$\Delta_{ALF} = \frac{0.04L\left(1+\tan^2\theta\right)}{\left(\tan\theta + P\left(\dfrac{s_h}{A_{sv}f_{yv}d_c\tan\theta}\right)\right)} \tag{10}$$

where $\theta$ is the angle of the shear crack, $P$ is the axial load, $A_{sv}$ is the area of transverse steel with yield strength $f_{yv}$ at spacing $s_h$, and $d_c$ is the depth of the core concrete measured to the centerlines of the transverse reinforcement. In the derivation, $\theta$ is assumed to be 65 degrees. The values of $\Delta_{f,f}$ and $\Delta_{s,f}$ in Equation 9 are determined according to the expected failure mode and classification of the column into categories as explained in subsequent subsection.

## 3.4 Total lateral response

In order to model the response of a column subjected to lateral loading, three deformation components should be combined together considering their interconnectedness. In this study, total column lateral response is modeled as a set of three springs in series; each spring representing lateral displacement component due to flexure, bar slip and shear. Each spring is subjected to same force and total displacement is sum of responses of each spring. The pre-peak total response is obtained by simply adding deformation components due to flexure, bar slip and shear mechanism as described above. After reaching the peak, the mechanism limiting the peak strength (flexure or shear) will dominate the behavior. The procedure for combining deformation components for post peak is explained below and the model is illustrated in Figure 6.

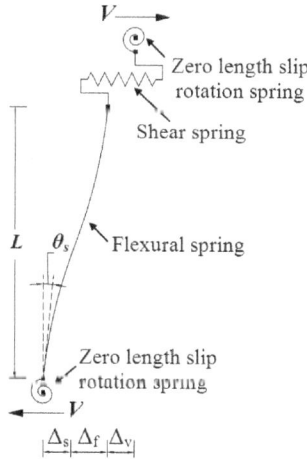

Fig. 6. Spring representation of total response model

For post-peak behavior, the column is classified into one of the five categories based on a comparison of its shear, yield and flexural strength and rules are specified for the combination of the deformation components for each category (Setzler & Sezen, 2008). The yield strength $V_y$ is defined as the lateral load corresponding to the first yielding of the tension bars in the column and flexural strength $V_p$ is the lateral load corresponding to the peak moment calculated from flexural analysis. Both of these loads are calculated from moment-curvature analysis of the fiber model as explained above. The shear strength $V_n$ for the columns failing in shear prior to the reaching flexural strength or failing close to flexural strength is determined from the proposed shear model, where $V_n = V_{peak}$ as discussed above. For other columns where peak strength by the proposed shear model is close to the flexural strength, shear strength is calculated as a function of displacement ductility (Sezen & Moehle, 2004).

$$V_n = k(V_c + V_s) = k\left[\left(\frac{6\sqrt{f_c'}}{a/d}\sqrt{1 + \frac{P}{6\sqrt{f_c'}A_g}}\right)0.80A_g + \frac{A_{sv}f_{yv}d}{s}\right] \qquad (11)$$

where $V_c$ is the concrete contribution to shear strength, $V_s$ is the steel contribution to shear strength, $A_g$ is gross cross-sectional area, $a/d$ is the aspect ratio and $k$ is a factor related to the displacement ductility which is the ratio of the maximum displacement to the yield displacement. The value for $k$ varies from 0.7 to 1.0 for displacement ductilites from less than 2 to grater than 6 respectively. In this study, the value for k is taken as 1.0 as classification of the columns is based on initial or low-ductility shear and flexural strengths.

The classification system and rules governing the post peak response in each category are described below and are illustrated in Figure 7. Peak response of the column is limited by the smaller of the shear strength and flexural strength, however post-peak response is assumed to be governed by the limiting mechanism (i.e., flexure or shear).

### 3.4.1 Category – I ( $V_n < V_y$ )

In this category of the columns, shear strength is less than the yield strength and column fails in shear while the flexural behavior remains elastic. The deformation at peak strength (i.e., shear strength) is the sum of deformations in each spring at the peak strength. After the peak strength is reached, the shear behavior dominates the response. As the shear strength degrades, the flexure and slip springs unload along their initial responses. The post-peak deformation at any lateral load level is the sum of the post-peak shear deformation and the pre-peak flexural and slip deformations corresponding to that load.

### 3.4.2 Category – II ( $V_y \leq V_n < 0.95V_p$ )

The shear strength is less than flexural strength and column fails in shear, however inelastic flexural deformation occurring prior to shear failure affects the post-peak behavior. The deformation at peak strength is the sum of the deformations in each spring at the peak strength. Shear deformations continue to increase after the peak shear strength is reached, but the flexure and shear springs are locked at their peak strength values. Hence, post-peak deformations at any lateral load level is the sum of flexural and slip deformations at the peak strength and post-peak shear deformation at that load.

### 3.4.3 Category – III ( $0.95V_p \leq V_n \leq 1.05V_p$ )

The shear and flexural strengths are nearly identical. Shear and flexural failure are assumed to occur "simultaneously," and both mechanisms contribute to the post-peak behavior. The post-peak deformation at any lateral load level is the sum of the post-peak flexure, slip, and shear deformations corresponding to that load.

### 3.4.4 Category – IV ( $1.05V_p < V_n \leq 1.4V_p$ )

The shear strength is greater than the flexural strength and the column may potentially fail in the flexure, however large shear deformations affect the post-peak behavior and shear failure may occur as the displacements increase. The deformation at peak strength is the sum of the deformations in each spring at the peak strength. After the peak strength is reached, flexural and slip deformations continue to increase according to their models, but the shear spring is locked at its value at peak strength. The post-peak deformation at any lateral load level is the sum of the post peak flexural and slip deformations corresponding to that load and the shear deformation at peak strength.

### 3.4.5 Category – V ( $V_n > 1.4V_p$ )

The shear strength is much greater than the flexural strength and column fails in flexure while shear behavior remains elastic. The peak strength of the column is the flexural strength calculated from the flexure model. If the column strength degrades, flexural and slip deformations continue to increase according to their models, while the shear spring unloads with an unloading stiffness equal to its initial stiffness. The post-peak deformation at any lateral load level is the sum of the post-peak flexural and slip deformations and the pre-peak shear deformation corresponding to that load.

For category-I columns, $\Delta_{f,f}$ and $\Delta_{s,f}$ values to be used in Equation 9 are assumed zero.

For the category-II columns, shear strength is lesser than flexural strength and these values are taken as the flexural and slip deformations at the load equal to the shear strength of the columns. For categories III, IV, and V specimens, $\Delta_{f,f}$ and $\Delta_{s,f}$ are the maximum calculated flexural and slip deformations.

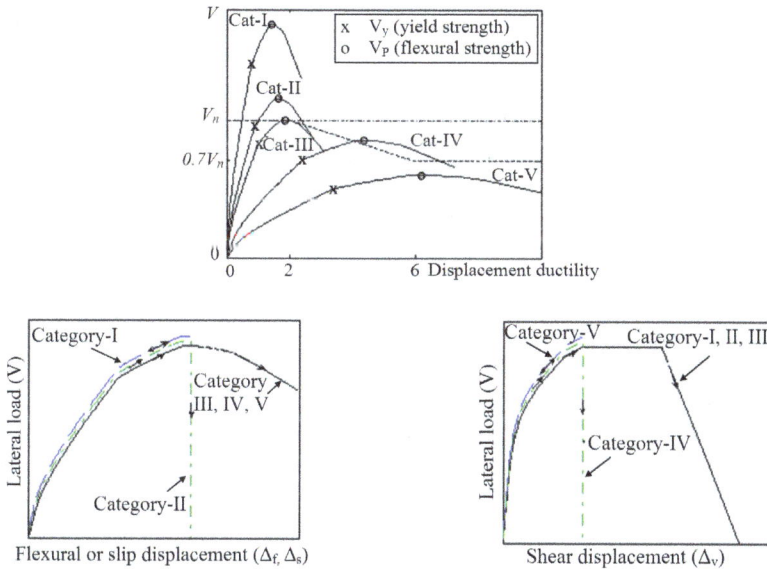

Fig. 7. Classification of columns into categories and rules governing combination of the deformation components (Setzler & Sezen, 2008)

## 4. Interaction between flexure and shear mechanisms

When a fixed-ended reinforced concrete column is subjected to lateral loading, such as during an earthquake, flexural and shear mechanisms interact with each other and affect overall response of the column. The interaction between flexural and shear deformations in the proposed analytical procedure is based on the ASFI approach (Mostafaei & Kabeyasawa, 2007). Interaction methodology in ASFI approach couples axial-flexure and axial-shear models with each other. Both mechanisms have to be evaluated simultaneously which makes ASFI approach relatively complicated and computationally intensive. The computational effort can be reduced significantly, if the analyses for flexural and shear behavior can be performed independently. Therefore, in the proposed procedure, the interaction of the shear deformations on flexural performance, and vice versa, are considered in a simplified manner that allows easy implementation and decoupled flexural and shear response evaluations.

## 4.1 Interaction of concrete compression softening

Cracked concrete behavior is considered in flexural section analysis to represent degradation in compressive strength of the concrete due to applied shear stresses. This requires determination of the compression softening factor $\beta$ to lower concrete stresses in uniaxial compression (Figure 2). The procedure for determining compression softening factor is adopted from the recently developed model called Uniaxial-Shear-Flexure Model (USFM) (Mostafaei & Vecchio, 2008). This is an approximate approach for response estimation of reinforced concrete elements which is derived after simplifying ASFI approach. USFM employs few fundamental equations of the MCFT and two assumptions on average principal compressive strain and average centroidal axial strain of the element to determine average principal tensile strain. The details of formulation, implementation and verification of USFM approach can be found in (Mostafaei & Vecchio, 2008).

Compression softening factor $\beta$, as defined in Equation 1, is a function of concrete principal tensile strain $\varepsilon_{c1}$ of the element being analyzed. The procedure to approximately determine principal tensile strain and subsequently compression softening factor for a fixed ended column subjected to in-plane lateral load is illustrated in Figure 8. For an element considered between inflection point and one of the end sections of the column, $\varepsilon_{c1}$ can be determined from the following MCFT equation.

$$\varepsilon_{c1} = \varepsilon_x + \varepsilon_{yv} - \varepsilon_{c2} \tag{12}$$

where, $\varepsilon_x$ is average axial strain at the centroid for the element and is obtained by averaging the values of centroidal axial stain at one of the end section $\varepsilon_o$ and axial strain of the inflection point $\varepsilon_{xa}$. Likewise $\varepsilon_{c2}$ is average concrete principal compressive strain for the element. Its value, as per USFM assumption (Mostafaei & Vecchio, 2008), can be taken as the average of the uniaxial concrete compressive strain corresponding to resultant compressive force of the stress block at end section $\varepsilon_c$ and axial strain at the inflection point $\varepsilon_{xa}$. Hence,

$$\varepsilon_x = \frac{\left(\varepsilon_o + \varepsilon_{xa}\right)}{2} \qquad \varepsilon_{c2} = \frac{\left(\varepsilon_c + \varepsilon_{xa}\right)}{2} \tag{13}$$

The other unknown quantity in Equation 12, strain of the transverse reinforcement $\varepsilon_{yv}$, can be determined from the following MCFT based relationship.

$$\varepsilon_{yv} = \sqrt{b^2 + c} - b \tag{14}$$

where,

$$b = \frac{f_{c1}}{2\rho_y E_{sy}} - \frac{\varepsilon_{c2}}{2} \;,\; c = \frac{\left(\varepsilon_x - \varepsilon_{c2}\right)\left(f_{c1} - f_{cx}\right) + f_{c1}\varepsilon_{c2}}{\rho_y E_{sy}} \;,\; f_{cx} = f_x - \rho_x f_{sx} \tag{15}$$

where $\rho_y$ is transverse reinforcement ratio, $E_{sy}$ is modulus of elasticity of transverse reinforcement, $f_{cx}$ is concrete stress in longitudinal axis of the column, $f_x$ is applied axial stress, $\rho_x$ is longitudinal reinforcement ratio, $f_{sx}$ is longitudinal steel stress obtained from section analysis based on average centroidal strain, $f_{c1} = 0.145\sqrt{f_c'}$ is concrete principal tensile stress (in MPa), $\varepsilon_x$ is normal strain at the centroid and $\varepsilon_{c2}$ is average concrete principal compressive strains, both determined from Equation 13.

After calculating concrete principal tensile strain $\varepsilon_{c1}$ from Equation 12, compression softening factor $\beta$ is determined from Equation 1 for a given curvature. This is the estimated value of $\beta$ which is employed in fiber model analysis to lower concrete stresses. In the proposed procedure, compression softening factor determined with the help of above mentioned procedure is employed till peak flexural strength and then a constant value equal to the last lowest is used for post peak flexural analysis.

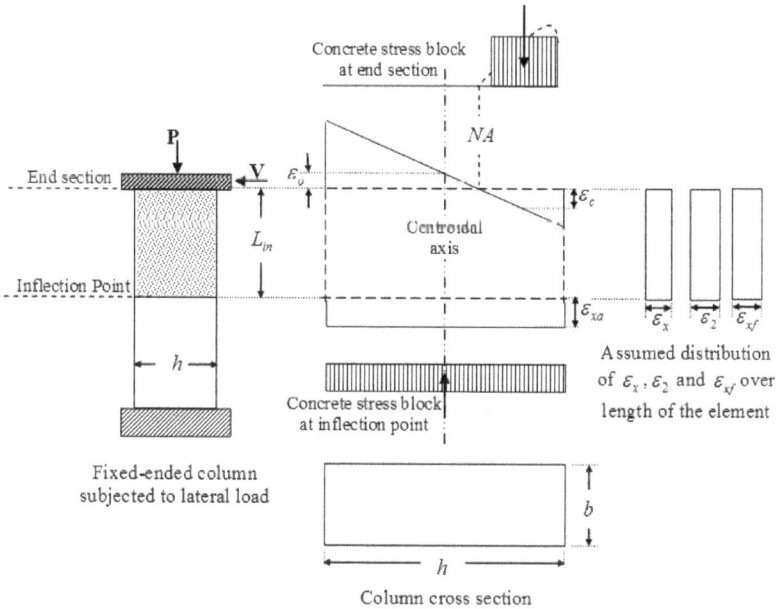

Fig. 8. Interaction of compression softening and axial strains (Mostafaei & Kabeyasawa, 2007; Mostafaei & Vecchio, 2008).

## 4.2 Interaction of axial strains

The effect of flexural deformations on shear behavior is considered by incorporating axial strain and shear stress due to flexure into in-plane analysis of the shear element based on axial strain interaction methodology of ASFI approach (Mostafaei & Kabeyasawa, 2007) and equilibrium of shear stresses in flexural and shear mechanisms. In this procedure, interaction of axial strain is taken into account by adding flexibility component of axial deformation due to flexure to the corresponding flexibility component of axial-shear model. By employing flexural shear stress to in-plane stress-strain relationship of the shear element, shear deformations are determined. The procedure for axial deformation interaction and determination of shear strain is described here for a fixed ended column subjected to lateral load.

The length of the column between inflection point and one of the end sections is considered as a shear element subjected to constant normal stress due to applied axial load and average shear stresses due to applied lateral load. Performing flexural analysis on fiber model of the end section and inflection point, average centroidal axial strain due to flexure $\varepsilon_{xf}$ (Figure 8) and corresponding flexibility component $f_{xf}$ can be determined with the help of following ASFI equations (Mostafaei & Kabeyasawa, 2007).

$$\varepsilon_{xf} = \frac{(\varepsilon_o - \varepsilon_{xa})}{2} \ , \quad f_{xf} = \frac{\varepsilon_{xf}}{\sigma_x} \ , \quad \sigma_x = \frac{P}{bd} \tag{16}$$

where, $\sigma_x$ is applied axial stress in longitudinal direction of the column and can be determined by dividing the applied axial load $P$ by the effective area of the cross section. A stress-strain relationship in terms of flexibility matrix for an in plane shear element (axial-shear model) can be defined as,

$$\begin{pmatrix} f_{11} & f_{12} & f_{13} \\ f_{21} & f_{22} & f_{23} \\ f_{31} & f_{32} & f_{33} \end{pmatrix} \begin{Bmatrix} \sigma_x \\ \sigma_y \\ \tau_{xy} \end{Bmatrix} = \begin{Bmatrix} \varepsilon_x \\ \varepsilon_y \\ \gamma_{xy} \end{Bmatrix} \tag{17}$$

where $f_{ij}$ $(i,j=1,2,3)$ are flexibility components of in plane shear model, $\sigma_x$ is normal applied stresses in longitudinal direction, $\sigma_y$ is normal stress in transverse direction, $\tau_{xy}$ is shear stress, $\varepsilon_x$ is normal strain in axial direction, $\varepsilon_y$ is normal strain in transverse direction, and $\gamma_{xy}$ is shear strain.

Axial strain due to flexure $\varepsilon_{xf}$ can be taken into account in the axial-shear model by adding flexibility component obtained from Equation 16 into Equation 17.

$$\begin{pmatrix} (f_{11} + f_{xf}) & f_{12} & f_{13} \\ f_{21} & f_{22} & f_{23} \\ f_{31} & f_{32} & f_{33} \end{pmatrix} \begin{Bmatrix} \sigma_x \\ \sigma_y \\ \tau_{xy} \end{Bmatrix} = \begin{Bmatrix} \varepsilon_x + \varepsilon_{xf} \\ \varepsilon_y \\ \gamma_{xy} \end{Bmatrix} \tag{18}$$

In Equation 18, stresses in transverse direction (clamping stresses) are zero due to inexistence of lateral external force along the column, i.e., $\sigma_y = 0$. In addition, the applied shear stress $\tau_{xy}$ of the element is taken from flexural section analysis (Equation 2) as $\tau_{xy} = \tau_f$. In Equation 18, knowing the applied stresses, corresponding strains can be calculated. The flexibility matrix is obtained by inverting material stiffness matrix of the shear element formulated using secant stiffness methodology of the MCFT approach (Vecchio & Collins, 1986).

## 5. Buckling of compression bars

Longitudinal reinforcing bars in columns may experience inelastic axial compression under severe loading and exhibit large lateral deformation known as buckling. The behavior in the compressive face of a concrete member at overload depends on a variety of factors such as, size and shape of the cross-section, the amount of longitudinal compression steel, the amount of transverse reinforcement providing confinement to the section, thickness of the cover concrete, and stress-strain properties for the steel and concrete (Potger et al., 2001). The tendency for the compressively loaded steel bars to buckle and deflect outwards is initially resisted by the lateral restraint provided by the surrounding cover concrete as well as the transverse steel ties. As the compressive loads increase and approach the section capacity, the concrete surrounding the compressive bars carries large longitudinal compressive stress, and eventually becomes prone to longitudinal cracking, and spalling. After the cover concrete spalls off, ties restrain lateral movement and buckling.

In this study, compression steel stresses were reduced to account for buckling using the model shown in Figure 9. According to this model, compression stresses in longitudinal bars start to decrease when unconfined cover concrete starts to spall. When this happens, corresponding strain in the relevant steel layer can be calculated from flexural strain distribution across the cross section depth. This strain is $\varepsilon_{sp}$ as shown in the figure. This point can fall anywhere on typical stress-strain relationship for steel depending upon the level of flexural strain. Steel stresses follow their usual constitutive stress-strain relationship until strain reaches this limit. Then compression stresses in reinforcement follow new path defined by a line joining peak stress point to residual strength point having a slope $m$, which is calculated from following relationship (Inoue & Shimizu, 1988).

$$m = 100\varepsilon_{yx}\left(\frac{1}{\sqrt{1+500\lambda^2}}-1\right)E_{sx} \;\; ; \;\; \lambda = \frac{\alpha s_h}{i_r} \tag{19}$$

where, $\varepsilon_{yx}$ is yield strain of longitudinal bars, $E_{sx}$ is modulus of elasticity for longitudinal steel, $\alpha$ is 1.0 for corner bars and 0.5 for intermediate bars, $s_h$ is stirrup or tie spacing, and $i_r$ is radius of gyration of longitudinal bar.

Diameter of the longitudinal bar and spacing of the transverse reinforcement is important parameters that affect the buckling of the compression bars (Monti & Nuti, 1992). Smaller diameter bars restrained by widely spaced ties are most likely to undergo lateral deformations and buckling much earlier during loading history than larger diameter bars confined by closely spaced transverse reinforcement. Therefore, in the proposed model, for tie spacing to bar diameter ratio $s_h/d_b$ of less than 5.00, no buckling is considered and compressive behavior of the reinforcement is similar to its tensile behavior. For $s_h/d_b$ ratio above 11.00, the bars are considered to buckle as soon as reinforcement yields. For $s_h/d_b$ ratios between 5.00 and 11.00, post-buckling softening is considered soon after spalling of the cover concrete with the proposed model.

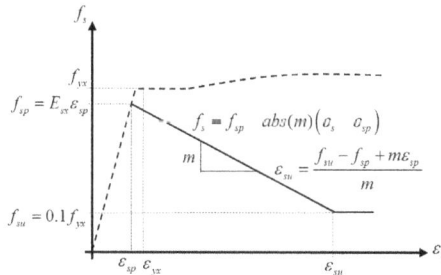

Fig. 9. Proposed compression bar buckling model

## 6. Analytical steps for implementation of the proposed procedure

Total lateral displacement of a concrete column can be calculated under lateral loads. Note that the procedure can be used to model the response prior to peak strength (under increasing loads) and also beyond peak response (possible decreasing lateral loads due to strength and stiffness degradation). The major steps of the proposed procedure are to:

- Define uniaxial material properties for unconfined and confined concrete, and reinforcing steel. Include the effect of compression softening of concrete (Equation 1 and Figure 2). Consider the effect of compression bar buckling under large axial deformations (Equation 19 and Figure 9).
- Define fiber cross section for flexural analysis and perform moment-curvature analysis of the cross section.
- Calculate lateral load versus flexural displacement by integrating curvatures over the height of column (Equations 2 and 3).
- Calculate lateral load versus reinforcement slip displacement (Equations 5 through 7).
- Perform MCFT analysis to lateral load versus shear displacement up to maximum shear strength (Figure 5) while considering the interaction of axial strains due to flexure (Mostafaei & Kabeyasawa, 2007). Alternatively, an approximate method (Sezen, 2008) can be used and this step can be skipped. Obtain lateral load versus shear displacement envelope (Figure 5).
- Classify the column and combine flexure, slip, and shear responses (Section 3.4, and Figure 7). During calculation of combined or total displacement, at each step, consider the interaction between axial-shear and flexure mechanisms (Section 4 and Figure 8).

## 7. Model verification

The proposed procedure is implemented to evaluate response of previously tested reinforced concrete columns (Sezen, 2002) and predicted responses are compared with experimental test data. The columns tested by (Sezen, 2002) are very useful as these provide experimental force-displacement data for each of the flexure, slip, and shear components individually in addition to the overall response. Hence, the experimental data from these columns are used to validate the component and total deformation models proposed in this study (Lodhi, 2010; Sezen & Moehle 2006).

These columns were lightly reinforced and had shear and flexural design strengths very close to each other. These are 18 in. (457 mm) square columns with fixed ends at top and bottom having height of 116 in. (2946 mm). The columns had eight No.9 bars and No. 3 column ties with 90-degree end hooks spaced at 12 in. (305 mm). Specimens-1 and -4 were tested with a constant axial load of 150 kip (667 KN), whereas, Specimen-2 was tested under a constant axial load of 600 kip (2670 KN). The columns were tested under unidirectional cyclic lateral loading, except for Specimen-4, which was tested under monotonically increasing load after few initial cycles of elastic loading. All of the test specimens are modeled with average concrete compressive strength of 3.08 psi (21.2 MP). The yield strength of longitudinal and transverse reinforcement are taken to be 63 ksi (434 MP) and 69 ksi (476 MP), respectively. Other details of test specimens, material properties used for the development of reinforcing steel and unconfined and confined concrete models can be found in (Setzler & Sezen, 2008; Sezen, 2002, 2008).

Lateral load-flexural displacement relationships for Specimen-1 and 4 are presented in Figure 10 (a) and 10(b), respectively. In this comparison, test specimens are also analyzed using another displacement component model (Setzler & Sezen, 2008). This model also treats deformations due to flexure, bar slip and shear individually, however, does not consider softening of concrete compression strength, concrete tensile behavior and buckling of compression bars in flexural analysis. It can be seen that both approaches predict identical pre-peak responses, which matches very well with the experimental data. Peak load and deformation at peak load is also estimated very well by both approaches. For post peak behavior, however, predicted responses are quite different. After reaching the strains corresponding to the start of compression bar buckling, response predicted by the proposed procedure gradually drops and generally follows stiffness of the measured response. The diverging near-peak and post-peak predicted responses by both approaches highlight the need to consider concrete softening and bar buckling effect in the analysis. Figure 10 (c) and 10(d) presents load-displacement relationships due to reinforcement slip for Specimen-1 and 2, respectively. The predicted responses by displacement component model and proposed method produce almost identical response up until peak load and then diverge in the post peak range. Again, this highlights the need for considering buckling of compression bars in the flexural analysis.

Lateral load-shear displacement relationships for Specimen-1 and 2 are presented in Figure 10(e) and 10(f), respectively. For comparison of the predicted shear responses, the columns are also analyzed with ASFI approach (Mostafaei & Kabeyasawa, 2007). The predicted responses by ASFI approach and proposed procedure are identical until observed peak and follows experimental data generally well. Peak shear strength is generally captured well by both approaches. For post-peak behavior, proposed model shows strength degradation as deformations increase. In ASFI approach, after reaching peak load, shear deformations are calculated from secant stiffness at peak strength, which is kept constant for post peak behavior. As a result, post-peak predicted shear response in ASFI approach does not show shear strength degradation.

Figure 11 shows the comparison of predicted and experimental lateral load-total displacement relationships for Specimen-1, 2 and 4. Shear strength of Specimen-1 (Figure 11(a)) is calculated as 69.0 kips and flexural strength from moment-curvature analysis is 70.0 kips. Hence, this specimen is classified as category-III column, for which total displacement at any point in the response is sum of flexural, slip and shear displacement at that load step. With the proposed procedure, initial response is predicted very well up to the peak strength. Peak strength and deformation at peak load and the post peak response are captured well. The Specimen-2 (Figure 11(b)) has shear and flexural strengths of 92.0 and 72.0 kips, respectively, and is classified as category-IV column. For this column, shear deformation is frozen at its value at peak strength (flexural strength, 72.0 kips) and added to flexural and slip displacements for post-peak response. Predicted response by the proposed approach slightly overestimates the pre-peak stiffness and peak load in the positive direction and follows post peak experimental response fairly well in both directions. Specimen-4 (Figure 11(c)) is identical to Specimen-1 except that it was tested under monotonically increasing lateral load after few initial elastic cycles. Comparison of shear and flexural strength classifies this column into category-III column. The predicted response by the proposed procedure follows the trend in experimental data but slightly overestimates the initial stiffness and peak strength.

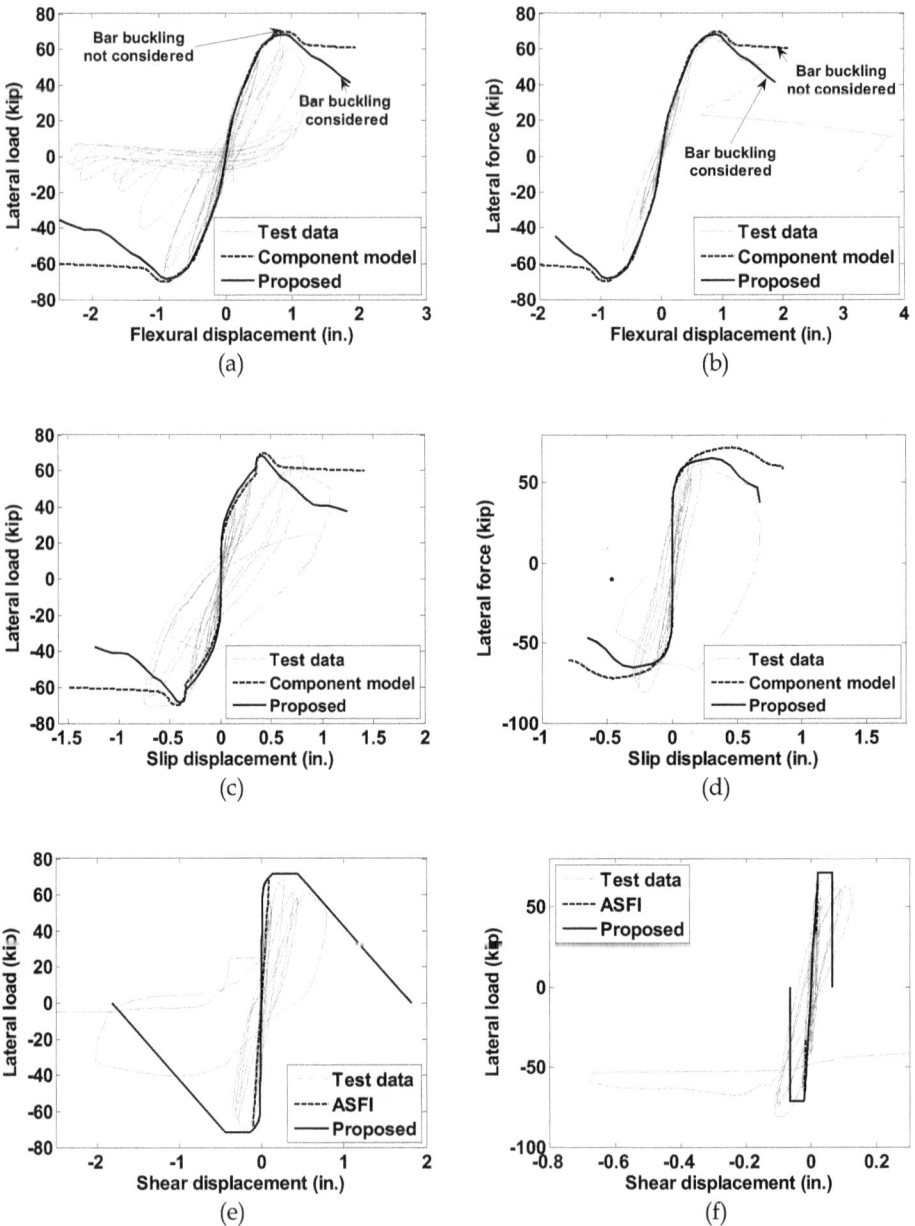

Fig. 10. Experimental and analytical results for: (a) flexural response of Specimen 1; (b) flexural response of Specimen 4; (c) reinforcement slip response of Specimen 1; (d) reinforcement slip response of Specimen 2; (e) shear response of Specimen 1; and (f) shear response of Specimen 2

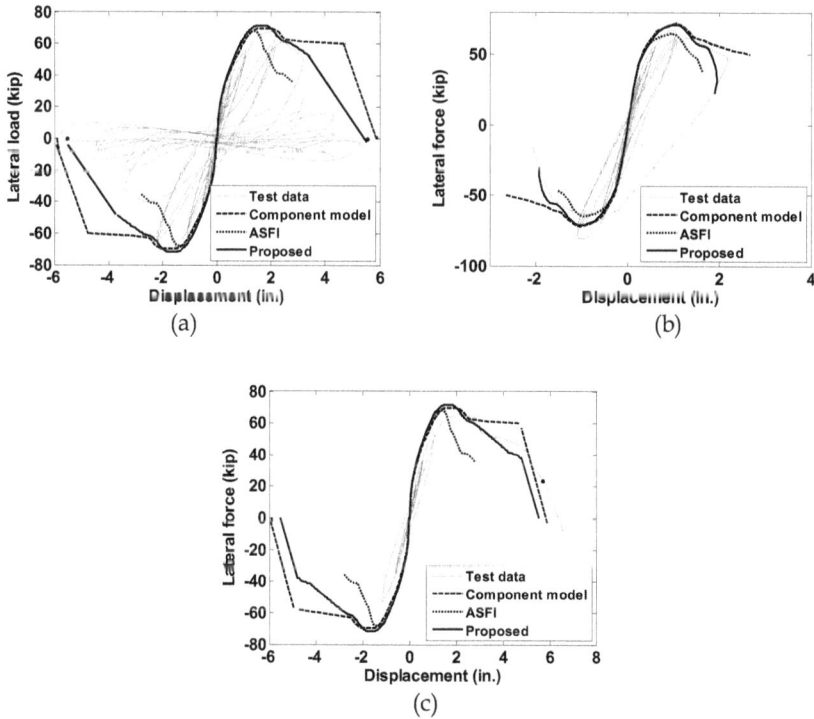

Fig. 11. Lateral load – displacement relationships : (a) Specimen 1; (b) Specimen 2; and (c) Specimen 4

## 8. Conclusions

A procedure is proposed for response estimation of reinforced concrete columns subjected to lateral loads. The procedure determines flexure, bar slip and shear deformations individually considering interaction between these mechanisms. The axial-flexure mechanism is decoupled from axial-shear model, allowing a relatively simpler analytical procedure. The flexure model in the proposed procedure incorporates concrete tensile behavior, interaction of compression softening and buckling of compression bars into the flexural analysis. The shear model includes the effect of flexural deformation on shear behavior. The pre-peak response is evaluated by employing MCFT, which cannot be used when the strength starts to degrade beyond peak strength. A post peak shear response envelope from displacement component model is adopted for predicting post peak shear behavior. All deformation components, i.e., flexural, bar slip and shear, are added together to get total response of the column. The total/combined peak response is limited by the smaller of the shear and flexural strength of the column and limiting mechanism governs the post peak response. The proposed procedure employs relatively simple calculations for the overall response estimation. The comparison of the predicted and observed responses indicates that the proposed procedure is a suitable displacement-based evaluation process that performs well in predicting the individual displacement components and total response.

# 9. References

ACI Committee 318. (2008). *Building Code Requirements for Structural Concrete (ACI 318-08) and Commentary (ACI 318R-08)*, American Concrete Institution, ISBN 0-87031-264-2, Farmington Hills, Michigan, USA

Arakawa, T. (1970). Allowable Unit Shearing Stress and Design Method of Shear Reinforcement for Reinforced Concrete Beams. *Concrete Journal*, Vol.8, No.7, pp. 11-20 (in Japanese)

Aydan, O. & Tano, H. (2011). Shaking-Induced Damage to Buildings by M 9.0 East Japan Mega Earthquake on March 11, 2011, In: *Tohoku Pacific Ocean Earthquake and Tsunami - Quick observations from the PEER/EERI/GEER/Tsunami Field Investigation Team*, 18.08.2011, Available from: http://www.jsce.or.jp/committee/eec2/eq_repo rt/201103tohoku/Aydan3.pdf

Bentz, E.C. (2000). Sectional Analysis of Reinforced Concrete Members. *PhD Thesis*, Department of Civil Engineering, University of Toronto, Toronto, Canada, 2000

Ceresa, P.; Petrini, L. & Pinho, R. (2007). Flexure-shear fiber beam-column elements for modeling frame structures under seismic loading-state of the art. *Journal of Earthquake Engineering*, Vol. 11, pp. 46–88, ISSN 1363-2469

Ceresa, P.; Petrini, L.; Pinho, R. & Sousa, R. (2009). A fiber flexure-shear model for seismic analysis of RC-framed structures. *Earthquake Engineering and Structural Dynamics*, Vol. 38, No. 5, pp. 565–586

Chao, S.H. & Loh, C.H. (2007). Inelastic response analysis of reinforced concrete structures using modified force analogy method. *Earthquake Engineering and Structural Dynamics*, Vol. 36, pp. 1659–1683

Earthquake Engineering Research Institute. (December 2005). First Report on the Kashmir Earthquake of October 8, 2005, *EERI Special Earthquake Report-December 2005/EERI's Learning from Earthquakes Program*, 18.08.2011, Available from: http://www.eeri.org /lfe/pdf/kashmir_eeri_1st_report.pdf

Inoue, K. & Shimizu, N. (1988). Plastic Collapse Load of Steel Braced Frames Subjected to Horizontal Force. *Journal of Structural and Construction Engineering*, Vol. 388, pp. 59-69 (in Japanese)

Lodhi M.S. (2010). Response estimation of reinforced concrete columns subjected to lateral loads. *M.S. Thesis*, The Ohio State University, Columbus, Ohio, 2010.

Mid-America Earthquake Center, University of Illinois Urbana-Champaign. (2005). The Kashmir Earthquake of October 8, 2005 - A Quick Look Report, *MAE Center Report No. 05-04*, 18.08.2011, Available from:   http://mae.cee.uiuc.edu/documents/cd_ro m_series/05-04/Report05-04.pdf

Mergos, P.E. & Kappos, A.J. (2008). A distributed shear and flexural flexibility model with shear–flexure interaction for R/C members subjected to seismic loading. *Earthquake Engineering and Structural Dynamics*, Vol. 37, PP. 1349–1370

Mergos, P.E. & Kappos, A.J. (2010). Seismic damage analysis including inelastic shear–flexure interaction. *Bulletin of Earthquake Engineering*, Vol. 8, pp. 27–46

Monti, G. & Nuti, C. (1992). Non-linear Cyclic Behavior of Reinforcing Bars Including Buckling. *Journal of Structural Engineering*, Vol. 118, No. 12, pp. 3269-3284

Morsch, E. (1922). *Der Eisenbetonbau-Seine Theorie und Anwendung*, (5th Edition, Vol. 1, Part 1), Wittwer, Stuttgart, Germany

Mostafaei, H. & Kabeyasawa, T. (2007). Axial-Shear-Flexure Interaction Approach for Reinforced Concrete Columns. *ACI Structural Journal* 2007, Vol. 104, No. 2, pp. 218-226

Mostafaei, H. & Vecchio, F.J. (2008). Uniaxial Shear-Flexure Model for Reinforced Concrete Elements. *Journal of Structural Engineering*, Vol 134, No. 9, pp. 1538-1547

Mullapudi, T.R. & Ayoub, A.S. Modeling of the seismic behavior of shear-critical reinforced concrete columns. *Engineering Structures* 2010; 32:3601–3615.

Mullapudi, T.R.; Ayoub, A.S. & Belarbi, A. (2008). A fiber beam element with axial, bending and shear interaction for seismic analysis of RC structures, *Proceedings of the 14th World Conference on Earthquake Engineering*, Beijing, China, 2008

Pacific Earthquake Engineering Research Center/Earthquake Engineering Research Institute/Geotechnical Extreme Event Reconnaissance/Tsunami Field Investigation Team. (2011). Tohoku Pacific Ocean Earthquake and Tsunami - Quick observations from the PEER/EERI/GEER/Tsunami Field Investigation Team, 18.08.2011, Available from: http://peer.berkeley.edu/news/wp-content/uploads/2011/04/Tohoku-short-interim-report.pdf

Patwardhan, C. (2005). Strength and Deformation Modeling of Reinforced Concrete Columns. *M.S.Thesis*, The Ohio State University, Columbus, Ohio, 2005

Potger, G.M.; Kawano, A.; Griffith, M.C. & Warner, R.F. (2001). Dynamic Analysis of RC Frames Including Buckling of Longitudinal Steel Reinforcement, *Proceedings of the NZSEE Conference*, 2001, Paper No. 4.12.01

Priestley, M.J.N.; Calvi, G.M. & Kowalsky, M.J. (2007). *Displacement-Based Seismic Design of Structures*, IUSS Press, Pavia, Italy

Ritter, W. (1899). Die Bauweise Hennebique. *Schweizerische Bbauzeitung*, Vol. 33, No. 7, pp 59-61

Schlaich, J.; Schafer, I. & Jennewein, M. (1987). Towards a Consistent Design of Structural Concrete. *Journal of the Prestressed Concrete Institute*, Vol. 32, No. 3, pp. 74-150

Setzler, E.J. & Sezen, H. (2008). Model for the lateral behavior of reinforced concrete columns including shear deformations. *Earthquake Spectra*, Vol. 24, No. 2, pp. 493-511

Sezen H., and Moehle J.P. (2006). Seismic tests of concrete columns with light transverse reinforcement. *ACI Structural Journal*, Vol. 103, No. 6, pp. 842-849

Sezen, H. (2002). Seismic Behavior and Modeling of Reinforced Concrete Building Columns. *Ph.D Thesis*, University of California, Berkeley, 2002

Sezen, H. (2008). Shear deformation model for reinforced concrete columns. *Structural Engineering and Mechanics*, Vol. 28, No. 1, pp. 39-52

Sezen, H., & Moehle, J.P. (2004). Strength and deformation capacity of reinforced concrete columns with limited ductility, *Proceedings of the 13th World Conference on Earthquake Engineering*, Vancouver, Canada, 2004

Sezen, H. & Moehle, J.P. (2004). Shear Strength Model for Lightly Reinforced Concrete Columns. *Journal of Structural Engineering*, Vol. 130, No. 11, pp. 1692-1703

Sezen, H. & Setzler, E.J. (2008). Reinforcement Slip in Reinforced Concrete Columns. *ACI Structural Journal*, Vol. 105, No. 3, pp. 280-289

Structural Engineers Association of California (SEAOC). (2002). Performance based seismic design engineering of buildings. *Vision 2000 Report*, Sacramento, CA, USA

Takewaki, I.; Murakami, S.; Fujita, K.; Yoshitomi, S. & Tsuji, M. (2011). The 2011 off the Pacific coast of Tohoku earthquake and response of high-rise buildings under long-period ground motions. *Soil Dynamics and Earthquake Engineering*, doi:10.1016/j.soil dyn.2011.06.001 (In press)

U.S Geological Survey/Earthquake Engineering Research Institute Advance Reconnaissance Team. (February 23, 2010). The $M_w$ 7.0 Haiti Earthquake of January 12, 2010, *USGS/EERI Team Report V.1.1*, 18.08.2011, Available from: http://www.eqclearingh ouse.org/20100112-haiti/wp-content/uploads/2010/02/USGS_EERI_HAITI_V1.1. pdf

Vecchio, F.J. (1989). Nonlinear finite element analysis of reinforced concrete membranes. *ACI Structural Journal*, Vol. 86, No. 1, pp. 26-35

Vecchio, F.J. (2000). Disturbed stress field model for reinforced concrete: formulation. *Journal of Structural Engineering*, Vol. 126, No. 9, pp. 1070-1077

Vecchio, F.J. & Collins, M.P. (1986). The modified compression field theory for reinforced concrete elements subjected to shear. *ACI Journal Proceedings*, Vol. 83, No. 2, pp. 219-231

Vecchio, F.J. & Collins, M.P. (1988). Predicting the Response of Reinforced Concrete Beams Subjected to Shear Using Modified Compression Field Theory. *ACI Structural Journal*, Vol. 85, No. 3, pp. 258-268

Williams, M.S. & Sexsmith, R.G. (1995). Seismic Damage Indices for Concrete Structures: A State-of-the-Art Review. *Earthquake Spectra*, Vol. 11, No. 2, pp. 319-349

Xu, S.Y. & Zhang, J. (2011). Hysteretic shear–flexure interaction model of reinforced concrete columns for seismic response assessment of bridges. *Earthquake Engineering and Structural Dynamics*, Vol. 40, pp. 315–337

Zhang, J.; Xu, S.Y. & Tang, Y. (2010). Inelastic displacement demand of bridge columns considering shear–flexure interaction. *Earthquake Engineering and Structural Dynamics*, DOI 10.1002/eqe.1056

# 10

# Seismic Behavior and Retrofit of Infilled Frames

Mohammad Reza Tabeshpour[1], Amir Azad[1] and Ali Akbar Golafshani[2]
*[1]Mechanical Engineering Department,*
*[2]Civil Engineering Department*
*Sharif University of Technology, Tehran,*
*Iran*

## 1. Introduction

The most important recent earthquakes showed the importance of seismic behavior of various types of structures such as infilled frames: Japan 2011 (Takewaki et al., 2011), Haiti 2010 (Eberhard et al.) and Newzealand 2010 (Ismail et al. 2011). There are many important questions in the field of infilled frames, such as:

1. What are the negative and positive effects of infill walls in various types of buildings (local and global)?
2. Where infills should be considered (in which type of structures, considering lateral resisting system)?
3. When it is better to have interaction between frame and infill? In what type of structures considering the number of stories?
4. When it is better to have a gap between frame and infill?
5. What is the state of the art in this field?
6. Comparing National and international codes, what is the state of the practice?
7. What are the suitable strategies to retrofit existing infilled frames?

The aim of this chapter is to answer some of these questions. A categorized discussion is presented here to classify the problems solved or to be solved.

An ideal model of structure (bare frame) is considered usually in order to analyze and design the structure, which undoubtedly has important differences with its actual model. The actual model has also some differences with the considered model such as effects of infill walls. Existence of the infill walls basically provides higher stiffness and strength for the frames, but their detrimental effects on the structure performance is ignored due to lack of adequate information about the behavior of frames and infill walls. Meanwhile, recent studies have shown that different arrangements of stiffness, mass and strength by each other can have significant effect on structure behavior and their response. One of the most common failure modes of structures in earthquakes is soft story failure which causes by discontinuity of lateral force resisting elements such as braces, shear walls or infill walls in the first story. In this case columns are imposed to large deformation and also plastic hinges are formed at top and bottom of the columns. This case usually is named as story mechanism. Due to eccentricity of the center of mass and stiffness, which causes by asymmetrical arrangement of infill walls, high torsional moment is produced. Other failure mode is short column that is a common mode in concrete structures.

Tabeshpour (2009d) has presented a comparative study of several building codes about masonry infill wall for design purposes. Tabeshpour et al. (2011d) investigated the lateral drift of concrete infilled frames to answer the following question: When and how is it better to separate infill from frame?

## 2. Masonry infill walls

Regarding to the combined behavior from the infill wall and structural frame which observed in many earthquakes, researchers predicted these events with modeling the masonry infill walls based on Fig. 1 as a compression strut elements. The existence of infill walls can change the structural behavior from flexural action into axial action. The advantages in the conversion of flexural action to axial action are:
- Reduce contribution of frame in lateral resisting;
- Reducing the lateral deformations.
The disadvantages of converting the flexural action to axial action:
- Increase of the axial load in the column and foundation,
- Creation of the concentrated shears at top and bottom of the column,
- Creation concentrated shears at beginning and end of the beam,
- Creation of huge shears on the foundation.
The equivalent struts of infill wall may be modeled in 3 different types: beam-to-beam (Fig. 2), column-to-column (Fig. 3) and node-to-node (Fig. 4). If it take places-to-column model like Fig. 3 then some forces would be exchanged between walls and column. This part of column is known as short column.
Because of significant stiffness and strength of the infill walls, they may cause severe irregularities in stiffness and strength in the building's elevation and plan. Various effects of masonry infill walls are summarized in table 1.

## 3. The history of modeling

Finite element modeling of masonry infill wall is a very complex and unreliable task due to several parameters such as: mortar characteristics, brick specifications, the interaction between brick and mortar, the interaction between masonry infill wall and frame. Possible discussions on this issue would be:
1. Modeling of infill walls categorized in two methods:
- Detailed models (Micro)
- Simple models (Macro)
The former is offered based on the finite element of the masonry infill wall which has utilized common methods in theories of elasticity and plasticity. The behavior of macro models are based on physical behavior of infill walls that can be modeled by using one or some structural elements.
2. The second important issue is related to the capacity of model to cover some/all of the related nonlinear phenomena. For example, actual stiffness in some models are not considered in the elastic limit. Some other models take into account its stiffness and decrease carefully as well as the decline in strength. That is how the structural behavior can be studied before reaching the fracture.
3. Another issue is to study the effects of one directional and cyclic loading on behavior and characteristics of the system.

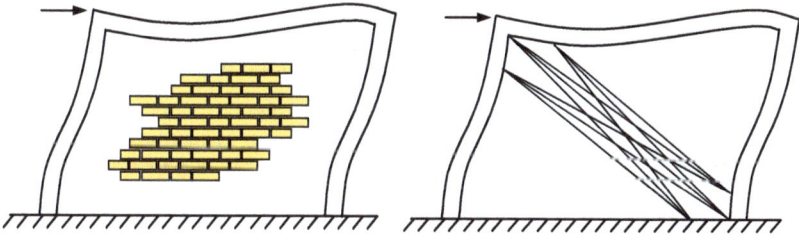

Fig. 1. Model with 9 struts

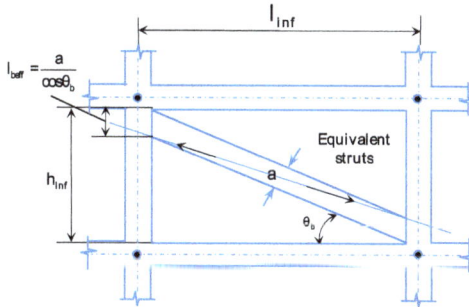

Fig. 2. Beam-to-beam model

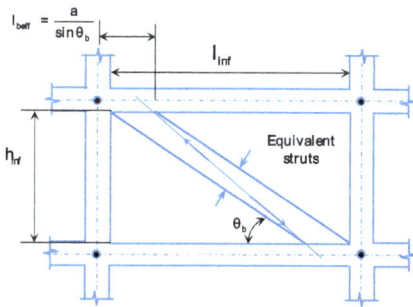

Fig. 3. Column-to-column model

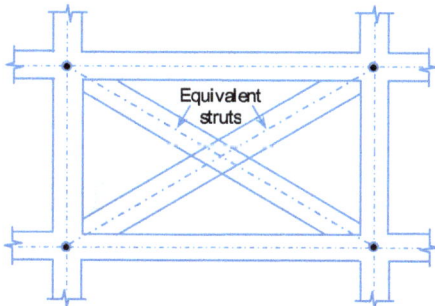

Fig. 4. Corner to corner model

| No. | Advantages | Disadvantages |
|-----|-----------|---------------|
| 1 | Higher stiffness and lower displacement | Stiffness irregularity in height (soft story) |
| 2 | Higher strength | Strength irregularity in height (weak story) |
| 3 | Lower ductility requirements | Stiffness irregularity in plan (torsion) |
| 4 | Higher base level in special conditions | Improper distribution of force between columns of a concrete frame |
| 5 | Ductile shear fracture in the steel short column | Improper distribution on force in plan (steel short column) |
| 6 | Frame design for small lateral loads | Increase in load design because of lower periods |
| 7 | Creation of couple system with axial action of frame | Increase on load design because of lower behavior factor for joint system |

Table 1. Advantages and disadvantages of masonry infill walls on steel or concrete frame

### 3.1 Micro modeling

All models discussed here are based on the finite element method which generally utilize 3 types of elements for providing the masonry infill wall, frame and the interaction between them. In most cases, special attention has been focused on the contacting elements between frame and masonry infill walls. Then, it has been clarified that numerical simulation of the infill wall is very important and their nonlinear phenamena must be modeled with great levels of accuracy. Research on this problem is closely related to develop elements used in masonry structures (Tabeshpour, 2009b).

Mallick and Severn (1967), have taken into account the contact of wall to frame particularly. The masonry infill walls were modeled as the rectangular elastic elements with 2 degree of freedom. The frame was also modeled using the elements without axial deformation. The slip of the frame against wall was also noticed along with the friction between them. They compared the results of analyses with experimental works and offered accurate presentation of the stiffness.

Goodman et al. (1968) developed an element in order to simulate the interaction between frame and wall. The rectangular element of the plain strain with 4 nodes and 2 transitional freedom degree in each node was modified to consider the contact condition properties. The shear strength of the element is dependent on adhesion and friction. This study has proposed a moderate mode between wall and frame with given length and zero initial width.

Researches from Malik and Garg (1971) modeled the effect of existence of shear slot between frame and masonry wall. They used the rectangular element of plain strain for the wall similar to the above mentioned technique. They also proposed that in the model of beam member for frame, the rotational degree of freedom must not be considered, which means that the frame must only be deformed under the shear and axial loads. This model was used to study 2 issues: the effect of openings available in the masonry infill wall in addition to the effect of shear slot available between frame and infill walls. Experimental work was also launched to evaluate the validity of the results.

Koset et al. (1974) observed that infill walls and frame cracks are occured and developed even in little lateral loads. This is attributed to the low tensile strength at the contact between infill walls and frame. Therefore, in order to simulate the system's response under the lateral load, opening/closing of gaps between masonry infill walls and frame must be considered.

King and Pandey (1978) used the element proposed by Goodman and his coworkers. Primary tests showed that the curves of shear stress were elasto-plastic in contact between the elements. Tangential stiffness characteristics ($K_n$ and $K_s$) of the moderate elements were defined as functions of these elements. That was how the frictional slip of connection/separation between the elements was studied. They achieved acceptable results from Mallick and Severn (1967) models.

Liauw and Kwan (1984) advanced a plastic theory which allowed three different fracture modes. Based on the relative strength of columns, beams and masonry infill walls behaviour, three type of failure are mention below.

- Corner crashing of infill walls and fracture in columns
- Corner crashing of infill walls and fracture in beams
- Diagonal crack of masonry infill walls

Rivero and Walker (1984) developed a nonlinear model for simulating the response of the frame system which is Infilled by walls and is under stimulus of earthquake. 2 types of element were developed base on surface between frame and walls. Gap element and the connection element. The former was used aiming to show the distance between frame and infill walls in no tangential conditions, while the later was utilized to model the contact mode. The process of crack formation and development was studied carefully.

Shing et al. (2002) simulated the nonlinear behavior of the masonry wall elements using the plasticity theory. The cracks through bricks and mortar as well as the cracks between mortar joint and members of reinforced concrete were modeled and studied. In this model, overall behavior of the combined system before cracking was considered homogeneously and isotropically. The behavior of materials was presumed elasto-plastic based on the Von Mises yield criterion with tensions of Rankeen type. After calibrating the model by using experimental studies, the finite element model was capable to simulate the actual behavior of the system properties.

### 3.2 Macro modeling

The idea of using a simple member for simulating infill walls inside the frame has always been attractive, and has several advantages in the process of analysis and design. At the beginning, it was explained that a diagonal strut with appropriate mechanical properties can be a suitable candidate for walls. By using the diagonal strut model, it will be possible to enter the following items to the model:

- Shear stiffness of the infill wall,
- Small shear and tensile stress of column at the contact between wall and frame,

Although this simple model cannot notice the following complexities in the model:

- Decreasing the stiffness and strength under cyclic loads,
- Out of plane behavior for masonry infill walls when diagonal crack occurred,
- Shear slip along joints which occurs at the middle height of infill walls.

These problems were solved to some extent in the equivalent strut model. For example, Klinger and Bertero (1976) modeled masonry infill walls with two equivalent struts and noted the effects of stiffness dimming. Polyakov (1956) studied the normal and shear stresses at the middle of infill walls, using the variation calculation method and offered a numerical technique to estimate the load which cause diagonal crash.

Holmz (1961) presented formula for a diagonal strut for the first time. He assumed that the width of equivalent strut is equal to one third of the diagonal length. After that, several studies were performed to define the width of the equivalent strut.

Stafford Smith (1968) observed that the equivalent diagonal strut has many simplifications and some modifications must be done on its equivalent width. He assumed that the distribution of the interactional forces between frame and infill walls is triangular. This idea has a very high accuracy and is still in use. Based on the interaction length between infill walls and frame, other proposals were introduced by Mainstone (1971) and Kadir (1971).

Klinger and Bertero (1976) provided the first diagonal member with cyclic behavior which was able to consider the stiffness dimming behavior through the modeling procedure.

Chrysostomou (1991) investigate the behavior of the frame and the infill wall system under the earthquake loading regarding the effects of decreasing of stiffness and wall strength. He modeled the wall in any diagonal direction with three bars based on Fig. 8. The The $\alpha L$ length is equal to the plastic hinge in column or beam. These members act compressively.

The effective width of equivalent strut in the infill wall proposed by different researchers has severe variation from 10 to 35%. Table 1 summarizes different relations for the effective width of equivalent brace in the masonry infill walls and Tabeshpour recommends some values for effective width in Hand book, part 18 (page 65) (2009). Modeling of infill wall using commercial softwares is needed for design purposes (Tabeshpour, 2009e).

| Researcher | | Effective Width ($b_w$) | $\lambda h$ | |
|---|---|---|---|---|
| Holmes (1961) | | $b_w = [0.33]d_w$ | - | 170 (max) |
| Mainstone (1971) | | $b_w = 0.16(\lambda h)^{-0.3} d_w$ | 5 | 50 |
| Klingner and Bertero (1978) | | $b_w = 0.175(\lambda.h)^{-0.4} d_w$ | 5 | 45(min) |
| Liauw and Kwan (1984) | | $b_w = 0.95 h_w \cos\theta (\lambda h)^{-0.5}$ | 5 | 90 |
| Paulay and Priestley (1992) | | $b_w = [0.25]d_w$ | - | 125 |
| Recommended | Upper band, Negative Effect | $b_w = [0.2]d_w$ | - | 100 |
| | Lower band, Positive Effect | $b_w = [0.1]d_w$ | - | 50 |

Table 2. Different formulae of equivalent masonry strut's effective width

## 3.3 Yung's modulus of masonry materials

The important point about Young's modulus of masonry materials is the range of values obtained from the relations proposed by different researchers which is attributed to the nature of masonry materials. Table 2 lists some relations presented by some researchers as well as the value of Yung's modulus for 15 Kg/cm² compressive stress (Tabeshpour Hand book, part 18, 2009). Shear strength of materials are usually demonstrated in codes by static friction relation as show in follow:

$$\tau = \tau_0 + \mu\sigma_y \tag{1}$$

$\tau_0$ : Joint shear strength

$\mu\sigma_y$ : Frictional strength component

$\mu$ : Internal frictional factor

$\sigma_y$ : Normal stress component along the horizontal direction

| Researcher | Module of Elasticity | $E_m$(kg/cm²) $f_m$=15 kg/cm² |
|---|---|---|
| Sahlin (1971) | $L_m$= 750 $f_m$ | 1100 |
| Paulay and Priestley (1992) | $E_m$= 750 $f_m$ | 1100 |
| Sanbartolome (1990) | $E_m$= 500 $f_m$ | 7500 (min) |
| Sinha&Pedreschi (1983) | $E_m$= 1180 $f_m$ | 16000 |
| Hendry (1990) | $E_m$= 2116 $f_m$ | 25000 (max) |
| Some others | $E_m$= 1000 $f_m$ | 15000 |

Table 3. Module of elasticity for equivalent masonry struts

| Researcher | Shear bond strengths, $\tau_o$ |
|---|---|
| Hendry (1990) | 0.3 to 0.6 MPa |
| Shrive (1991) | 0.1 to 0.7MPa |
| Paulay and Priestley (1992) | 0.1 to 1.5MPa |

Table 4. Shear bond strength for equivalent masonry struts

| Researcher | $\mu$ |
|---|---|
| Sahlin (1971) | 0.1 to 1.2 |
| Stöckl and Hofmann (1988) | 0.1 to 1.2 |
| Atkinson et al. (1989) | 0.7 and 0.85 |
| Hendry (1990) | 0.1 to 1.2 |
| Paulay and Priestley (1992) | 0.3 for design purposes |

Table 5. Ductility for equivalent masonry struts

## 4. Failure modes

### 4.1 Soft Story

The base floors of the existing buildings are generally arranged as garages or offices. No walls are built in these floors due to its prescribed usage and comfort problems. But upper floors have walls separating rooms from each other for the residential usage. In these arrangements, the upper floors of most buildings are more rigid than their base floors. As a result, the seismic behaviors of the base and the upper floors are significantly different from each other. This phenomenon is called as the weak-story irregularity. Weak stories are subjected to larger lateral loads during earthquakes and under lateral loads their lateral deformations are greater than those of other floors so the design of structural members of weak stories is critical and it should be different from the upper floors.

Sattar and Liel (2000) shown in their results of pushover analysis that infill walls were increased the initial stiffness, strength, and energy dissipation of the infilled frame, compared to the bare frame, despite wall's brittle failure modes. Vulnerability and damage analysis of existing buildings using damage indices have been presented by Golafshani et al. (2005). Such studies can be used for quantitative investigation of existing buildings. A case-study structure that collapsed because of a soft-story mechanism during the 2009 L'Aquila earthquake was studied with Verderame et al. (2009). Their study presented some peculiar details and results, but it could not be stated. It represents a common practice in the L'Aquila building stock.

Haque and Amanat (2009) shows that, when RC framed buildings having brick masonry infill on upper floor with soft ground floor is subjected to earthquake loading, base shear can be more than twice to that predicted by equivalent earthquake force method with or without infill or even by response spectrum method when no infill in the analysis model.

Tena-Colunga (2010) evaluated how the soft first story irregularity condition should be defined: (a) as a significant reduction of the lateral shear stiffness of all resisting frames within a given story, as established in the seismic provisions of Mexican building codes or, (b) as a substantial reduction of the lateral shear stiffness of one or more resisting frames within a given story, as proposed by the author.

Kirac et al. (2010) studied the seismic behavior of weak-story. Calculations were carried out for the building models which are consisting of various stories with different storey heights and spans. Some weak-story models were structural systems of existing buildings which were damaged during earthquakes. It was observed that negative effects of this irregularity could be reduced by some precautions during the construction stage. Also some recommendations were presented for the existing buildings with weak-story irregularity.

A conceptual and numerical analysis for investigating the effect of masonry infills on seismic behavior of concrete frames considering various type of infill arrangements was presented by Tabeshpour et al (2004). It was found that a large drift is concentrated in soft story (the story with no infill). Design of columns in soft stories is an important problem to have an acceptable mechanism in severe earthquakes. In order to avoid from soft story failure, columns should be designed for increased loads, Tabeshpour (2009f) has investigated increasing design load in specific columns and presented a simple formula for this purpose.

| Researcher | $\delta_{cr}(\%)$ URM | $\delta_{max}(\%)$ URM | $\delta_{max}(\%)$ Frame + Infill | $\delta_{max}(\%)$ Bare Frame | Other note |
|---|---|---|---|---|---|
| Fiorato (1970) | | | 1.1 | | |
| Zarnic & Tomazevic (1984) | 0.2 | | 1 | 3 | |
| Govindan et al. (1986) | | | 3 | 1.5 | |
| Valiasis & Stylianidis (1989) | | | 0.6 | 1 | $\tau_u \approx 0.25 - 0.3$ Mpa |
| Carydis et al. (1992) | | | | | Good system behaviour up to 0.14% drift; steel frame with infill |
| Pires&Carvalho (1992) | 0.1 | | 0.5 | | $\tau_u \approx 0.27 - 0.51$ Mpa |
| Shing et al. (1992) | | | | | $\tau_u \approx 0.34$ Mpa |
| Valiasis et al. (1993) | 0.2 to 0.3 | | | | |
| Fardis & Calvi (1995) | | | | | $\delta_{max}$ for URM is 0.1 $\delta_{max}$ for frame |
| Zarnic (1995) | 0.1 | 0.3 | 0.6 | 2 | |
| Pires et al. (1995) | | | 0.3 | 2 | $V = 0.8, V_{max}$ at 6% drift |
| Manos et al. (1995) | 0.15 | | 0.3 | 1 | $\tau_u \approx 0.3$ Mpa $V = 0.8$, $V_{max}$ at 2% drift for infill frame |
| Michailidis et al. (1995) | 0.1 | 0.25-0.35 | | | |
| Mehrabi et al. (1996) | 0.3 | | 0.6 | 3.1 | $\tau_u \approx 0.5$ Mpa, $V = 0.8$, $V_{max}$ at 1.5% drift for infilled frame, 6.8% for bare frame |
| Negro & Verzeletti (1996) | <0.3 | | 1.1 | 2.4 | $V_{max} = 0.4W$ for bare frame. $V_{max} = 0.62W$ for infilled frame |
| Aguilar (1997) | | | 1.3 | | |
| Marjani (1997) | 0.5 | | | | |
| Zarnic & Gostic (1997) | | 0.2 | 1 | >1 | |
| Zarnic (1998) | | | 0.3 | | $V = V_{max}$ for 0.3% $< \delta <$ 2% Used in mathematical model of URM infill |
| Kappos et al. (1998) | 0.3 | | 3 | | |
| Schneider et al. (1998) | 0.2 | | 2 | | |
| Mosalam, White and Ayala (1998) | ± 0.1% and ± 0.2% | | 0.5 | | |
| Mehrabi & shing (1998) | 0.013% to 0.037% | | 2% to 4.3% | | |
| Lili Anne Akin (2004) | 0.3 | | 0.7 | | |
| Al-Chaar et al. (2002) | 0.25% to 1% | | | | |
| Santiago et al. (2008) | 1 | | | | $K_{infill} = 5K_{bare}$ $f_{infill} = 2f_{bare}$ |

Table 6. Deformation limitations

### 4.1.1 Conceptual discussion

One of the main reasons of failure of structures due to earthquakes is discontinuity of lateral force resisting elements like bracing, shear wall or infill in the first story as show conceptually in Fig. 5. So first story act as soft story, in this case columns are imposed to large deformation and plastic hinges are formed at top and bottom of the element. Conceptual figure is obtained from actual earthquake observation as shown in fig. 6. This phenomena is so-called story mechanism (severe drift of the story). Most of these buildings have collapsed. The upper stories have infills and consequently their stiffness is much more than the first story.

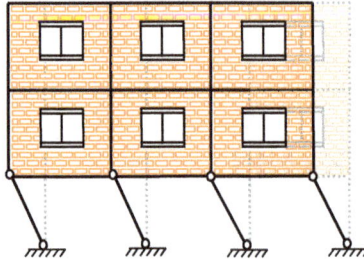

Fig. 5. Schematic view of soft story mechanism

Fig. 6. Soft story failure in a building during earthquake (Italy 1976)

The performance of a building in earthquake is shown in Fig. 6. This building is RC structure and has parking in the first story; there is no infill in the parking story. Deformations are localized in the first story and the columns of this story undergo large deformation, passing collapsed limit (4% of height).

### 4.2 Torsion

The effects of masonry infill walls in the structures are significant and very important in seismic responses of structures due to the experiences of the previous earthquakes. Many existing buildings are irregular in plan or elevation because of asymmetric placement of masonry infills. This kind of torsion should be considered by engineers.

The inelastic seismic response of a class of one-way torsionally unbalanced structures is presented By Bozorgnia et al. (1986). Tso (1988) shown that much better correlation exists between inelastic torsional responses and strength eccentricity than the traditionally used stiffness eccentricity parameter. Tso and Ying (1990) used a single mass three-element model, a study was made on the effect of strength distribution among elements on the inelastic seismic responses of eccentric systems. Additional ductility demands on elements and additional edge displacements are taken as response parameters of interest in optimizing the strength distribution.

Yoon and Stafford Smith (1995) presented a method to predict the degree of translational-torsional coupling of mixed-bent-type multistory building structures subject to dynamic loading.

Chopra and De la Llera (1996) focused on the description of two recently developed procedures to incorporate the effects of accidental and natural torsion in earthquake analysis and design of asymmetric buildings. Basu and Jain (2004) presented the definition of center of rigidity for rigid floor diaphragm buildings has been extended to unsymmetrical buildings with flexible floors.

Stefano et al. (2007) presents an overview of the progress in research regarding seismic response of plan and vertically irregular building structures. Dai Junw et al. (2009) shown some analytical results from 3D temporal characteristics of the responses of an RC frame building subjected to both a large aftershock and the main shock of Wenchuan $M_s$ =8.0 earthquake.

### 4.2.1 Conceptual discussion

Fig. 6 is a sample of previous earthquakes that shown the response of plan asymmetric structure. As a result, design codes incorporate procedures to account for such irregular plan-wise displacement distribution, leading to different stiffness's and capacities of resisting planes. Several researchers have carried out numerical and experimental investigations to understand these effects. The stiffness of masonry infill is a considerable value relating to that of the structure. Because of architectural and structural considerations, sometimes there is an eccentricity between center of mass and center of rigidity and the structure is irregular in plan called asymmetric building. The structure is also might be asymmetric as an irregular arrangement of infills in plan, which leads to unbalance distribution of stiffness. Produced torsion from eccentricity because of infill stiffness leads to extra forces and deformations in structural members and diaphragms. An appropriate alternative to solve this problem especially in existing buildings is using dampers.

a) Tortional failure of structure          b) Intraction between          c) Intraction between
(Kobe, 1995)                               infill walls and frame         infills wall and frame

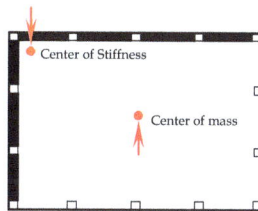

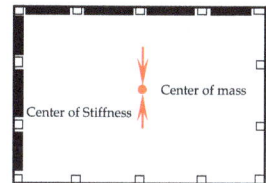

Fig. 7. Torsion of building

## 4.3 Short column

Shear failure is a critical kind of concrete column failure that occurs in short columns as repeatedly demonstrated during recent earthquake. Due to high brittle behavior and low ductility of these types of columns, it is important to investigate the behavior of short columns.

Moehle et al. (2000) examined loss of lateral and vertical load capacities by a study of columns tested in the laboratory. Correlations with geometric, materials, and loading characteristics were identified. They gathered test data to understand the effects of materials, geometry, and loading on failure mechanisms. Sezen and Moehle (2006) shown that columns with inadequate transverse reinforcement were vulnerable to damage including shear and axial load failure by earthquakes and laboratory experience. To study this behavior, they tested four full scale columns with light transfers reinforcement under unidirectional lateral load with either constant or varying axial load. Their tests shown that response of columns with nominally identical properties varied considerably with magnitude and historical and lateral loads. Kwon (2007) presented back-analysis result of a RC building in Ica, Peru which was severely damaged during the Pisco-Chincha Earthquake. Kwon confirmed in analysis results that shear force demand on columns with infill walls was significantly higher than those without infill walls. Turel Gur et al. (2007) made three surveys of damage to concrete structures following the 1999 Marmara and Düzce earthquakes. They observed that the sole severely damaged structure was damaged not by failure in the ground story, as all the other school buildings, but by failure of captive columns at basement level as a result of discontinuity of the foundation walls in height. The structural walls of the building, which were not damaged at all, prevented the collapse of the building by providing sufficient lateral strength and enhancing the gravity load capacity. Their observation was that the presence of structural walls improved the behavior of reinforced concrete systems drastically. Tabeshpour et al. (2005) presented numerical study of short column failure using IDARC. Non-ductile behavior of short columns was modeled for nonlinear damage analysis. Tabeshpour and Mousavy (2011b) presented plastic hinge properties for short column surveying because of masonry infill wall using nonlinear static analysis.

### 4.3.1 Conceptual discussion

#### 4.3.1.1 Local short column

*4.3.1.1.1 Flexural failure*

Flexural failure in columns depends on the shear span ratio that is:

$$a_s = M / (VH) \qquad (2)$$

Slender columns ($a_s > 3.5$) are characterized by a flexural type of failure. This type of damage consists of spalling of the concrete cover and then crushing of the compression zone, buckling of longitudinal bars and possible fracture of hoops due to the expansion of the core (Elnashai and Sarno, 2008).

*4.3.1.1.2 Shear failure*

Short columns ($a_s < 2$) are characterized by a shear failure presented a brittle failure. This type of failure occur when columns have conventional reinforcement (hoops and longitudinal bars) and high axial load, when subjected to cyclic loading results in cross-inclined shear cracks. This behavior may be improved if cross-inclined reinforcement is utilized, and particularly if multiple cross-inclined reinforcement (forming a truss) is used. Some columns in RC frames may be considerably shorter in height than the other columns in the same story. Short columns are stiffer, and require a larger force to deform by the same amount than taller columns that are more flexible. Regarding to earthquake damage photos due to short column phenomena increased force generally incurs extensive damage on these columns. The upper portion of the column next to the window behaves as a short column due to the presence of the infill wall, which limits the movement of the lower portion of the column. In many cases, the heights of the columns in each story are the same, as there are no walls adjoining them. When the floor slab moves horizontally during an earthquake, the upper ends of all columns undergo the same displacement. Effective height of columns is shorter when masonry infills added during construction. Consequently these columns attract a larger force as compared to a regular column. The damage in these short columns as shown in fig. 7 is often in the form of X-shaped cracks, which is characteristic for shear failure. In new buildings, the short column effect should be avoided during the architectural design process.

### 4.3.1.2 Global short column (very stiff non-ductile story)

Earthquake observation indicates that, total collapse of the story occurs, due to shear failure of the columns in the story. If the negative effects of infill walls are not consider in design procedure, brittle failure in low drift ratio may occur and it makes detritions drop in strength.

### 4.4 Interaction

Interaction between infill walls and concrete columns cause the brittle failure as repeatedly demonstrated during recent earthquake. Existing of infill walls and adjacent to concrete frames is the most important and determinant effects in behavior of concrete structures during earthquake. During recent earthquake great damage occur because of interaction phenomena.

Smith & Coull (1991) presented a design method for infilled frame based on diagonally braced frame criteria. They proposed a method that considered three possible failure modes of infill: shear along the masonry, diagonal crashing of infill walls and corner crushing of infill. Paulay & Priestley (1992) proposed a theory about the seismic behavior of masonry infilled frame and a design method for infilled frames. They said that although masonry infill may increase the overall lateral load capacity, it can result in altering structural response and attracting forces to different or undesired part of structure with asymmetric arrangement. This means that masonry infill may affect on structural behavior in earthquake.

Bell and Davidson (2001) reported on the evaluation of a reinforced concrete frame building with brick infill walls. They used in their evaluation an equivalent strut approach for modeling the infill walls. Their results indicated that infill walls, where presented in a regular arrangement, had a significant beneficial influence on the behavior of RC buildings that contrasted with New Zealand guidelines which gave an impression that infill masonry walls had a detrimental influence on the behavior of buildings due to interaction effects.

Naseer et al. (2007) overviewed buildings damaged during October 08, 2005 Kashmir earthquake. They understood that most of the buildings in the earthquake affected area were non-engineered. The following conclusions were drawn from the analysis of data collected in the post-earthquake damage assessment surveys that most of the buildings are either non-engineered or semi-engineered before October 08, 2005 earthquake. Mohyeddin-Kermani et al. (2009) focused specifically on observations made on concrete construction with masonry infill walls during the Sichuan earthquake with identification of damage and key of failure modes. This will be related to the damage and failure modes observed in past earthquakes because of interaction between masonry infill walls and concrete frame. Baran and Sevil (2010) studied on behavior of infilled frames under seismic loading. They considered hollow brick infills as "structural" members during the structural design process. They emphasized that since the behavior is nonlinear and closely related to the interaction conditions between frame and infill, analytical studies should be revised and supported by experimental data.

### 4.4.1 Conceptual dissection
Earthquake reports indicate the negative effect of infill wall as shown in fig. 7-9. Due to observation, damage of structure because of interaction effects categorize in two groups:
a.   Interaction between masonry infill walls and concrete frame (Fig. 8)
b.   Interaction effects in confined masonry structure (Fig. 9)
Design procedure of infilled frame structure in most codes is base on the bare frame. Earthquake observation indicate that infill frame have effective role in response of structure. Local damage of infill walls in recent earthquake indicate that the actual behavior of structure adjust to the material strength basis.
Confined masonry building is commonly used structure both in small cities and rural areas. This type of structure is very similar to infilled frames with two differences: masonry walls carry vertical loads and tie beams are not moment frames and don't carry vertical loads. Fig. 10 shows the interaction effect between tie beams and masonry wall. Because of concentrated shear force, the corner part of the beams is very vulnerable and shear cracks occur in earthquakes (Fig. 9).
In order to have a deep view on the structural behavior of infilled frames, simplified models of infills and frames are presented in table 7.

Fig. 8.  shear failures in columns (California 1994)

Fig. 9. Shear failure of column (Italia 2002)

Fig. 10. interaction between infill wall and vertical tie beam

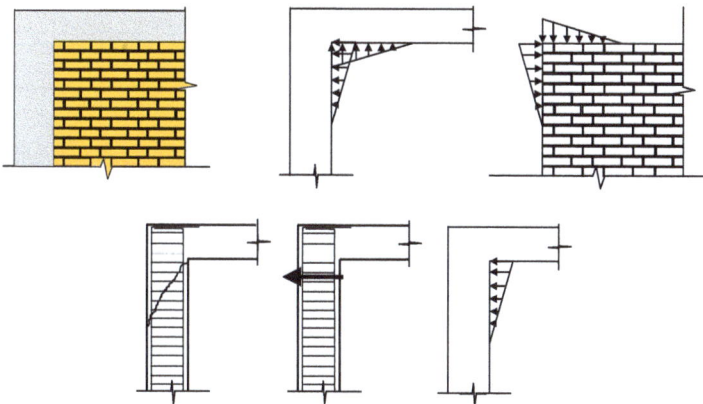

Fig. 11. Impose force from masonry wall into vertical tie beam

| Capacity Curve | Strength ($F_y$) | Stiffness (k) | System |
|---|---|---|---|
| | 0 | 0 | |
| | $\dfrac{A_m}{l} f_m \cos\theta =$ $\dfrac{t \times a}{l} f_m \cos\theta =$ $t \times a \times f_m \cos\theta$ | $\dfrac{A_m E_m}{l} \cos^2\theta =$ $\dfrac{t \times a \times E_m}{l} \cos^2\theta =$ $t \times a \times E_m \cos^2\theta$ | |
| | $\dfrac{4M_p}{h}$ | $\dfrac{24E_s I}{h^3}$ | |
| | $\dfrac{4M_p}{h} + t \times a \times f_m \cos\theta$ | $\dfrac{24E_s I}{h^3} +$ $t \times a \times E_m \cos^2\theta$ | |
| | $\dfrac{0.5\sqrt{f_c'}}{h'/d}\sqrt{1+\dfrac{P}{0.5\sqrt{f_c'}A_g}}\,0.8A_g$ $+A_m f_m \cos\theta$ | $\dfrac{24E_s I}{h'^3}$ | |
| | $\dfrac{0.5\sqrt{f_c'}}{h'/d}\sqrt{1+\dfrac{P}{0.5\sqrt{f_c'}A_g}}\,0.8A_g$ $+A_m f_m \cos\theta$ | $\dfrac{24E_s I}{h'^3}$ | |
| | $2M_p\left(\dfrac{1}{h}+\dfrac{1}{h'}\right)$ $+A_m f_m \cos\theta$ | $\dfrac{24E_s I}{h'^3}$ | |

Table 7. Simplified models for capacity curves

## 5. Case study

Tabeshpour et al. (2011c) showed that the infill walls can lead to severe torsion increase through the frame which can be solved by using friction damper device. We can say that in the irregularities and changes in structural properties because of infill walls will not considered, the structural design may be inefficient and the seismic response of the structures may not be acceptable. Tabeshpour and Ebrahimian, (2010) have presented design of friction/yielding damping devices. Considering the infill walls leads to determine the period of the structure in high accuracy and therefore, the seismic responses will be reliable and the design of a friction damper will be performing in a correct way. Many control devices have been developed to achieve the first purpose, and they have been applied to high-rise buildings and towers such as friction damper device (FDD) (Mualla and Belev, 2002). Friction damper is the simplest kind of dampers and easy to construct and install. The second purpose of vibration control is to prevent of imparting damage to the main elements of a structure during severe earthquakes. In seismic design of friction dampers, the structural stiffness and fundamental period directly affect the damper properties.

### 5.1 Modeling of masonry infill walls

From experimental observations, it is evident this type of structure exhibits a highly nonlinear inelastic behavior, even at low-level loading. The nonlinear effects mentioned above introduce analytical complexities, which require sophisticated computational techniques in order to be properly considered in the modeling. Due to the stiffness and strength degradation occurring under cyclic loading, the infilled frame structures cannot be modeled as elasto-plastic systems, while models that are more realistic should be used to obtain valid results, especially in the dynamic analysis of short period structures, such as infilled frames. The aim of this chapter is to introduce the modeling of a masonry infill walls, which will implement in the analysis in the following chapter. The elastic in-plane stiffness of a solid unreinforced masonry infill walls prior to cracking shall be represented with an equivalent diagonal compression strut of width, $a$, given by the following equation:

$$a = 0.25 \left( \lambda h_{col} \right)^{-0.4} r_{inf} \tag{3}$$

$$\lambda_1 = \left( \frac{10 E_{me} t_{inf} \sin 2\theta}{E_{fe} I_{col} h_{inf}} \right)^{0.25} \tag{4}$$

where:

$h_{Col}$ = Column height between centerlines of beams, cm

$h_{inf}$ = Height of infill walls, cm.

$E_{fe}$ = Expected modulus of elasticity of frame material, kg/cm

$E_{me}$ = Expected modulus of elasticity of infill material, kg/cm

$I_{col}$ = Moment of inertia of column, cm

$L_{inf}$ = Length of infill walls, cm.

$r_{inf}$ = Diagonal length of walls panel, cm.

$t_{inf}$ = Thickness of infill walls and equivalent strut, cm

$\theta$ = Angle (it's tangent is the infill height-to length, radians)

$\lambda_1$ = Coefficient used to determine equivalent width of infill strut

The equivalent strut shall have the same thickness and modulus of elasticity which is represented in fig. 4.

| $E_{me}\ (kg\,/\,cm^2)$ | $t_{inf}\ (cm)$ | $\lambda_1$ | $a\ (cm)$ |
|---|---|---|---|
| 12000 | 20 | 0.009043 | 71.95 |

Table 8. Summary of calculated masonry parameters

## 5.2 Parameters of compression equivalent strut

In this research equivalent compression strut used instead of masonry infill walls. This strut is in a diagonal and node-to-node manner whose length is equal to the diameter of the frame and its effective width is 0.2 of the diameter of the frame. The thickness of strut is same as the wall's thickness.

In order to obtain the masonry materials of strut, Australia's building code was used concerning the conventional compression bricks and mortars which produce stress-strain curves below relating to the mode of a 23cm one. The equation of stress-strain of masonry materials (bricks) in compression is considered as a parabolic function up to the maximum stress ($f_{m0}$) based on Table 9. Then with increase in strain, the value of stress decreases linearly, therefore it remains constant. These values are described in details in appendix.

| Parameter | Value |
|---|---|
| Thickness | 23 (Cm) |
| $f_{mo}$ | 4 (Mpa) |
| $\varepsilon_{mo}$ | 0.0014 |
| $f_{mu}$ | 0.8 (Mpa) |
| $\varepsilon_{mu}$ | 0.0028 |

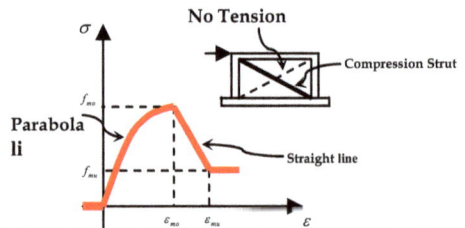

Table 9. Equivalent masonry strut's material properties

## 5.3 Damper description and principle of action

Friction dampers have often been employed as a component of these systems because they present high energy-dissipation potential at relatively low cost, easy to install and maintain. A friction damper is usually classified as one of the displacement-dependent energy dissipation devices, because its damper force is independent from the velocity and frequency-content of excitations. A friction damper is activated and starts to dissipate energy only if the friction force exerted on its friction interface exceeds the maximum friction force (slip force); otherwise, an inactivated damper is no different from a regular bracing. This devise used to dissipate the energy not only in the usual structure (building) but also it used in platforms and jackets (offshore structure) as well (Komachi et al., 2011).

The damper main parts are the central (vertical) plate, two side (horizontal) plates and two circular friction pad discs placed in between the steel plates as shown in Fig. 11. The central plate has length $h$ and is attached to the girder mid span in a frame structure by a hinge. The hinge connection is meant to increase the amount of relative rotation between the central and side plates, which in turn enhances the energy dissipation in the system. The ends of the two side plates are connected to the members of inverted V brace at a distance r from the FDD center. The bracing makes use of pretension bars in order to avoid compression stresses and subsequent buckling. The bracing bars are pin-connected at both ends to the damper and to the column bases. The combination of two side plates and one central plate increases the frictional surface area and provides symmetry needed for obtaining plane action of the device. When a lateral force excites a frame structure, the girder tends to displace horizontally. The bracing system and the forces of friction developed at the interface of the steel plates and friction pads will resist the horizontal motion. Fig. 11 explain the functioning of the FDD under excitation. As is shown, the device is very simple in its components and can be arranged within different bracing configurations to obtain a complete damping system.

### 5.4 Numerical study

A 3-story frame with 3 bays has been investigated in this study. Fig. 12 show plan and elevation of the building. Frame A has been filed by masonry walls with thickness of 23 cm. Lateral force resisting system is intermediate steel moment frame and the type II of soil according to Iranian seismic code of practice (Standard No. 2800). Since investigating the effects of masonry infill walls is the main goal of this research, the considered frames are designed according to last version of Iranian building codes without considering infill walls. Dimensions of the elements have been shown in Table 10. Dead and live loads of stories are considered 600 (kg/m^2) and 200 (kg/m^2) respectively. These parameters are considered 550 (kg/m^2) and 150 (kg/m^2) respectively for roof story. Dead load is considered 133 (kg/m^2) for 23 cm thick walls respectively.

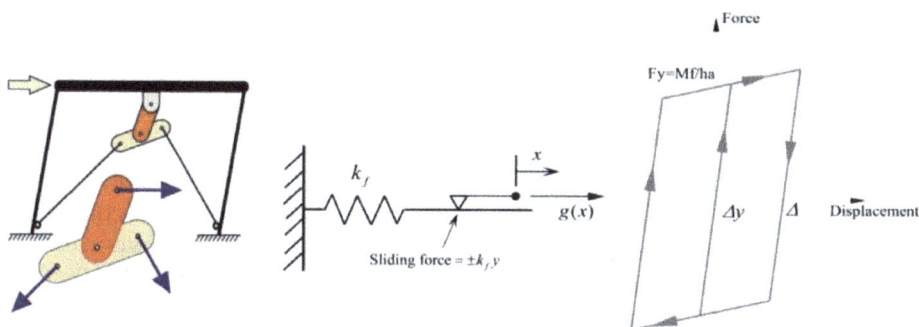

Fig. 12. Component of FDD

In order to compare the behavior of the original structure and equipped structure, the pushover curves of three cases are shown in Fig. 13. Infill walls lead to increase stiffness and strength of buildings compared to a building without considering infill walls. Changing the slope in pushover curves shows this phenomenon. In the push over curves with friction damper, stiffness and strength of buildings in the elastic part of analysis are increased. Since infill walls are brittle material and have a high stiffness, these walls absorb a large amount of lateral load until they fail. After failure of infill walls, we have a drop of stiffness (slope) and strength in curves. As it can be seen in the figure after failure of the infill walls, the slope of the curve will be the same as bare frame. The local interaction between frame A and infill walls is not considered. These results are achieved when the shear strength of columns are sufficient. This can be supposed in steel structures. In order to have a clear sense of the effect of infill and friction damper on the structural behavior, Fig.14 shows the scaled deformations of 3 cases named on Fig.13 as a, b and c.

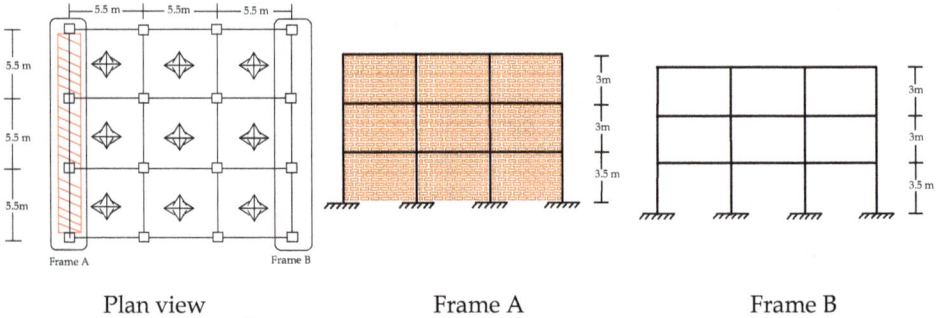

|   Plan view   |   Frame A   |   Frame B   |

Fig. 13. Building elevation

| Story | Column | | Exterior Beam | | Interior Beam | |
|-------|--------|--|---------------|--|---------------|--|
|       | Dimension (cm) | Thickness (cm) | Web (cm) | Flange (cm) | Web (cm) | Flange (cm) |
| 1 | 25×25 | 1 | 30×1 | 12×1 | 30×1 | 15×1 |
| 2 | 20×20 | 1 | 30×1 | 12×1 | 30×1 | 15×1 |
| 3 | 20×20 | 1 | 30×1 | 12×1 | 30×1 | 15×1 |

Table 10. Details of element sections

As shown in Fig. 14 by adding infill walls to the bare frame they lead to increase the stiffness of the system and the torsional problems occurs. This torsion leads to structural failure because of concentration of stress in one side and concentration of deformations in the other side. By using friction damper, eccentricity can be omitted and the distance between the center of mass and center of stiffness will be controlled to satisfy code requirement. In region I with increasing lateral deformation, rotation is increased both in infilled frame and equipped frame. However in this region the rotation at equipped frame is considerably less than infilled frame. This reduction of rotation is because of transforming asymmetric system (infilled frame) to a symmetric system (equipped frame). When infill walls start to fail, rotation starts to reduce Region II in the case of equipped frame. But for asymmetric infilled frame, the rotation increases with increasing lateral drift. In region III failing the infill walls cause to reduce the eccentricity and torsional rotation. Therefore the rotation decreases with increasing lateral displacement. In the case of equipped frame, failing the infill walls leads to increase rotation clockwise.

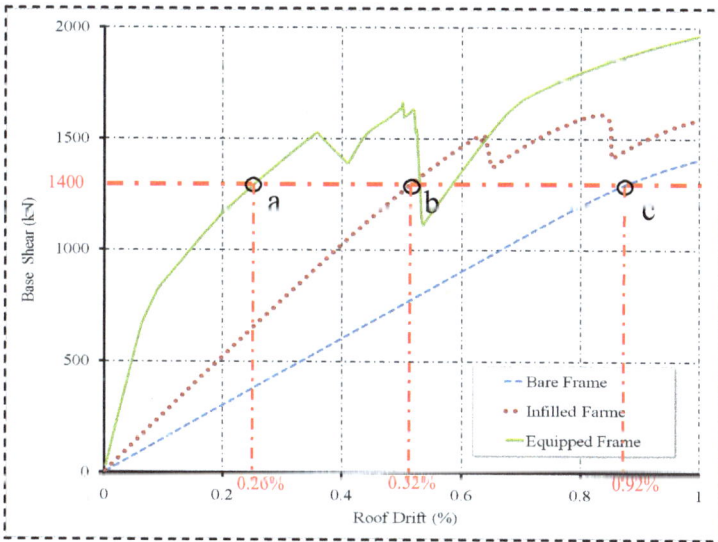

Fig. 14. Pushover curves

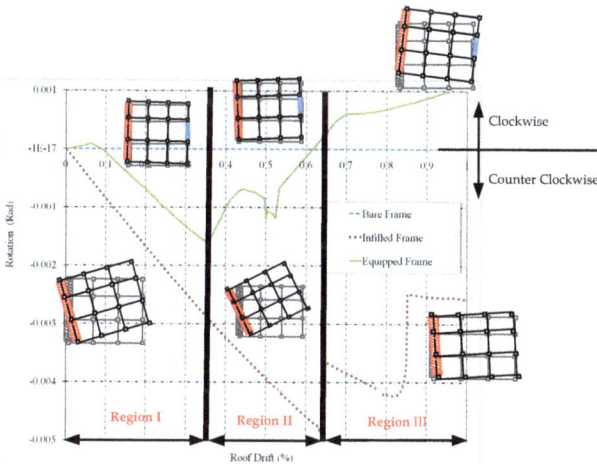

Fig. 15. Rotation curves for center of mass

After record selection procedure the maximum response of the structure with this arrangement are selected. In this type of structure, in Y direction infill wall is applied in one side and it changes the center of mass and stiffness in the structure. This eccentricity makes torsion in the plan. By applying the FDD in the perpendicular directions decreasing the eccentricity is possible.

Acceleration time history of Loma Perita earthquake has been shown in Fig. 15. Displacement time history response for bare and infilled frame under Loma Perita earthquake has been shown in Figs. 16 and 17 respectively. Comparing Figs. 17 and 18 it is seen a considerable reduction in response when it is equipped by FDD.

| Earthquake | Data | Site | Station and Component | M | R(km) | PGA (g) | PGV (cm/s) | PGA/PGV (s) |
|---|---|---|---|---|---|---|---|---|
| Loma perieta | 24.4.1989 | Soil | Holister – South & Pine | 6.9 | 27.9 | 1.298 | 37.1 | 15.8 |

Table 11. Details of records

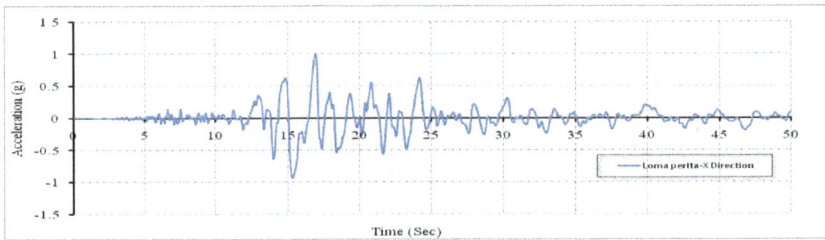

Fig. 16. Acceleration Time History of Loma Perita Earthquake

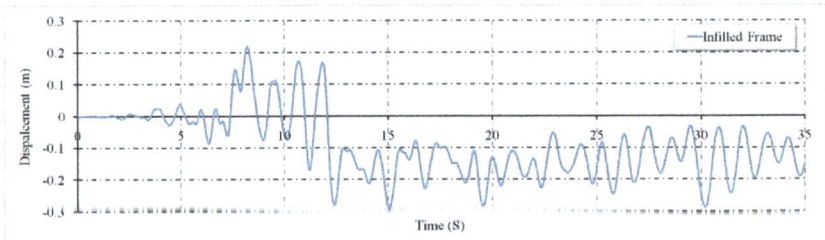

Fig. 17. Time history response for bare frame under Loma perita earthquake

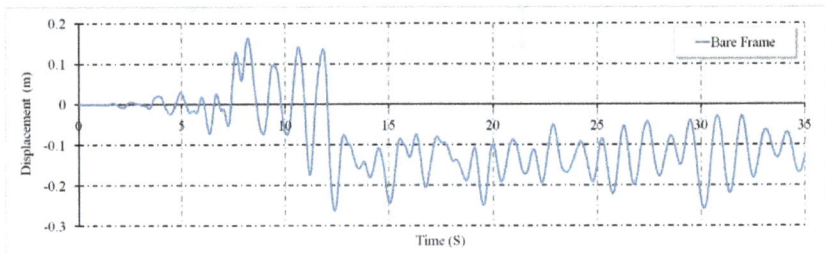

Fig. 18. Time history response for Infilled frame under Loma perita earthquake

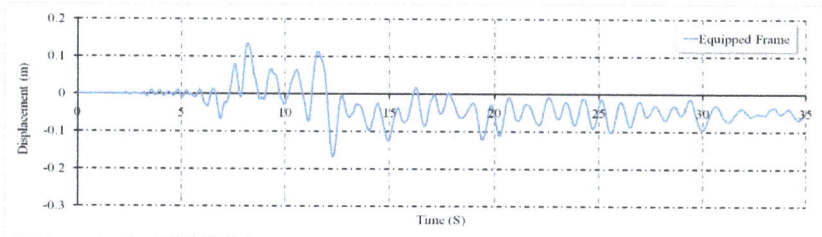

Fig. 19. Time history response for equipped frame under Loma perita earthquake

## 5.5 Sumarry and conclusion

Because of high stiffness of the infill walls, considering them as structural elements leads the initial stiffness of structures to increase. Such elements show high strength at the first step of seismic loading, but by reaching to the maximum strength, the infill walls fail and high loss of strength occurs in small drifts. This drop down of strength can be seen in push over curves of the structures. A relatively complete review on the positive and negative effects of masonry infills were presented in a categorized manner. As an example for numerical study torsion produced by infills were discussed as an engineering problem. Existing of these walls causes high differences between a center of mass and center of stiffness. Therefore by applying the lateral forces in center of mass, high torsional torque is generated in the diaphragm. For solving this problem, the FDD is used. Sensitivity analysis on effective variables on the FDD behavior shows that increasing sliding force causes decreasing the differences between the center of mass and center of stiffness, so the problem would be solved. It can be seen that, the FDD modifies structural torsion under earthquake excitation. By increasing PGA the positive effect of FDD in structural behavior is reduced, but equipped structure has better performance related to other structures without FDD. Seismic code requirements are considered. A detailed structural model has been produced using OpenSees. Both static and dynamic nonlinear analyses have been carried out. Because of sensitivity of the friction damper to pulse type excitation, near filed input motion has been considered as excitation force.

## 6. Acknowledgement

The authors gratefully acknowledge the contribution of Afsaneh Mousavi, graduated student of earthquake engineering, for helping in this research.

## 7. References

Aguilar, R. G. (1997). *Efecto del Refuerzo Horizontal en el Comportamiento de Muros de Mampostería Confinada Ante Cargas Laterales.* MSCE. Thesis, Universidad Nacional Autónoma de México.

Al-Chaar, G. Issa, M. & Sweeney, S. (2002). Behavior of masonry-infilled nonductile reinforced concrete frames, *J. Struct. Eng., ASCE* 128, 1055–1063.

Arturo Tena-Colunga, (2010), Review of the Soft First Story Irregularity Condition of Buildings for Seismic Design, *The Open Civil Engineering Journal*, 4, 1-15.

Atkinson, R.H., Amadei, B.P., Saeb, S. and Sture, S. (1989). Response of Masonry Bed Joints in Direct Shear, *ASCE Journal of Structural Division*, Vol. 115, No. 9, pp. 2276-2296

Australian Standard, AS 3700 – 2001, *"Masonry Structures "* , prepared by Committee BD-004, published on 27 November 2001

Baran, M. & Sevil, T. (2010), Analytical and experimental studies on infilled RC Frames , *International Journal of the Physical Sciences* Vol. 5(13), pp. 1981-1998, 18 October

Basu ,D. & Jain, S. K. (2004), Seismic Analysis of Asymmetric Buildings with Flexible Floor Diaphragms, *Journal of structural engineering*, AUGUST

Bell, D.K. & Davidson, B.J. (2001), Evaluation of Earthquake Risk Buildings with Masonry Infill Panels, *NZSEE Conference, 2001*

Bozorgnia1, Y., Tso, W. K. & M. ASCE. (1986), Inelastic earthquake rresponse of asymmetric structures, *Journal of Structural Engineering*, Vol. 112, No. 2, February

Carydis, P.G. Mouzakis, H.P. Taflambas, J.M. and Vougioukas, E.A. (1992). Response of infilled frames with brickwalls to earthquake motions. *Proceedings of the 10th World Conference on Earthquake Engineering*, Madrid, A.A. Balkema, Rotterdam, Vol. 5, pp. 2829-2834.

Chrysostomou, C.Z. (1991). *Effect of Degrading Infill Walls on the Nonlinear Seismic Response of Two-Dimensional Steel Frames*, PhD dissertation, Cornell University, Ithaca, N.Y

Chopra, A.K.& Juan C. De la Llera. (1996 ). Accidental and natural torsion in earthquake response and design of building, *eleventh word conference on earthquake Engineering*

Eberhard, M.O., Baldridge, S., Marshall, J. Mooney, W. & Rix, G.J. (2010). The Mw7.0 Haiti earthquake of january 12, 2010: USGS/EERI Advance Reconnaissance Team

Elnashai, A. S. & Sarno L. D. (2008), *Fundamentals of Earthquake Engineering*, John Wiley & Sons, Ltd. ISBN: 978-0-470-02483-6

Fardis, M.N. and Calvi, M.C., 1995. Effects of infills on the global response o reinforced concrete frames, *Proceedings of the 10th European Conference on Earthquake Engineering*, Vol. 4, pp.2893-2898.A.A. Balkema, Rotterdam

Fiorato, A. E., Sozen, M. A. and Gamble, W. L.(1970), An Investigation of the Interaction of Reinforced Frames with Masonry Filler Walls, *Structural Research Series*, No. 370, November ,University of Illinois, Urbana, Illinois, Civil Engineering Studies.

Gerardo M. Verderame, Flavia De Luca, Paolo Ricci and Gaetano Manfredi. (2010), Preliminary analysis of a soft-storey mechanism after the 2009 L'Aquila earthquake, *Earthquake Engng Struct. Dyn*

Golafshani, A.A., Bakhshi, A. & Tabeshpour, M.R. (2005). Vulnerability and damage analysis of existin buildings, *Asian Journal of Civil Engineering*, No. 1, 6, 85-100.

Goodman, R.E.; Taylor, R.L.; Brekke, T.L., (1968). *A model for the mechanics of jointed rock*. ASCE, Div.,94(3), pp. 637-659, USA

Govindan, P., Lakshmipathy, M., and Sanathkumar, A.R. (1986). Ductility of infilled frames. *American Concrete Institute Journal*, 83: 567 - 576.

Gur, T., Cihan Pay, A., Ramirez, J. A., Sozen, M. A., Johnson, A. M. & Ayhan Irfanoglu, (2007), Performance of School Buildings in Turkey during the 1999 Düzce and the 2003 Bingöl Earthquakes, *Submitted to Earthquake Spectra*, EERI

Hendry, A.W. (1990) *Structural Masonry*, Macmillan Education Ltd, London, England.

Holmes M. (1963). Combined Loading on Infilled Frames, *Proceeding Of The Institution Of Civil Engineers*, Volume 25: 31-38

Iranian Code of Practice for Seismic Design of Buildings, Standard No. 2800, 3rd Edition, Building and housing research center, 2006

Ismail, N., Griffith, M.& Ingham J.M. (2011) Performance of masonry buildings during the 2010 Darfield (New Zealand), 11th NAMC

Junwu, D., Yanru ,W., Qingsong ,G., Xiaozhai, Q., Xueshan, Y., Mai, T. & Lee, G. C. (2009), Three dimensional temporal characteristics of ground motions and building responses in Wenchuan earthquake, *Earthq Eng & Eng Vib*, 8:287-299

Kadir M.R.A. (1974). *The structural behavior of masonry infill panels framed structures*, University of Edinburgh, PhD thesis

Kappos, A.J. Stylianidis, K.C. and Michailidis, C.N. (1998). Analytical models for brick masonry infilled r/c frames under lateral loading. *Journal of Earthquake Engineering*, Vol. 2(1),pp. 59-87

King G.J.W. & Pandey P.C. (1978). The analysis of infilled frames using finite elements. Proc. Instn Civ. Engrs, Part 2, 65, 749-760

Klingner, R.E. Rubiano, N.R. Bashandy, T.R. & Sweeney, S.C. (1996). Evaluation and Analytical Verification of Shaking Table Data from Infilled Frames, Part 2: Out-of-Plane Behavior, *Proceedings of the 7th North American Masonry Conference*, Notre Dame.

Klinger, R.E. & Bertero, V.V. (1976). *Infilled frames in earthquake resistant construction*. Earthquake Engineering Research Center, University of California, Berkeley, CA, Rep. EERC 76-32, Dec

Komachi, Y. Tabeshpour, M.R. & Mualla, I. & Golafhshani, A.A. (2011). Retrofit of Ressalat Jacket Platform (Persian Gulf) using Friction Damper Device. *Journal of Zhejiang University*, in press

Koset, E.G. et al. (1974). Non-*linear dynamic analysis of frames with filler panels*, Am. Soc.Civ.Engrs, J.struct. Div.,100, 743-757

Kwon, O. (2008). Damaging Effects of the Pisco-Chincha (Peru) Earthquake on an Irregular RC Building, *The 14th World Conference on Earthquake Engineering* October 12-17, Beijing, China 2008

Lili Anne Akin. (2004). behavior of reinforced concrete frames with masonry infills in seismic regions. PHD Thesis in Purdue University

Liauw, T.C. & Kawn, K.H. (1984). New Development in Research of Infilled Frames, *Proc. 8th World Conf. on Earthq.Engng*, San Francisko, 4, 623-630

Mainstone, R. J. (1971). *On the stiffness and strengths of infilled frames*, Proceedings, Institution of Civil Engineers, Supplement IV, 57–90

Mallick D.V. & Severn R.T. (1967). *The Behavior of Infilled Frames under Static Loading*, The Institution of Civil Engineers, Proceedings, 39, 639-656

Mallick, D. V. & Garg, R. P. (1971). *Effect of openings on the lateral stiffness of infilled frames*, Proceedings of the Institution of Civil Engineers 49, 193–209

Manos, G.C. Triamataki, M. & Yasin, B. (1995). Experimental and numerical simulation of the influence of masonry infills on the seismic response of reinforced concrete framed structures.*Proceedings of the 10th European Conference on Earthquake Engineering*, A.A. Balkema, Rotterdam, Vol. 3, pp. 1513-1518.

Marjani, F. & Ersoy, U. Behavior of Brick Infilled Reinforced Concrete Frames Under Reversed Cyclic Loading. 06/23/2004.

Mehrabi, A. B., Shing, P. B., Schuller, M.P. and Noland, J.L. (1996). Experimental Evaluation of Masonry-Infilled RC Frames. *ASCE Journal of Structural Engineering*, Vol. 122, No. 3, pp. 228-237.

Mehrabi, A. B., and Shing, P. B. (1998). Seismic Resistance of Masonry- Infilled RC Frames. *Proceedings of the 8th North American Masonry Conference*, The Masonry Society, USA, pp. 465-476.

Michailidis, K., Sklavounos, S., and Plimer, I. (1995). Chromian dravite from the chromite ores of Vavdos area, Chalkidiki peninsula, Northern Greece. Neues Jahrbuch fur Mineralogie. Monatshefte, 1995, 513–528.

Moehle, J. P., Elwood, K. J. & Sezen, H., Shear strength and axial load collapse of existing reinforced concrete columns

Moghadam, Hasan. (2001). Earthquake Engineering, Basic and Application Seismic design of buildings-2800 code. 3. S.l:Building and housing research center

Mohyeddin-Kermani, A. et al. (n.d).The Behaviour of RC Frames with Masonry Infill in Wenchuan Earthquake

Mosalam, K.M., White, R.N., & Gergely, P. (1997). Static Response of Infilled Frames using Quasistatic Experimentation. *ASCE Journal of Structural Engineering*, Vol. 123, No. 11, pp. 1462-1469.

Mualla, IH, Belev B. Performance of steel frames with a new friction damper device under earthquake excitation, *Engineering Structures*, 24 (2002) 365-371

Naseer, A. et al. (2007). Building code of Pakistan: before and after the October o8, 2005 Kashmir earthquake, *Sixth National Conference on Earthquake Engineering*, 16-20 October, Istanbul, Turkey.

Negro P. and Verzeletti G. (1996). Effect of Infills on the Global Behaviour of R/C Frames: Energy Considerations from Pseudodynamic Tests. *Earthquake Engineering and Structural Dynamics*, 25, 753-773

Nevzat, Kirac. Mizam, Dogan, Hakan, Ozbasaran. (2010). Failure of weak-storey during earthquakes, *Engineering Failure Analysis*

Paulay, T. & Priestley, M.J.N. (1992). *Seismic Design of Reinforced Concrete and Masonry Buildings*, John Wiley & Sons, New York, United States

Paulay, T. & Priestley, M. J. N. (1992). *Seismic Design of Reinforced Concrete and Masonry Buildings"*, John Wiley & Sons, Inc

Polyakov, S. V. (1956). *Masonry in Framed Buildings (An investigation into the strength and stiffness of masonry infilling)*, Moscow.

Pires, F. & Carvalho, E.C. (1992). The behaviour of infilled reinforced concrete frames under horizontal cyclic loading. *Proceedings of the 10th World Conference on Earthquake Engineering*, Madrid, A.A. Balkema, Rotterdam, Vol. 6, pp. 3419-3422.

Rivero C.E. & Walker W.H. (1984), An Analytical Study of The Interaction of Frames and Infill Masonry Walls, *Proc. 8th World Conf. on Earthq. Engng.*, San Francisko, 4, 591-598.

Sahlin, S. (1971). *Structural Masonry*, Prentice-Hall Inc., New Jersey, United States.

San Bartolomé, A. (1990). *Collección del Ingeniero Civil*, (in Spanish), Libro No. 4, Colegio de Ingenierios del Peru.E

Santiago, P. Amadeo, B.C. Mario, E. R. and Paul Smith-Pardo, J. masonry infill walls: an effective alternative for seismic strengthening of low-rise reinforced concrete building. *The 14th World Conference on Earthquake Engineering* October 12-17, 2008, Beijing, China

Schneider, S.P. Zagers, B.R. and Abrams, D.P. (1998). Lateral strength of steel frames with masonry infills having large openings. *ASCE Journal of Structural Engineering*, Vol. 124(8), pp. 896-904.

Sezen, H. Moehle, J. P, (2006), Seismic Test of Concrete Columns With Light Transverse Reinforcement, *ACI Structural Journal*, V.103, No.6, November-December

Sharany Haque & Khan Mahmud Amanat. (2009), Strength and drift demand of columns of RC framed buildings with soft ground story, *Journal of Civil Engineering (IEB)*, 37 (2) (2009) 99-110

Shing, P. B. & Mehrabi, A. B. (2002). Behaviour and analysis of masonry-infilled frames, *Progress in Structural Engineering and Materials*, 4(3), 320-331.

Shrive, N.G. (1991). Materials and Material Properties, *Reinforced and Prestressed Masonry*, A.W.Hendry, Longman Scientific and Technical, London, England

Siamak Sattar & Abbie B. Liel, (n.d), Seismic performance of reinforced concrete frame structures with and without masonry infill walls

Sinha, B.P. & Pedreschi, R. (1983). Compressive Strength and Some Elastic Properties of Brickwork, *International Journal of Masonry Construction*, Vol. 3, No. 1, pp. 19-27.E

Smith, B. S. & Coull, A. (1991), Tall Building Structures: Analysis and Design, John Wiley & Sons, Inc

Stafford-Smith B.S. (1968). Model Test Results of Vertical and Horizontal Loading of Infilled Frames, *ACI Journal*, Volume 65, No. 8: 618-625

Stefano, Ma. D. & Pintucchi, B. (2007), A *review of research on seismic behaviour of irregular building structures since 2002*, Springer Science+Business Media B.V. 2007.

Stöckl, S. & Hofmann, P. (1988). Tests on the Shear Bond Behaviour in the Bed-joints of Masonry, *Proceedings of the 8th International Brick and Block Masonry Conference*, Dublin, Ireland.

Tabeshpour, M. R. Ghanad, M. A. Bakhshi, A. & Golafshani, A. A. (2004). Effect of masonry infills on seismic behavior of concrete frames, *1th national conference of civil engineering*, sheriff University, Tehran, Iran.

Tabeshpour, M. R. (2005). Numerical study of short column failure, *Earthquake research letters*, IISEE, Tehran, Iran.

Tabeshpour, M.R. (2009a). *Masonry infills in structural frames (Handbook, Part 18)* , FadakIsatis Publisher, Tehran, Iran.

Tabeshpour, M.R. (2009b). *Seismic retrofit of infilled frames (Handbook, Part 19)*, Fadak Issatis Publisher, Tehran, Iran.

Tabeshpour, M. R. (2009c). *Requirements of infills in Iranian seismic code of practice, Standard No. 2800 (Handbook, Part. 20)*, Fadak Issatis Publisher, Tehran, Iran.

Tabeshpour, M. R. (2009d) A comparative study of several codes about masonry infill wall, *4th national conference of seismic design code of buildings (standard No. 2800)*, Tehran, Iran.

Tabeshpour, M. R. (2009e) Modeling of infill wall using commercial software, *4th national conference of seismic design code of buildings (standard No. 2800)*, Tehran, Iran.

Tabeshpour, M. R. (2009f) Increasing design load in specific columns, *4th national conference of seismic design code of buildings (standard No. 2800)*, Tehran, Iran.

Tabeshpour, M.R. Ebrahimian, H. (2010). Seismic retrofit of existing structures using friction dampers, *Asian Journal of Civil Engineering*, No. 4, 11, 509-520.

Tabeshpour, M. R. Azad, A., & Golafshani, A. A. (2011a). Nonlinear static analysis of steel frame with asymmetric arrangement of masonry infill wall, *3rd national conference of urban management and retrofitting*, Tehran, Iran.

Tabeshpour, M. R. Mousavy, A. (2011b). Short column surveying because of masonry infill wall existence using nonlinear static analysis, *6th national conference of civil engineering*, semnan, Iran.

Tabeshpour, M.R. Azad, A. Golafshani, A.A. (2011c). Seismic Retrofitting by using Friction damper in Horizontally Irregular Infilled Structures, 6th National Congress on Civil Engineering, Semnan University, Semnan, Iran.

Tabeshpour, M. R. Kalatjari, V.R. & Karimi, K. (2011d), Investigation of lateral drift of concrete infilled frames, 1st *National conference on earthquake and lifelines*, Tehran, Iran.

Takewaki, I.; Murakami, S. & Fujita, K. & Yoshitomi, S. & Tsuji, M. (2011). The 2011 off the Pacific coast of Tohoku earthquake and response of high-rise buildings under long-period ground motions, *Soil Dynamics and Earthquake Engineering*, In Press

Tso, W.K. (1988), Strength eccentricity concept for inelastic analysis of asymmetrical structures, *Eng. Struct.* 1989, Vol. 11

Tso, W. K. & Ying, H. (1992), Lateral strength distribution specification to limit additional inelastic deformation of torsionally unbalanced structures, *Eng. Struct.* Vol. 14, No 4, July

Valiasis, T. Stylianidis, K. (1989). Masonry Infilled R/C Frames under Horizontal Loading experimental Results. *European Earthquake Engineering*, Vol. 3, No. 3, pp. 10-20

Valiasis, T.N. Stylianidis, K.C. and Penelis, G.G. (1993). Hysteresis model for weak brick masonry infills in R/C frames under lateral reversals. *European Earthquake Engineering*, Vol. 7(1), pp. 3-9.

Yoon, Y. & Stafford Smith, B. (1995), Assessment of translational-torsional coupling in asymmetric uniform wall-frame structures, *Journal of structural engineering*, October

Zarnic, R. (1994). Inelastic Model of R/C Frame with Masonry Infill – Analytical Approach. *Engineering Modelling*, 1-2, 47-54.

Zarnic, R. & Tomazevic, M. (1985). Study of Behaviour of Masonry Infilled Reinforced Concrete Frames Subjected to Seismic Loading. *Proc. 7th International Brick Masonry Conference*, Brick Dev. Res. Inst. & Dept. of Arch. And Bldg., University of Melbourne, 2, 1315-1325.

Zarnic, R. Gostic, S. (1998). Non Linear Modelling of Infilled Frames. *11th European Conference Earthquake Engineering*, Paris, Balkema: Rotterdam.

# 11

# Seismic Vulnerability Analysis of RC Buildings in Western China

Zhu Jian
*Ning Xia University*
*China*

## 1. Introduction

Kobe earthquake in Japan (Ms6.9) happened in 1995 Jan 17 at five o'clock in the morning, depth of the seismic focus was 20km, the seismic characteristic was shallow vertical focus earthquake, peak acceleration was 813gal, predominant period was 0.3~0.5s, The main sediments of the region under 20~30 meters were sandy silt, and Generally to moderately weathered rock. 18 thousand buildings were been damaged and 1.2 thousand buildings were collapsed, half timber houses which built before 1980'were destroyed seriously, at the same time middle-storey reinforced-concrete (RC) buildings which were 7~8 layer buildings damaged also seriously, the mainly damage showing was integral overthrow of buildings, in addition columns between third to forth layer of RC buildings were squashed in the earthquake, and were collapsed in the middle of structures.

Tohoku Earthquake (Ms9.0) in Japan occurred on March,11 2011, it is regarded as the most devastating killer earthquake after the 1923 great kanto earthquake in Japan, in which almost 30000 people were killed or missed in the earthquake and the subsequent monster tsunami. The maximum height of the tsunami is reported to have been almost 40 m. The recorded maximum peak ground acceleration was 2933 gal and large long-period wave components were recorded in Tokyo during the 2011 off the Pacific coast of Tohoku Earthquake. It is remarkable in this earthquake that the number of collapsed or damaged buildings and houses remains unclear because most of the damage resulted from the tsunami (Takewaki,2011).

Northridge earthquake(Ms 6.7) happened in 1994 Jan 17 at five o'clock in the morning, this is terrible earthquake with tremendous horizontal and vertical acceleration evenly to reach 1.0g, 2500 houses were been collapsed and 4000 houses were been damaged seriously by powerful natural energy, the main characteristic of buildings' damage was shear failure of columns in old second to third layer timberwork apartment blocks, and RC frames were damaged very small include several 40 layer high-rise RC buildings.

Taiwan Chichi earthquake (Ms7.6) happened in 1999 Sep 21 at one o'clock in the morning, depth of the seismic focus was 8km, and the earthquake was powerful shallow focus earthquake, damaged data of 8733 buildings were collected post earthquake, one to three layer RC buildings were damaged seriously, or about 52.5 percent, among which old RC buildings built before 1982 based insufficient seismic design and outdated seismic code were destroyed deeply, and occupied about 59.4 percent. Small high-rise shear-wall RC buildings were damaged tiny, and which occupied 6.4 percent.

China Wenchuan earthquake(Ms8.0) happened in 2008 May 12 at two o'clock in the afternoon, in the seismic region are three main geological fault zones, and the Wenchuan earthquake occurred in Longmenshan (LMS) geological fault zone, total length of LMS is 530 kilometer(km), width of which is 40~50km, presenting northeast-southwest running, and leaning to northwest about 30~70 angles, LMS divided into two parts by Jiangyou(JY) city in Sichuan province, northeast part of which is geological fault zone during Early Pleistocene-middle Middle Pleistocene, southwest part of which is fault zone during Holocene, the earthquake was happened in southwest part, depth of the seismic focus was 14km, and fracture length was 240km, the process of fracture consisted of several continuous events, every event was a earthquake which magnitude was Ms7.2 to Ms7.6, all type of buildings were damaged overlying in the continuous vibration, the peak acceleration recorded was 957.8gal, waves diffused all directions in three-dimensional, at the same time vibration along northeast of fracture zone was more powerfully, and continued for about 100 seconds, spreading to 16 provinces of China, specialists from State Seismological Bureau of China surveyed 500 thousand square kilometer, and in where 2419 square kilometer earthquake intensity reached 11 scale, far surpass design intensity in the region where is 7 scale.

In the Wenchuan earthquake, 5.46 million buildings were collapsed, number of serous damaged buildings were 5.93 million, amount to total damaged buildings exceeded 15 million, old brick masonry structures built 70s~80s damaged most seriously, in the next place low-rise RC buildings destroyed also severely, which were integral overthrow of base layer columns yielded because of wrong site, low material strength and fault layout. With Indian plate moving to north continuously and squeezing Asian plate, in the near future there is still high risk of major earthquake happen again in western China region, so how to evaluate reliability and vulnerability of the lifeline systems for future earthquake in the area and search reasonable design practice of seismic strengthening of these buildings is urgent mission.

## 2. Summarize methods of structural seismic vulnerability analysis

Damage from earthquake is comprehensive, there are different ways for various damage of engineering, seismic technician classify the vulnerability assessment as four sorts: empirical, judgmental, experimental and analytical according to whether the damage data used in their generation derives mainly from observed post-earthquake surveys, expert opinion, analytical simulations or combinations of these respectively.

As described below, the seismic vulnerability assessment of buildings at large geographical scales has been first carried out in the early 70's, through the employment of empirical methods initially developed and calibrated as a function of macro-seismic intensities. This came as a result of the fact that, at the time, hazard maps were, in their vast majority, defined in terms of these discrete damage scales (earlier attempts to correlate intensity to physical quantities, such as PGA, led to unacceptably large scatter). Therefore these empirical approaches constituted the only reasonable and possible approaches that could be initially employed in seismic risk analyses at a large scale.

(Whitman, 1973) first proposed the use of damage probability matrices for the probabilistic prediction of damage to buildings from earthquakes. The concept of a DPM is that a given structural typology will have the same probability of being in a given damage state for a given earthquake intensity. The format of the DPM was suggested by (Whitman, 1973), where example proportions of buildings with a given level of structural and non-structural damage are provided as a function of intensity (note that the damage ratio represents the ratio of cost of repair to cost of replacement). (Whitman, 1973) compiled DPMs for various structural typologies according to the damaged sustained in over 1600 buildings after the 1971 San Fernando earthquake.

One of the first European versions of a damage probability matrix was produced by (Braga et al., 1982), which was based on the damage data of Italian buildings after the 1980 Irpinia earthquake, and this introduced the binomial distribution to describe the damage distributions of any class for different seismic intensities. The binomial distribution has the advantage of needing one parameter only which ranges between 0 and 1. On the other hand it has the disadvantage of having both mean and standard deviation depending on this unique parameter. The buildings were separated into three vulnerability classes (A, B and C) and a DPM based on the MSK scale was evaluated for each class. This type of method has also been termed 'direct' (Corsanego & Petrini, 1990) because there is a direct relationship between the building typology and observed damage. The use of DPMs is still popular in Italy and proposals have recently been made to update the original DPMs of Braga. (Di Pasquale, 2005) have changed the DPMs from the MSK scale to the MCS (Mercalli-Cancani-Sieberg) scale because the Italian seismic catalogue is mainly based on this intensity, and the number of buildings has been replaced by the number of dwellings so that the matrices could be used in conjunction with the 1991 Italian National Statistical Office (ISTAT) data. (Dolce,2003) have also adapted the original matrices as part of the ENSeRVES (European Network on Seismic Risk, Vulnerability and Earthquake Scenarios) project for the town of Potenza, Italy. An additional vulnerability class D has been included, (Grüntal, 1998) using the EMS98 scale to account for the buildings that have been constructed since 1980. These buildings should have a lower vulnerability as they have either been retrofitted or designed to comply with recent seismic codes.

Judgmental method was based structural damage data can be considered the greatest cause of life and monetary loss in the majority of seismic events, vulnerability curves were been received under-predict the damage observed in buildings after earthquake, for example: Miyakoshi et al. [6] used damage data observed in RC buildings after the Kobe earthquake (Japan, 1995) in constructing damage parameter equation, (Yamazaki & Murao, 2000) also made up of empirical vulnerability curve though analyzing damage data observed in buildings after the Kobe earthquake, (Orsini et al.,1999) used the Parameterless Scale of Intensity (PSI) ground-motion parameter to derive vulnerability curves for apartment units in Italy. Both studies subsequently converted the PSI to PGA using empirical correlation functions, such that the input and the response were not defined using the same parameter.

The use of observed damage data to predict the future effects of earthquakes also has the advantage that when the damage probability matrices are applied to regions with similar characteristics, a realistic indication of the expected damage should result and many uncertainties are inherently accounted for. However, there are various disadvantages associated with the continued use of empirical methods:

1.  A macro-seismic intensity scale is defined by considering the observed damage of the building stock and thus in a loss model both the ground motion input and the vulnerability are based on the observed damage due to earthquakes.

2.  The derivation of empirical vulnerability functions requires the collection of post-earthquake building damage statistics at sites with similar ground conditions for a wide range of ground motions: this will often mean that the statistics from multiple earthquake events need to be combined. In addition, large magnitude earthquakes occur relatively infrequently near densely populated areas and so the data available tends to be clustered around the low damage/ground motion end of the matrix thus limiting the statistical validity of the high damage/ground motion end of the matrix.

3.  The use of empirical vulnerability definitions in evaluating retrofit options or in accounting for construction changes (that take place after the earthquakes on which those are based) cannot be explicitly modeled; however simplifications are possible, such as upgrading the building stock to a lower vulnerability class.

4.  Seismic hazard maps are now defined in terms of PGA (or spectral ordinates) and thus PGA needs to be related to intensity; however, the uncertainty in this equation is frequently ignored. When the vulnerability is to be defined directly in terms of PGA, where recordings of the level of the ground shaking at the site of damage are not available, it might be necessary to predict the ground shaking at the site using a ground motion prediction equation; however, again the uncertainty in this equation needs to be accounted for in some way, especially the component related to spatial variability.

5.  When PGA is used in the derivation of empirically-defined vulnerability, the relationship between the frequency content of the ground motions and the period of vibration of the buildings is not taken into account.

Laboratory testing represents the third alternative tool for vulnerability assessment. The main advantage of this method is the freedom in selecting model to suite the application. However, this method is hampered by the limitations of scale, laboratory and equipments capacities. Another factor affecting the reliability of this tool is the effect of loading type and routine on the response. Furthermore, the deficiency in modeling soil-structure interactions represents one of the main disadvantages of using the laboratory testing as a seismic vulnerability assessment tool. Pseudo dynamic testing has been used in several studies. However, the slow rate of testing represents a main obstacle in using this type of testing for vulnerability assessment where large sets of data points are required to make a comprehensive assessment. Recently sub-structuring and distributed testing have been used in different studies.

The last tool is analysis vulnerability method. The method has received much attention from researchers in recent years. Many researchers of countries such as (M.A.Erberik, 2004), (Rossetto. T, 2005), (S. Kircil,2006), (Jun Ji,2007) and (Barbara Borzi,2008) have studied seismic characteristic and vulnerability of many kinds of structures like reinforced concrete buildings, masonry buildings, bridges, museums and dams. Analysis offers two main advantages: the ease of controlling the level of refinement and feasibility of parametric studies. However, results of recent studies emphasized several issues that need to be resolved in order to enhance the use of analysis in realistic vulnerability assessment, including influence level between different models, ground vibration importing selecting, comprehensive structures analysis and computing consume. So analysis method is selected based all four tools characteristic, and the analysis tool will be used in classic RC buildings of western region in China in connecting fielding damage data observed after Wenchuan earthquake (2008) in China.

## 3. Response spectrum fitting stochastic artificial waves

Probability response spectrum is an elastic response spectrum, which is enough as external load for most sample and general layout buildings of seismic design, but which is not suitable for comprehensive and lifeline structures, so nonlinear dynamic analysis is very necessary and importable to the lifeline structures. In fact, actual earthquake recorded waves are not easy for attainable in engineering research, and why is the reason how to use spectrum density and design response spectrum for fitting artificial stochastic seismic waves.

### 3.1 Analysis procedure

Artificial seismic waves analysis methods could be divided into engineering simulation method and seismology simulation method. Earthquake focus also be divided into many element focus in seismology simulation method, and which element focus excite site vibration in using theoretical or experimental ways, that in the end gathering site vibration by total element focus and to move forward to construct site seismic function. Engineering simulation method is made up appropriate seismic function in accordance with seismic Fourier frequency spectrum, energy spectrum density and response spectrum, engineering method has been accepted by engineering technician because of well academic foundation. So engineering simulation method is been selected in Matlab7.0 programming and simulate and fit artificial seismic stochastic waves according with design response spectrum. Five steps are following:

1. initial setting artificial seismic waves;
2. compute response spectrum of artificial seismic waves;
3. compare initial response spectrum with target response spectrum and compute both ratio;
4. regulate artificial seismic waves on the base of response ratio;
5. repeat step 2 and degree of fitting satisfy demand.

### 3.2 Artificial seismic waves simulation

There are three methods for artificial waves: tri-angle series method, stochastic impulse method and natural regression method, the third ways raised in 1970s, but the first method is accepted widely.

We suppose that an zero mean value $(E[x(t)=0])$ steady state diffuse scattering wave shape time history is $x(t)$, which may be showed by Fourier analysis, as shown in Eq.(1):

$$x(t) = \sum_k A_k \exp[i(\omega_k t + \phi_k)] \tag{1}$$

Where $\omega_k = k\Delta\omega$, $\Delta\omega = 2\pi/T_d$, $T_d$ is seismic time history, $A_k = A(\omega_k)$ is Fourier amplitude spectrum value, $\phi_k$ is phase angle, so we can decided $A_k$, $\phi_k$ according with vibration characteristic value, in other word, we can create steady vibration time history based above function. In which $\phi_k$ is created by random phase way and random range from 0 to $2\pi$, certainly it my also take from real earthquake recorded waves. Frequency of vibration could shown by spectrum density function $S_{\ddot{x}_g}(\omega)$, and $S(\omega)$ has relation with $A(\omega_k)$,

$$E[x(t)^2] = \int_{-\infty}^{\infty} S(\omega)d\omega = \sum_k A_k^2 \text{, mean } S(\omega_k)\Delta\omega = \frac{A_k^2}{2} \text{, mean also } A_k = [2S(\omega_k)\Delta\omega]^{\frac{1}{2}} .$$

In the mean time (Kaul, 1978) computed relation energy spectrum density function with response spectrum on the base of diffuse scattering theory shown as Eq. (2):

$$S(\omega_k) = \frac{\left[\zeta\middle/\pi\omega_k\right]S_a^2(\omega_k,\zeta)}{In\left[\omega_k T_d\middle/{-\pi In(1-p)}\right]}$$ (2)

Where $S_a$ is accelerate response spectrum, $\zeta$ is system damping ratio and initial setting as 5%, $p$ is surpass probability and setting as 10%, $T_d$ is time history 30 seconds, initial steady artificial seismic time history are shown by programmed with Matlab 7.0 in Fig.1:

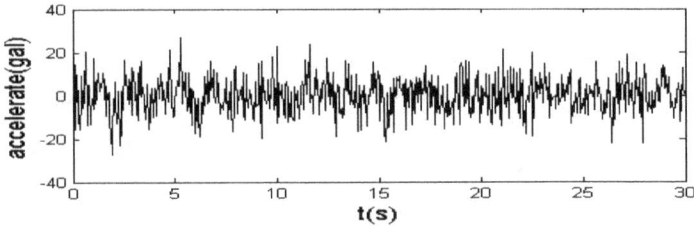

Fig. 1. Initial steady artificial vibration time history

Real earthquake wave is not an sample steady vibration, in consideration of initial vibration phase, main vibration phase and decline phase, so artificial waves only shown more accurately as stationary time history, in present, common way is steady state $x(t)$ multiply by decided time varying envelop function $g(t)$, that is to say, change stationary state into steady form.
Seismic acceleration could be shown as:

$$A_g(t) = g(t)\ddot{X}_g(t)$$ (3)

Where $\ddot{X}_g(t)$ is $x(t)$, $g(t)$ is envelop curve, so structure of spectrum frequency is not vary with time but variance is changing with time.
Generally envelop function is decided in according with seismic total duration and strong vibration duration. Seismic total duration means earthquake time history from begin time to end time, we can know that seismic duration change from several seconds to dozens of seconds and at most one to two minutes, (Murphy and O'Brien,1977) from American analysis about 400 recorded earthquake waves and found the vast majority of recorded waves duration within 25~40s when the total duration within 2~100s, so 30s was decided as standard artificial wave duration. After engineering technician analyze vast actual seismic recorded acceleration $a(t)$, they found an common characteristic: stationary of intensity and frequency spectrum, vibration intensity divided as ascent stage, strong vibration sustain stage and weakening stage, meanwhile stationary in every stage. In engineering we often pay attention to strong vibration stage called strong vibration duration $T_\alpha$, because importance of strong vibration duration has been realized by engineering technician, but definition of which is not definite, so in the article relative duration based

energy was been considered, and (Husid,1974), (Trifunac & Bardy,1975) studied that they thought using $\int_0^t a^2(t)dt$ as time-vary characteristic of earthquake vibration energy, and shown as Eq.(4):

$$I(t) = \frac{\int_0^t a^2(t)dt}{\int_0^T a^2(t)dt} \tag{4}$$

Where T is total duration of seismic wave, $I(t)$ is function within 0~1. Definition of strong vibration duration is :

$$T_\alpha = t_2 - t_1 \tag{5}$$

The relation of $t_2$ and $t_1$ is :

$$I(t_2) - I(t_1) = \alpha\% \tag{6}$$

Japanese and American researchers (Takjzawa,1980), (Jennings & Housner,1968) compared 90%, 80%, 70%, 60% and 50% of energy duration for destructive effect of buildings and decided 70% of energy duration was more fitting to actual situation, the article author also analysed and computed representative seismic record in Tab.(1), in the same time considered standard deviation ±5% , and energy in ascent stage occupy 10% of total energy , moreover $I(t_1)$ is 10%, $I(t_2)$ is 80%.

| Earthquake (T) | 10% $t_1$ | 80% $t_2$ | Strong duration $T_\alpha$ | Earthquake (T) | 10% $t_1$ | 80% $t_2$ | Strong duration $T_\alpha$ |
|---|---|---|---|---|---|---|---|
| El-centro (37.03s) | 6.12s | 10.18s | 4.06s | Northridge (15.01s) | 1.90s | 3.80s | 1.90s |
| Taftew (27.11s) | 1.87s | 7.73s | 5.86s | Chichi long (39.98s) | 10.26s | 15.00s | 4.74s |
| Kobe (46.38s) | 4.00s | 8.50s | 4.50s | Nanjin (16.10s) | 3.00s | 8.90s | 5.90s |
| Tianjinew (19.08s) | 7.26s | 10.01s | 2.75s | Qiananns (22.10s) | 1.79s | 4.74s | 2.95s |
| Tangshanew (49.20s) | 20.9s | 39.4s | 18.5s | Wenchuan (500s) | 11.92s | 91.7s | 79.78s |

Table 1. Statistics and analysis of strong vibration duration in earthquake record

Tri-stage curve and exponent curve have been accepted widely now by engineering because of simple, visualized and physical significance（Ohsaki,1978;Kaul,1978;Amin & Ang, 1968）, tri-stage curve could be shown as:

$$g(t) = \begin{cases} \left(\dfrac{t}{t_1}\right)^2 & t \le t_1 \\ 1 & t_1 < t \le t_2 \\ \exp\left[-c(t - t_2)\right] & t > t_2 \end{cases} \tag{7}$$

Where $t_1$, $t_2$ and $c$ are model parameters, $t_1$ and $t_2$ are begin moment and end moment, $c$ present weakening velocity in descent stage, $T_s$ is vibration duration in strong vibration duration for 70% energy of total seismic energy, in which $t_1 = 0.5T_s$, $t_2 = 1.2T_s$, $c = 2.5/T_s$. Exponent curve is shown as $g(t) = \beta t e^{-\alpha t}$, in which $\alpha$ and $\beta$ is model parameter, $\alpha = 0.01$, $\beta = 0.028$.

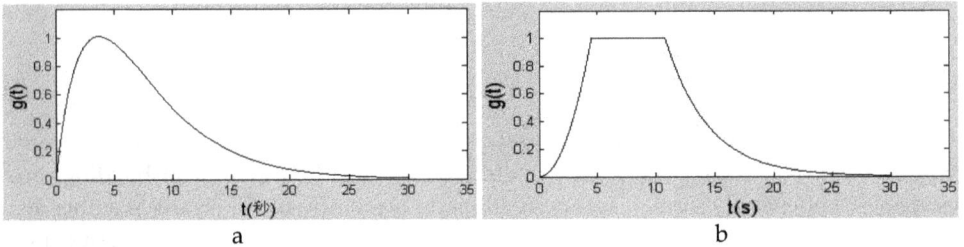

a                                          b

Fig. 2. Envelop curve (a: exponent form, b: tri-stage form)

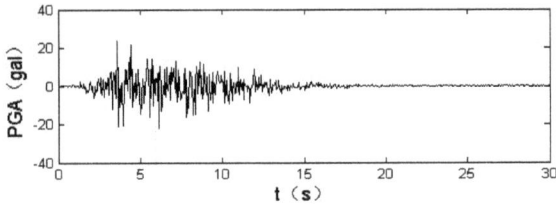

Fig. 3. Initial stationary artificial stochastic acceleration time history

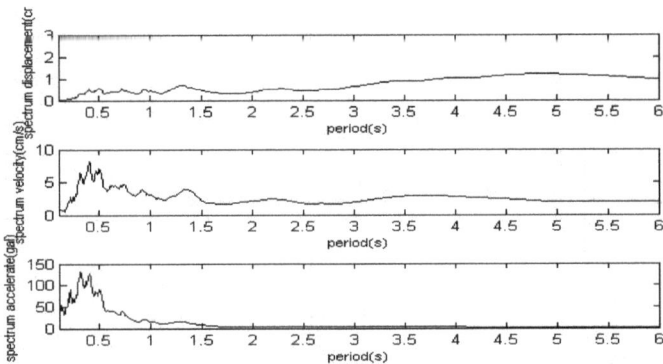

Fig. 4. Initial artificial stochastic response spectrum(from top to bottom: spectrum displacement, spectrum velocity, spectrum acceleration)

## 3.3 Artificial seismic record update

If initial artificial seismic response spectrum from record $a_0(t)$ isn't meet degree of fitting with specify earthquake response spectrum or code response spectrum, there has need to update for more precision. Generally two ways now have been used for updating. The article used the second method.

First method：

$$A_i'(t) = \frac{S_V}{S_V'} A_i(t) \tag{8}$$

Where $A_i'(t)$ is updated Fourier spectrum, $A_i(t)$ is Fourier spectrum of initial artificial record, $S_V$ is target response spectrum velocity, $S_V'$ is spectrum velocity of initial record.

Second method：

$$S_g'(\omega) = S_g(\omega) \left[ \frac{TRSA(\omega)}{SRSA(\omega)} \right]^2 \tag{9}$$

Where $S_g'(\omega)$ is updated power spectrum density, $S_g(\omega)$ is initial power spectrum density, $TRSA(\omega)$ is intend to compared target response spectrum, $SRSA(\omega)$ is computed stationary artificial seismic response spectrum, next step compute new steady artificial wave and multiply seismic envelop curve for updating new stationary artificial seismic wave, the updating could be repeated several times until degree of fitting is satisfy with accuracy.

It is aware of that we need to control frequency in updating process, at first spectrum acceleration value of control points need be computed out, secondly comparing with control points of frequency, adjust ratio between two points need linear interpolation method to compute. As shown in Tab. 2.

| The number of control points | Frequency Range （Hz） | |
|:---:|:---:|:---:|
| 5 | 0.10 | 0.18 |
| 29 | 0.20 | 3.00 |
| 4 | 3.15 | 3.60 |
| 7 | 3.80 | 5.00 |
| 11 | 5.25 | 8.00 |
| 14 | 8.50 | 15.00 |
| 4 | 16.00 | 30.00 |

Table 2. Frequency adjust control points for artificial wave

Fig. 5-6 are updated samples for intensity 9 from (Chinese Seismic Code, 2001). The program iterative computed 5~8 times and result meet degree of fitting with Chinese code response spectrum, now artificial wave after updated was fitting to code response spectrum, 120 seismic record fitting to major earthquake and code response spectrum were created in this way and become applied load for next computation.

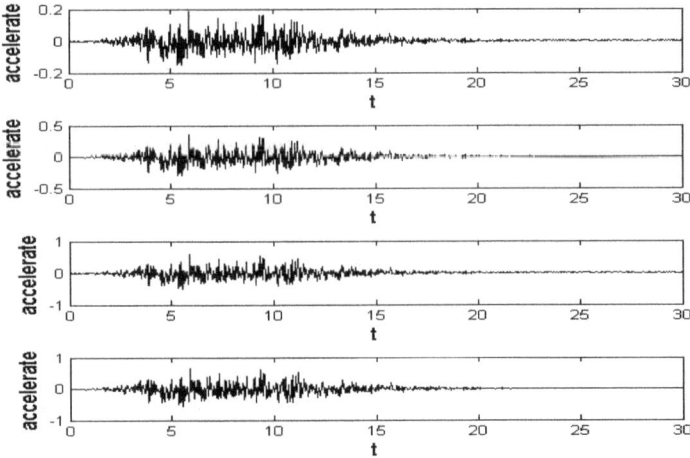

Fig. 5. Updating process of artificial seismic acceleration

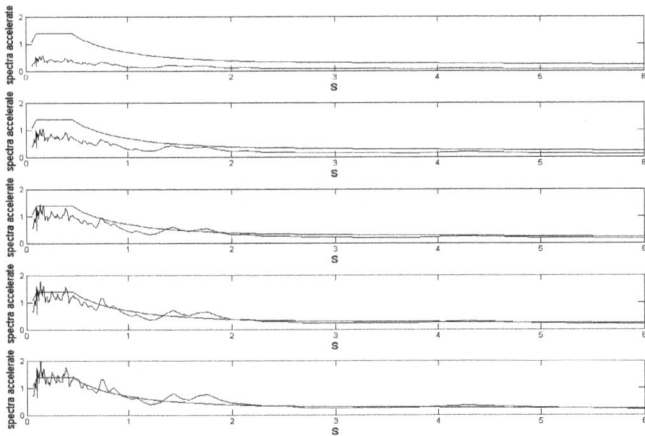

Fig. 6. Updating process of artificial response spectrum

## 4. Proposed damage index of frame buildings

At the moment seismic theory of China is more emphasis in prevent buildings collapse instead of prevent structural damage, first aim of buildings design is prevent collapse in major earthquake but obvious plastic deformation and damage is evitable, and key point is how to assure structural damage within acceptable degree based moment code, especially many lifeline constructions for example: nuclear energy station, bridge, tunnel, gymnasium and important factory buildings, so design based performance now have been approved widely, and content of structural vulnerability index is significant part of design based performance.

Damage index in early research mainly emphasis in member ductility, ductility ratio could be define as curvature $\mu_\phi$, rotation $\mu_\theta$ and deformation $\mu_\delta$,

$$\mu_\phi = \frac{\phi_m}{\phi_y} \quad \mu_\theta = \frac{\theta_m}{\theta_y} \quad \mu_\delta = \frac{\delta_m}{\delta_y} \tag{10}$$

Where $\phi_m$, $\theta_m$ and $\delta_m$ are respectively maximum curvature, maximum rotation and maximum deformation of member end, $\phi_y$, $\theta_y$ and $\delta_y$ are respectively yielding curvature, yielding rotation and yielding deformation.

(Banon,1982) also proposed a correctional ductility parameter that considering structural rigid and strength degeneration in same time,

$$FDR = \frac{K_o}{K_m} \tag{11}$$

Where $K_o$ is initial tangent stiffness, $K_m$ is equivalent tangent stiffness after maximum reaction. But in fact whether ductility ratio or correctional ductility ratio all could not give accurate judgment for structural damage and invalid.

Another damage index is relative storey displacement or relative storey rotation, in which relative storey rotation is widely accepted as estimate damage index at present in fragility research, in many research articles the index have taken as fragility standard in reinforced concrete buildings, and that relate to compatibility and effective in judging structural damage, that is to say that the index has characteristic in revealing both whole damage and local damage, and more better express structural whole damage compare with building bearing capacity, HAZUS99 series damage evaluation manual of US also adopted relative storey rotation as seismic vulnerability index.

Above both index could not reflect material accumulative damage effect, so other scholars lately proposed some index with accumulative damage effect, in which Park-Ang damage index is widely influence, the index is consist of two parts as shown in Eq. (12) , and one is transformation effect and second is absorbing energy effect.

$$D = \frac{\delta_m}{\delta_u} + \beta_e \frac{\int dE}{F_y \delta_u} \tag{12}$$

Where $D$ damage parameter, $\delta_m$ is maximum seismic transformation, $\int dE$ is accumulative absorbing hysteresis energy, $\delta_u$ is ultimate deformation under monotonic loading, $F_y$ is yielding strength with longitudinal steel bars, $\beta_e$ is constant damage parameter under considering hysteresis loading, which is connect with ratio of shear span to effective depth of section, axial-load ratio, ratio of longitudinal reinforcement and stirrup ratio.

Damage parameter $D$ is function with structural maximum deformation $\delta_m$ and whole hysteresis energy $\int dE$ , which is connect to loading time history, while the quantities $\beta_e$, $\delta_u$ and $F_y$ are independent of the loading history and are determined from experimental tests (Moustafa,2011), $D \geq 1$ means that buildings damage completely and could not bear loading.

$$\beta_e = \left[ 0.37 n_0 + 0.36 \left( k_p - 0.2 \right)^2 \right] 0.9^{100 \rho_c} \tag{13}$$

Where $n_0 = N/(bdf'_c)$ is standard axial loading, which is 0.05 when it less than 0.05, $\rho_c$ is stirrup ratio, which is stirrup volume compare to core concrete volume, and is 0.004 when less than 0.004, $k_p = \rho_p f_{yp}/(0.85 f'_c)$ is standard reinforcement ratio.

$$\delta_u = 0.0052\left(\frac{l_s}{d}\right)^{0.93}\left(0.85k_p\right)^{-0.27}\left(\frac{100\rho_c}{n_0}\right)^{0.48}\left(\frac{f'_c}{0.6895}\right)^{-0.15}l_c \tag{14}$$

Where $l_s$ is span length with shearing, $l_c$ is respective member length or height, $f'_c$ is uniaxial concrete compression strength, unit is $kN/cm^2$, $d$ is effective height, units is $cm$.

In 1985, Park suggested that set $D = 0.4$ as limit line between repairable and irreparable, in 1987, he proposed more detailed qualitative classified chart of concrete buildings' damage, as shown in Tab. 3.

| $D < 0.1$ | Undamaged or localized trivial crack |
|---|---|
| $0.1 \le D < 0.25$ | Light damage – trivial crack throughout |
| $0.25 \le D < 0.4$ | Moderate damage-severe crack and localized buckling |
| $0.4 \le D < 1$ | Severe damage-concrete crashing and reinforcing bars exposure |
| $D \ge 1$ | collapse |

Table 3. Qualitative damage description for concrete buildings (Park & Ang, 1985)

Strength and stiffness degeneration is main characteristic of material and structure damage, because of high variability of stiffness degeneration in cylinder-load lead to difficult in actual application, in 2005 Colombo-Negro proposed modified Park-Ang damage parameter model corresponding to strength degeneration, and which is defined:

$$\begin{aligned} D &= 1 - \frac{M_{ac}}{M_{yo}} \\ &= 1 - \left(\left(1-\frac{\mu_{max}}{\mu_u}\right)^{1/\beta_1} \cdot 0.5\left(1-\tanh\left(\beta_2\frac{\int dE}{E_u^*} - \pi\right)\right) \cdot \exp\left(-\beta_3\frac{\int dE}{E_u^*}\right)\right) \end{aligned} \tag{15}$$

Where $M_{ac}$ is yield force or moment actual degradation value, $M_{yo}$ is force or moment of theoretical yielding point on skeleton envelop curve, $\mu_{max}$ is attainable deformation ductility, $\mu_u$ is ultimate ductility, $E_u^*$ is ultimate cylinder energy, $\beta_1$ is coefficient of harden slope or soften slope in stress-strain curve, $\beta_2$ is consuming energy point in structure damage model when resistance force sloping, $\beta_3$ is strength dropping ratio in fragility damage model, as shown in Tab. 4.

| coefficient | (a) | (b) | (c) | (d) | (e) |
|---|---|---|---|---|---|
| Ductility-based strength decay parameter $\beta_1$ | 0.10 | 0.10 | 0.10 | 0.15 | 0.15 |
| Energy-based ductility strength decay parameter $\beta_2$ | 2.40 | 3.20 | 7.00 | 0.10 | 0.10 |
| Energy-based brittle strength decay parameter $\beta_3$ | 0.10 | 0.10 | 0.10 | 9.00 | 2.20 |

(a) Damage behavior of well confined reinforced concrete columns. (b) Damage behavior of concrete-filled steel rectangular columns. (c) Damage behavior of welded steel joints. (d) Damage behavior of poorly confined reinforced concrete columns. (e) Damage behavior of shear-deficient reinforced concrete walls.

Table 4. Damage model coefficient suggested value（Colombo & Negro,2005）

Where assuming system is double linear hysteresis model, post-yield stiffness is equal 3% of elastic yielding stiffness, so when ductility $= \mu$, hysteresis consumed energy is shown as Eq.(16).

$$\int dE = \frac{1}{2}\mu_y f_e - \mu_y f_y (\mu - \frac{1}{2}) \tag{16}$$

Where $\mu_y$ is yielding deformation, $f_e$ is yielding strength of equivalent linear system, $\frac{1}{2}\mu_y f$ is energy of equivalent linear system, $\mu_y f_y(\mu - \frac{1}{2})$ is energy of plastic system, ultimate hysteresis energy is

$$E^* = \mu_y f_e (\frac{1}{2} - \mu_u \overline{f}_{yu} + \frac{1}{2}\overline{f}_{yu}) \tag{17}$$

Where $\overline{f}_{yu}$ ultimate standard yielding strength, and assumed as 0.8, in according with above equations that could compute Colombo-Negro adjusted damage coefficient of specific structure

| Damage Scale | Value Range |
|---|---|
| Light damage | 0.01-0.10 |
| Moderate damage | 0.10-0.40 |
| Severe damage | 0.40-0.70 |
| collapse | >0.70 |

Table 5. Colombo-Negro suggested revised damage index for different limit states

One of primary object for calculate damage parameters is decided level of limit-value of different damage coefficient, and provide vulnerability scale for later seismic vulnerability analysis.

## 5. Finite element and frame modeling

In order to assess the adequacy of results obtained from Finite Element (FE) analyses have been carried out. The reinforced concrete models have been conducted with SeismoStruct

V5.0, a fibre-element based professional 3D program for seismic analysis of framed structures, which can be freely downloaded from the Internet. The program is capable of predicting the large displacement behavior and the collapse load of framed structures under static or dynamic loading, duly accounting for geometric nonlinearities and material inelasticity. Section with fibre-element of column or beam is shown as Fig.7, and 200 fibre elements fulfill calculated requirement.

Distributed fibre inelasticity elements are becoming widely employed in earthquake engineering applications, Whilst their advantages in relation to nonlinear constant-confined concrete model theory proposed by (Madas,1992), (Martinez-Rueda J.E. & Elnashai A.S.,1997) in 1990s, which uniaxial stain-stress relationship shown as Fig.8a, skeleton envelop curve could reflect confinement effect and hysteresis-stiffness degenerated characteristic. Steel stress-strain relationship proposed by (Menegotto and Pinto,1973), coupled with the isotropic hardening rules proposed by (Filippou,1983). The current implementation follows that carried out by (Monti & Nuti, 1992). Its employment should be confined to the modelling of reinforced concrete structures, particularly those subjected to complex loading histories, where significant load reversals might occur. As discussed by (Prota et al.,2009) , with the correct calibration, this model, initially developed with ribbed reinforcement bars in mind, can also be employed for the modelling of smooth rebars, often found in existing structures, as shown in Fig.8b.

In the article frame models proposed by (Ghobarah et al.,1999), who had been calculated on the based above fibre element, results coincide with results from general finite element program Drain-2D, in the same time, its accuracy in predicting the seismic response of reinforced concrete structures has been demonstrated through comparisons with experimental results derived from pseudo dynamic tests carried out on large-scale models (Casarotti et al.,2005).

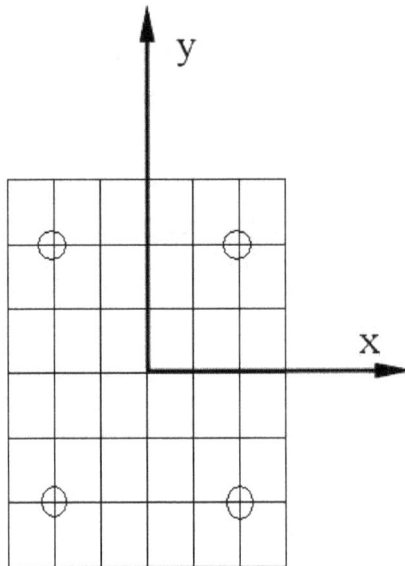

Fig. 7. Fibre in Section

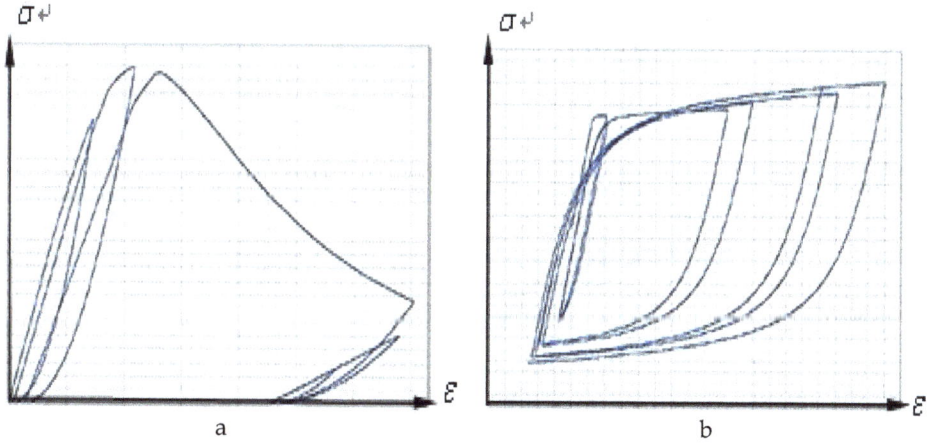

Fig. 8. Material stress-strain model (a: concrete b: steel bar)

## 6. Seismic vulnerability analysis of RC shear wall frame buildings

At present, there are still many RC buildings in Western China regions. Most of them are small high-rise (12~16 storey) RC frames in recent years. The sample buildings were designed according to the prescriptions for loading, material, member dimensioning and detailing of the seismic design and gravity load design codes in place in China in 2002. The full design of the sample RC frame in Western China is presented in Fig.9

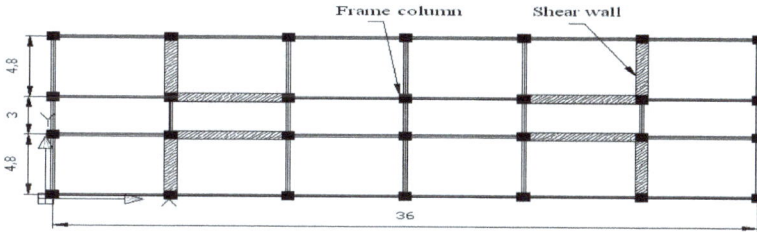

Fig. 9. Plan of typical frame-shear wall structure of 12-story

The structure consists of seven frames with bay width of 4.8m ,3m ,4.8m respectively and frame spacing of 6m. It is symmetrical in plan and elevation, and RC beam section is 0.60m×0.30m, at the same time RC column is 0.60m×0.60m, which connected beams around the building at the corner and intersect with inner columns of building. Thickness of RC shear-wall is 0.3m and floors consist of cast-in place reinforced concrete slabs is 120 mm thick.

Diameter longitudinal reinforcements of columns and   beams' section   are revealed in Fig.10, 8mm diameter stirrups must be spaced 100mm apart at the extremes and 200mm at the centre of the elements. The stress method used for the design is according to China seismic code.

The gravity load scenario consists of dead load and live load. When calculating the dead load, the weight of the structural members and the infill walls was included. The live load used was 2.5 kN/m², which is typical for school building. Other types of loading, such as wind and snow, were not considered.

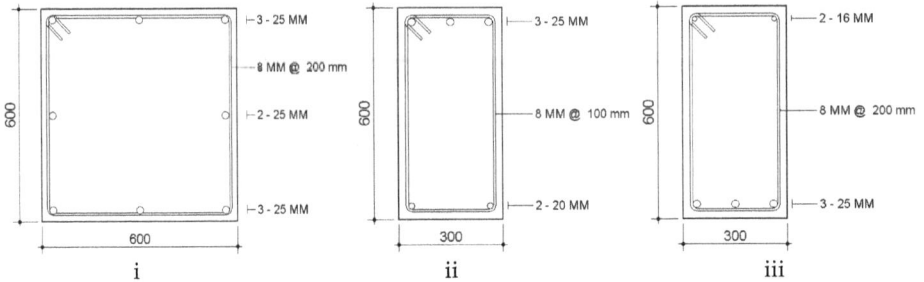

Fig. 10. Reinforced bars of section (i: column ii: side-beam iii:mid-beam )

Because of complicated climate in China all of year, the concrete strength must at least C30 according to China seismic code, reinforced bar (HPB235 HRB335) strength is 235 and 335 Mpa respectively, concrete strength is 30 Mpa, the stress-stain relationship are illustrated in Fig.8.

Ground motion characteristics have a significant effect on the vulnerability curves and special attention is required during the record selection phase in Fig.11.

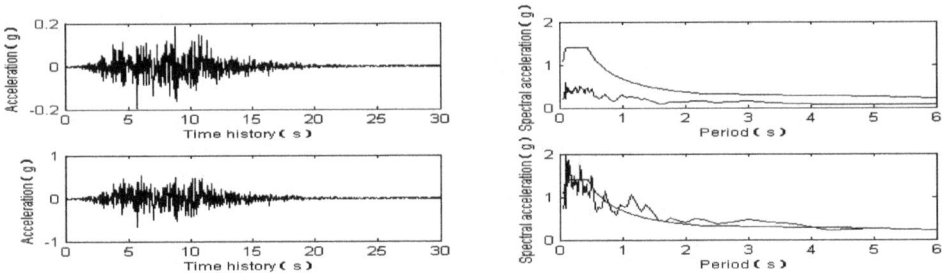

Fig. 11. Stochastic record compatible with the target spectrum of China code

## 6.1 Determination of limit states

The aim of this analysis is to evaluate the building's potential seismic performance. Four limit state conditions have been taken into account: light damage, moderate damage, extensive damage and complete damage. So if a building deformation beyond the extensive damage limit state it might not be economically advantageous to repair the building because many of the school buildings of Sichuan region were being set up without carefully thought of structural seismic codes of China.

| Limit state | Inter-storey drift (mm) | Inter-storey drift ratio(%) | Limit state | Inter-storey drift (mm) | Inter-storey drift ratio(%) |
|---|---|---|---|---|---|
| Light | 6.6 | 0.20 | Extensive | 33.0 | 1.00 |
| Moderate | 13.2 | 0.40 | Complete | 82.5 | 2.00 |

Table 6. Limit states and corresponding inter-storey drifts ratios(ISD%)

## 6.2 Vulnerability curve

Dynamic time-history analysis is used to evaluate the seismic response and to derive the vulnerability curve. This approach is the most tedious but it is also the more accurate way to assess the vulnerability of RC buildings in China. The selected frame was subject to each group of the stochastic artificial records.

The stochastic damage scatter diagram and damage versus hazard relationship of the typical RC frame is illustrated in Fig.12. The damage axis (y-axis) described as the hazard axis (x-axis) is described as spectral acceleration pga. Each vertical line of scattered data corresponds to an intensity level. The horizontal lines in the figure represent the limit states used in this study and described in terms of ISD%.

A statistical distribution is fitted to the data for each intensity level on each vertical line. The normal parameters, the mean $M_{ds}$ and standard deviation $\beta_{ds}$ of the damage state are calculated for each of these $S_k$ intensity levels. At each intensity level, the probability of exceeding each limit state is calculated. LS1, LS2, LS3 and LS4 represent the limit states for light, moderate, extensive and complete damage, respectively, as mentioned above. The mean and standard deviation values of the response data are also given in the Fig.13.

The probability of exceedance of a certain limit state is obtained by calculating the area of the standard normal distribution over the horizontal line of that limit state.

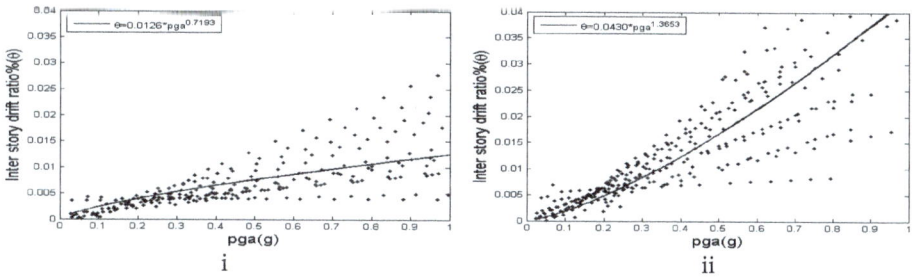

Fig. 12. Damage scatter dots relationship with PGA-ISD%$_{max}$  (i:X axis; ii:Y axis)

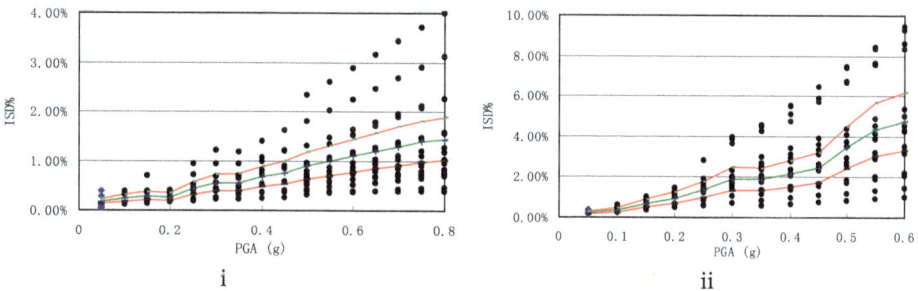

Fig. 13. Damage column dots plot with PGA-ISD%$_{max}$  (i:X axis; ii:Y axis)

After calculating the probability of exceedance of the limit state for each intensity level, the vulnerability curve can be constructed by plotting the calculated data versus spectral acceleration. In this study, a standard deviation fit is assumed as 0.3.

Figure.14 represents the vulnerability curves of typical frame-shear wall buildings in China with different spectral characteristic parameter. The curves become flatter as the nature of the statistical distribution of the response data. Vertical curves would represent deterministic response.

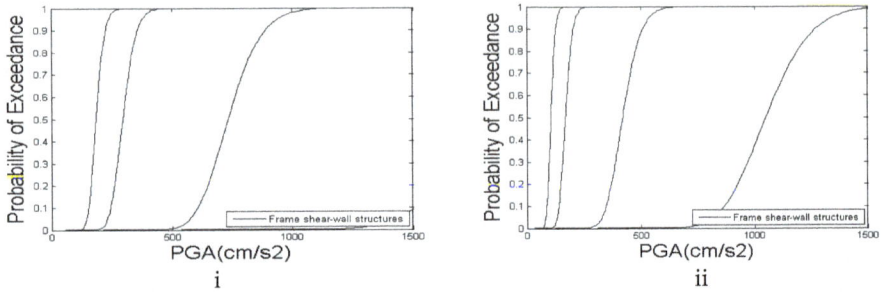

Fig. 14. Vulnerability curves with respected PGA (i : X axis; ii : Y axis)

## 7. Seismic vulnerability analysis of structures with mid-story seismic isolation and reduction (MIRS)

At present, there are many new built RC buildings in China large cities because of rapid developing economy. The buildings are designed according to the prescriptions for loading, material, member dimensioning and detailing of the seismic design and gravity load design codes in place in China in 2002.

The full design of the typical 12 stories MIRS in China is presented in Fig.15.

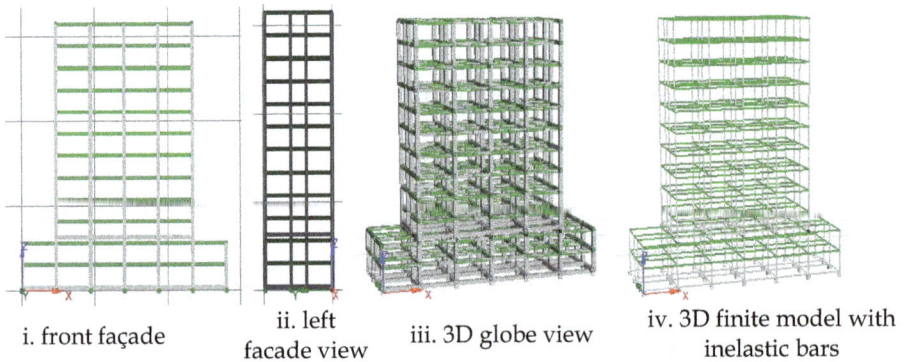

i. front façade     ii. left façade view     iii. 3D globe view     iv. 3D finite model with inelastic bars

Fig. 15. The full design of the typical MIRS in China

The structure consists of seven frames with bay width of 4.8m ,3m ,4.8m respectively and frame spacing of 6m. It is symmetrical in plan and elevation, and RC beam (0.6×0.35m2) around the exterior perimeter and along the top of interior longitudinal and transverse columns in all the floors of the building including ground base level according China code. At the same time RC column (0.6×0.6m2) which connected beams around the building at the corner and intersect with inner columns of building. Story height is 3.3m, floors consist of cast-in place reinforced concrete slabs is 120 mm thick. Reinforcement of section of MIRS is shown in Fig.16.

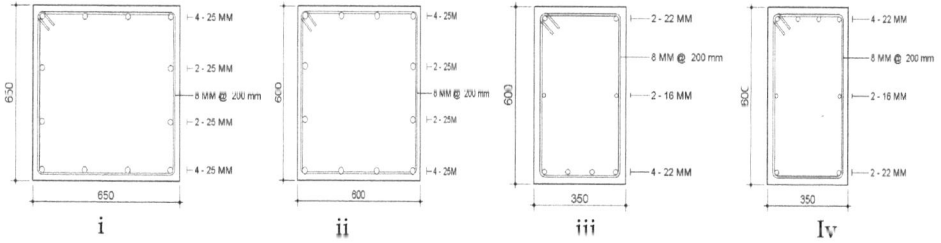

Fig. 16. Reinforcement of section of MIRS (i: column section of subway platform ii:column section on isolation layer iii: mid-beam section iv: side-beam section)

Columns have twelve 20mm diameter longitudinal reinforcements, 8mm diameter stirrups must be spaced 100mm apart at the extremes and 200mm at the centre of the elements, beams have four tension bars and two compressive bars. The stress method used for the design is according to China seismic code.

The gravity load scenario consists of dead load and live load. When calculating the dead load, the weight of the structural members and the infill walls was included. The live load used was 2.5 kN/m², which is typical for city tall building. Other types of loading, such as wind and snow were not considered.

## 7.1 Material and member property

Because of moist climate in Sichuan district all of year, the concrete strength must at least C30 according to China seismic code, reinforced bar strength is 235 and 335 Mpa respectively, concrete strength is 30 Mpa.

Design of laminated rubber bearing is chosen LRB-G4-850-180 based China isolation design code(CECS126,2001), the key parameters of bearing are: equivalent damping ratio is 0.27, secant stiffness ratio is 0.128,initial stiffness $k_0$ is 18100 KN/m , yielding force $F_y$ is 203 KN and post yielding ratio is 0.1, laminated rubber bearing stress-strain relationship is shown in Fig.17.

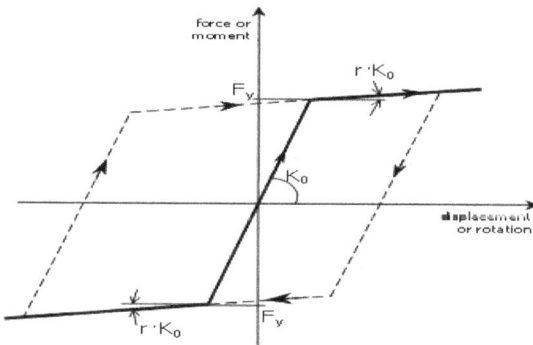

Fig. 17. Rubber bearing stress-strain model

## 7.2 Analytical model

In order to evaluating seismic vulnerability of MIRS accurately, the building is modeled as a 3-D pole frame with lumped masses the study when assessing seismic response. Inelastic frame elements means 3D beam-column elements capable of modeling members of space frames with geometric and material nonlinearities with 5% global damp coefficient.

Ground motion characteristics have a significant effect on the fragility curves and special attention is required during the record selection phase. Considering this fact, 30 corrected stochastic artificial ground motions have been used in this study from different PGA covering a wide range of characteristics with a magnitude range between 5.1 and 7.8. The motions are generally recorded on soft to medium sites according to Sichuan region's circumstance where basis prominent period range from 0.35s to 0.45s according China seismic code and crossed over the foundation in random orientation within x-direction or y-direction or both direction, as shown in Fig.11.

## 7.3 Damage level definition

The aim of this analysis is to evaluate the building's potential seismic performance, establishing a relation between the intensity of the seismic action and different damage states up to collapse. so damage level is defined according to the cracking, yielding or collapse of a set of elements or connections in the structure, as presented following Table.7 .

In view of being prone to brittle shear injury of inner brick masonry wall under strong motion, four limit state conditions have been taken into account: light damage, moderate damage, extensive damage and complete damage. So if a building deforms beyond the extensive damage limit state it might not be economically advantageous to repair the building because many of the school buildings of Sichuan region were being set up without carefully thought of structural seismic codes of China.

| Limit state | Inter-story drift (mm) | ISD ratio(%) |
|:---:|:---:|:---:|
| Light damage | 6.6 | 0.20 |
| Moderate damage | 16.5 | 0.50 |
| Extensive damage | 33.0 | 1.00 |
| Collapse | 99.0 | 3.00 |

Table 7. Limit States and corresponding inter-story drifts ratios(ISD%)

## 7.4 Fragility curve

Dynamic time-history analysis is used to evaluate the seismic response and to derive the fragility curve. This approach is the most tedious but it is also the more accurate way to assess the vulnerability of MIRS in China. The selected frame was subject to each group of the stochastic artificial records. The stochastic damage scatter diagram of the typical MIRS is illustrated in Fig.18.

A statistical distribution is fitted to the data for each intensity level on each vertical line. The normal parameters, the mean $M_{ds}$ and standard deviation $\beta_{ds}$ of the damage state are calculated for each of these Sa intensity levels. At each intensity level, the probability of exceeding each limit state is calculated. LS1, LS2, LS3 and LS4 represent the limit states for light, moderate, extensive and complete damage, respectively, as mentioned above. The mean and standard deviation values of the response data are also given lately.

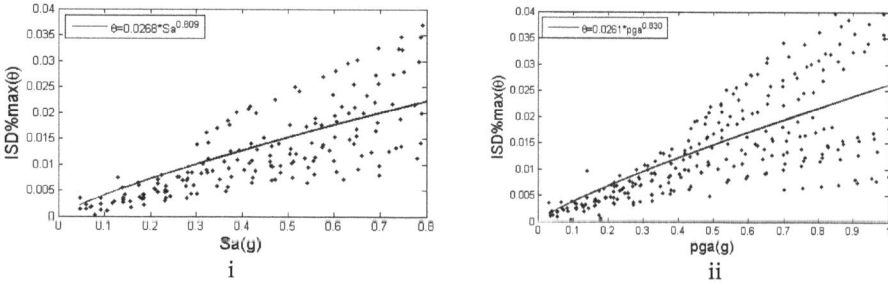

Fig. 18. Stochastic damage scatter diagram for MIRS(i : Sa; ii: PGA)

The probability of exceedance of a certain limit state is obtained by calculating the area of the standard normal distribution over the horizontal line of that limit state.

After calculating the probability of exceedance of the limit state for each intensity level, the vulnerability curve can be constructed by plotting the calculated data versus spectral acceleration. In this study, a standard deviation fit is assumed as 0.3.

Fig.19 represents the fragility curves of typical MIRS in China with different spectral characteristic parameter. The curves become flatter as the nature of the statistical distribution of the response data. Vertical curves would represent deterministic response.

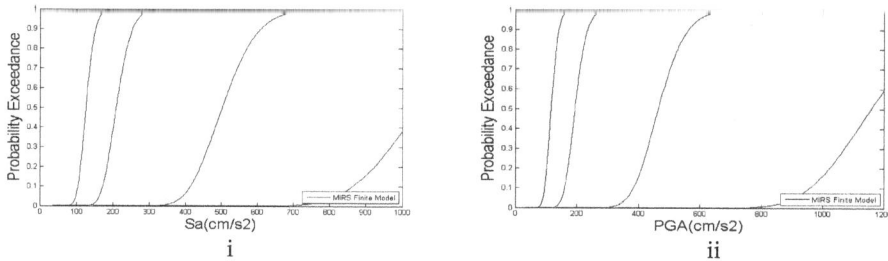

Fig. 19. Vulnerability curve for typical MIRS of China (i : Spectral accelerate; ii: PGA)

## 8. Seismic vulnerability analysis of RC industrial buildings

At present, there are many new built and old RC industrial buildings in Western China because of rapid developing economy. The buildings are designed according to the prescriptions for loading, material, member dimensioning and detailing of the seismic design and gravity load design codes of China.

The full design of the typical 12 stories MIRS in China is presented in Fig.20.

The structure consists of two frames with general configuration of bent widths and bay widths of 6m and 24m respectively, so the whole building have 66 meters long with 12 columns and 24m width, It is symmetrical in plan and elevation, and rectangular reinforced concrete ring beam (0.30  0.40m2) on the bracket of two side longitudinal columns of the building. At the same time RC column (0.40  0.60m2) which connected ring beam as confined frame element array along the exterior side of the building at the intersect with inner confined brick wall between columns. The roof of building which height selected is 9.6 m consist in steel fibres truss, the truss depth is changed from 2.4m in centre to 1.5m of two sides, there are four kind of circular hollow rod being defined to using with diameter from 0.03m to 0.05m, and thickness of bar' section is also verify from 2 to 3mm.

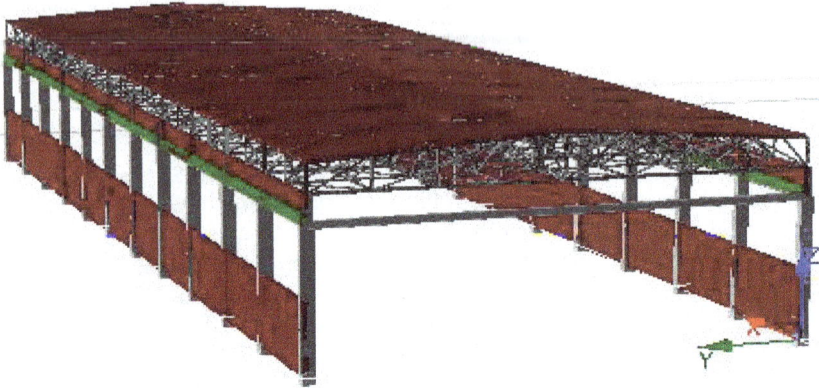

Fig. 20. The full design of the typical industrial buildings in China

The building's wall between columns generally consist of load-bearing infill masonry walls commonly made of clay bricks in Sichuan district of western China , confined by reinforced concrete vertical columns and width and thickness of wall are changed from 0.24m to 0.37m according China masonry code. Columns must have four 24mm and 28mm diameter longitudinal reinforcements, 8mm diameter stirrups must be spaced 100mm apart at the extremes and 200mm at the centre of the elements. The stress method used for the design is according to China code.

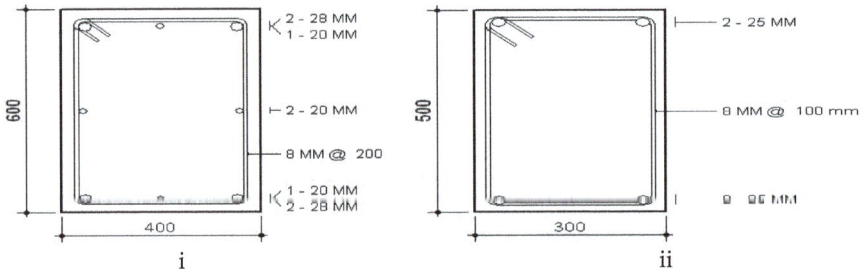

Fig. 21. RC industrial buildings' section (i: column ii: beam)

The industrial building was designed according to the China code for design of structure load for both gravity and seismic loads. The gravity load scenario consists of dead load and live load. When calculating the dead load, the weight of the structural members and the masonry infill walls was included. The roof live load used was 0.5 kN/m², which is typical for an industrial building. Other types of loading, such as wind and snow, were not considered. At the same time lifting capacity of crane is considered randomly from 22 to 66 ton base on crane span and working condition.

## 8.1 Material and member property
Because of moist climate in Sichuan district all of year, the clay brick strength must at least MU15 and the mortar strength must at least M10 according to China masonry code, so typical masonry shear strength is 0.27-1 Mpa. Bilinear stress-strain relationships with strain hardening were used for reinforced members which yield strength is 200Mpa and 300 Mpa, concrete strength is 30 Mpa in considering of that many industrial buildings in Sichuan district have been built two decades ago, some respective buildings among those built even without any consideration of horizontal seismic loads. And coefficient of variation of 30% have been considered for steel and concrete respectively. Uniaxial nonlinear constant confinement concrete model that constant confining pressure is assumed throughout the entire stress-strain range is proposed by (Madas,1992) to apply to element of concrete.

## 8.2 Generation of ground stochastic motion input
Ground motion characteristics have a significant effect on the fragility curves and special attention is required during the record selection phase. Considering this fact, 90 corrected stochastic artificial ground motions have been used in this study from different PGA covering a wide range of characteristics with a magnitude range between 5.1 and 7.8. The motions are generally recorded on soft to medium sites according to Sichuan region's circumstance where basis prominent period range from 0.35s to 0.45s according China seismic code and crossed over the foundation in random orientation within x-direction or y-direction or both direction. as shown in Fig.11.

## 8.3 Damage level definition
The aim of this analysis is to evaluate the building's potential seismic performance, establishing a relation between the intensity of the seismic action and different damage states up to collapse. so damage level is defined according to the cracking, yielding or collapse of a set of elements or connections in the structure, as presented following Table.8 .
Four limit state conditions have been taken into account: light damage, moderate damage, extensive damage and complete damage as table1 So if a building deforms beyond the extensive damage limit state it might not be economically advantageous to repair the building because many of the industrial buildings of Sichuan region were be set up without carefully thought of structural seismic codes of China.

| Limit state | Inter-story drift (mm) | ISD ratio(%) |
|---|---|---|
| Light damage | 24 | 0.25 |
| Moderate damage | 38.5 | 0.40 |
| Extensive damage | 96.0 | 1.00 |
| Collapse | 240 | 2.50 |

Table 8. Limit States and corresponding inter-story drifts ratios (ISD%)

## 8.4 Fragility curve
Dynamic time-history analysis is used to evaluate the seismic response and to derive the fragility curve. This approach is the most tedious but it is also the more direct and accurate way to assess the fragility of Sichuan industrial buildings. The selected frame with confined

masonry shear wall was subject to each group of the stochastic artificial records. Each group records were consisted of three stochastic ground motions.

There have some peak-displacement-history of original frame to be shown in Fig.22.

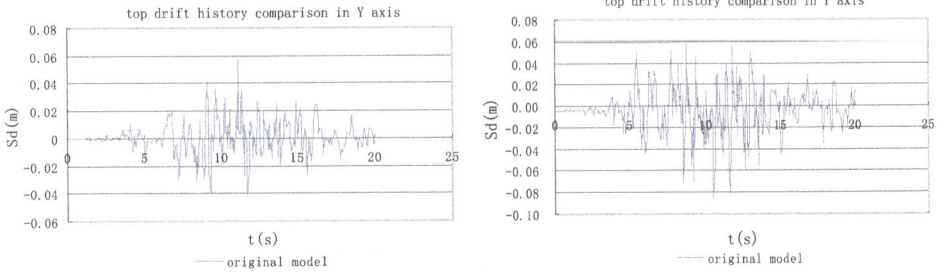

Fig. 22. Top drift time history with pga=300 gal in X and Y direction

The damage versus motion relationship is illustrated in Fig.23, The damage axis (Y-axis) described in terms of maximum inter storey drift ratio ($ISD_{max}$%) when the hazard axis (X-axis) is described as spectral acceleration (Sa). Each vertical line of scattered data corresponds to an intensity level. From bottom to top, these are four limit states as light damage to complete damage respectively. Moreover the average value of each vertical data are connect by red line when ±0.3 variance have been considered as brown lines. The results also revealed that difference of seismic capacity of structure at two directions of industrial buildings, a standard deviation fit is assumed as 0.3.

The probability of exceedance of a certain limit state is obtained by calculating the area of the standard normal distribution over the horizontal line of that limit state.

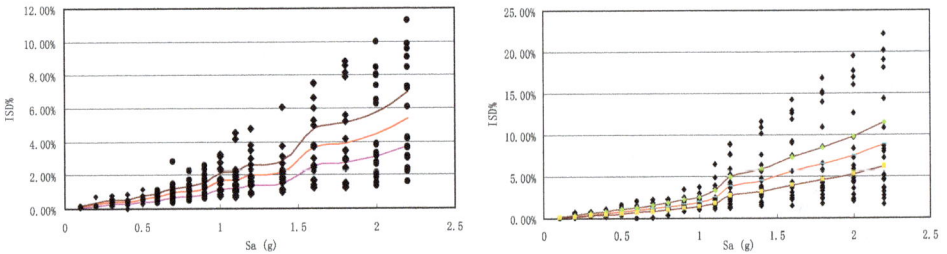

Fig. 23. Damage versus motion relationship in X and Y direction

After calculating the probability of exceedance of the limit state for each intensity level, the vulnerability curve can be constructed by plotting the calculated data versus Sa.

Fig.24 represents the fragility curves of typical industrial buildings in Western China with different spectral characteristic parameter. The curves become flatter as the nature of the statistical distribution of the response data. Vertical curves would represent deterministic response.

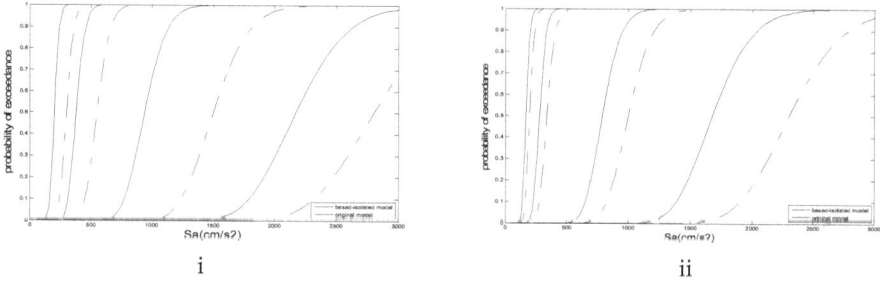

i                                                    ii

Fig. 24. Vulnerability curve with spectral acceleration (i : X direction; ii: Y direction)

## 9. Conclusion

The following conclusions have been obtained.

1. The seismic vulnerability research is actual important part of perform based seismic design (PBSD), and the research work is inevitably lasting along perform based seismic design theory. The representative response spectrum of West China was been built and fit unsteady initial stochastic record in consideration of characteristic of frequency and random angle of phase position. The actual applied stochastic wave was been created after initial wave optimized with China code spectrum.

2. The representative small high rise RC shear wall buildings were been selected and seismic vulnerability analysis been made based dynamic time-history method, the analytical vulnerability curves were been obtained firstly in homeland. Probability of moderate damage was much seldom when PGA<0.1g under foundation stability precondition, in contrary probability of moderate damage was started to increasing when PGA>0.3g. The seismic capability of longitudinal and transversal orientation of small high-rise RC-shear wall  buildings was different, probability of slight damage was coherent under minor earthquake, vulnerability of longitudinal orientation (X axis) was less than transverse orientation (Y axis) of frame under major earthquake obviously.

3. The typical MIRS in China were been modeled and their seismic fragility characteristic were studied on the base of dynamic nonlinear method. The results display better response features than general type of RC frames. The steep light damage curve reflects the roles of the infill brick panels that dominate the response in the vicinity of the light damage limit state. When the confined infill column are damaged and laminated rubber bearings began to play a key role in seicmic energy dissipation, the building reach interstory drift more flexible than before.

4. The seismic performance and vulnerability of industrial buildings were also been analyzed, seismic capacity and vulnerability of based-isolated models and original models were been compared. The seismic capability of two orientations was also different and longitudinal orientation of industrial buildings (X axis) was more capacity in seismic resistance than transverse orientation (Y axis) of buildings. Based-isolated model minimized 30%-50% of $ISD_{max}$% comparing to original model under major earthquake, so the based-isolated industrial buildings are more high seismic resistance than original industrial buildings in major earthquake.

## 10. Acknowledgement

The author was funded by the State key Program of National Natural Science Foundation of China (Grant No. 90815027), National Natural Science Foundation of China (Grant No. 51078097) and Natural Science Foundation of Ningxia Province (Grant No. NZ1156).

## 11. References

I. Takewaki (2011). Preliminary report of the 2011 off the Pacific coast of Tohoku Earthquake. *Journal of Zhejiang Univ-Sci A (Appl Phys & Eng)*, Vol.12, No.5, pp. (327-334), ISSN: 1862-1775

Whitman, R.V.; Reed, J.W. and Hong, S.T. (1973). Earthquake Damage Probability Matrices, *Proceedings of the Fifth World Conference on Earthquake Engineering*, Rome, Italy, 12,1973.

Braga, F.; Dolce, M. and Liberatore, D.A. (1982). Statistical Study on Damaged Buildings and an Ensuing Review of the MSK-76 Scale. *Proceedings of the Seventh European Conference on Earthquake Engineering*. Athens, Greece.1982.

Corsanego, A. & Petrini, V. (1990). Seismic Vulnerability of Buildings – Work in Progress. *Proceedings of the Workshop II on Seismic Risk Vulnerability and Risk Assessment*, Trieste, Italy, 1990.

Di Pasquale, G.; Orsini, G. and Romeo, R.W. (2005). New Developments in Seismic Risk Assessment in Italy, *Bulletin of Earthquake Engineering*, Vol. 3, No. 1, (2005), pp. (101-128), ISSN:1573-1456

Dolce, M.; Masi, A.; Marino, M. and Vona, M. (2003). Earthquake Damage Scenarios of the Building Stock of Potenza (Southern Italy) Including Site Effects, *Bulletin of Earthquake Engineering*, Vol. 1, No. 1, (2003), pp.(115-140), ISSN:1573-1456

Grünthal, G. (editor) (1998). *Cahiers du Centre Européen de Géodynamique et de Séismologie: Volume 15 – European Macroseismic Scale 1998*, European Center for Geodynamics and Seismology, ISBN : 2-87977-008-4,Luxembourg.

Yamazaki, F. & Murao, O. (2000). *Vulnerability functions for Japanese buildings based on damage data from the 1995 Kobe earthquake. Implication of recent earthquakes on seismic risk: Series on Innovation and Construction*, Vol. 2, Imperial College Press, ISBN: 978-1-86094-233-4, London

Orsini, G. (1999). A Model for Buildings' Vulnerability Assessment Using the Parameterless Scale of Seismic Intensity (PSI). *Earthquake Spectra*, Vol. 15, No. 3, (1999), pp. (463-483), ISSN: 8755-2930

M. A. Erberik & Amr S. Elnashai. (2004). Fragilityanaly sis of flat-slab structures. *Engineering Structures*, Vol.26, pp. (937–948), ISSN: 0141-0296

T. Rossetto & Amr Elnashai. (2005).A new analytical procedure for the derivation of displacement-based vulnerability curves for populations of RC structures. *Engineer Structures*, Vol.27, pp.(397-409), ISSN: 0141-0296

M. S. Kircil & Z. Polat. (2006), Fragility analysis of mid-rise R/C frame buildings. *Engineering Structures*, Vol.28, pp.(1335–1345), ISSN: 0141-0296

Jun Ji ; Amr S. & Elnashai. (2007). Daniel A. Kuchma, An analytical framework for seismic fragility analysis of RC high-rise buildings. *Engineering Structures*, Vol.29, pp.(3197-3209), ISSN: 0141-0296

B. Borzi.; R. Pinho & H. Crowley.(2008). Simplified pushover-based vulnerability analysis for large-scale assessment of RC buildings. *Engineer Structures*, Vol.30, pp.(804-820), ISSN: 0141-0296

Kaul, M.K. (1978). Stochastic Characterization of Earthquake through Their Response Spectrum. *Earthquake Engineering and Structural Dynamics*, Vol.6, No.5, pp.(497-509), ISSN: 0098-8847

Murphy JR & O'Brien LJ. (1977).The correlation of peak ground acceleration amplitude with seismic intensity and other physical parameters. *Bull Seismol Soc Am*, Vol.67,pp.(877–915), ISSN: 0037-1106

Husid,R.L. (1969)Analisis de Terremotos: Analisis General, Revista de IDIEM. *Santiago Chile*, Vol.8, No.1,pp.(21-42), ISSN:0716-1832

Trifunacc,M.D & Brady,A.G. (1975). On the Correlation of Peak Accelaration of Strong Motion with Earthquake Magnitude Epicentral Distance and Site Condition Proc.US Nat.Conf. *Earthquake Engineering*, pp.(43-52), ISSN: 1570-761X

Takizawa,H.et.al. (1980).*Collapse of a Modal for Ductile Reinforced Concrete Frames under Extreme Earthquake Motion*. EESD, Vol.8, No.2, ISSN: 1521-334X

Jennings PC. & Housner GW. (1968). *Tsai NC Simulated earthquake motions. Technical Report*. Earthquake Engineering Research Laboratory, California Institute of Technology

Ministry of Construction P.R. China.(2002). *Code for seismic design of buildings (GB50011-2001)*. Beijing: China Construction Press

China Association for Engineer construction standardization.(2001).*Technical specification for seismic-isolation with laminated rubber beating isolators CECS126:2001*. Beijing: China Construction Press.

Banon,H. & Veneziano,D. (1982). Seismic Safety of Reinforced Concrete Members and Structures. *Earthquake Engineering and Structural Dynamics*, Vol.10, pp.(179-193), ISSN: 0098-8847

Park, Y.J. & Ang, A.H.S. (1985). Mechanistic Seismic Damage Model for Reinforced Concrete. *Journal of Structural Engineering*, Vol. 111, No. 4, pp.(722-739), ISSN: 0970-0137

Colombo,A & Negro,P. (2005). A damage index of generalised applicability. *Engineer Structures*, Vol.27, pp.(1164-1174), ISSN: 0141-0296

Madas, P. and Elnashai, A.S. (1992). A new passive confinement model for transient analysis of reinforced concrete structures. *Earthquake Engineering and Structural Dynamics*, Vol. 21, pp. (409-431), ISSN: 0098-8847

Martinez-Rueda, J.E. & Elnashai, A.S. (1997). Confined concrete model under cyclic load. *Materials and Structures*, Vol. 30, No. 197, pp. (139-147), ISSN: 1359-5997

Menegotto, M., Pinto, P.E. (1973). Method of analysis for cyclically loaded RC plane frames including changes in geometry and non-elastic behaviour of elements under combined normal force and bending, *Symposium on the Resistance and Ultimate Deformability of Structures Acted on by well defined Repeated Loads*, International Association for Bridge and Structural Engineering, Zurich, Switzerland, pp. (15-22)

Filippou, F.C.& Popov, E.P., Bertero, V.V. (1983) Modelling of R/C joints under cyclic excitations, *ASCE Journal of Structural Engineering*, Vol. 109, No. 11, pp. (2666-2684), ISSN: 0970-0137

Monti, G. & Nuti, C. (1992). Nonlinear cyclic behaviour of reinforcing bars including buckling, *Journal of Structural Engineering*, Vol. 118, No. 12, pp.( 3268-3284), ISSN: 0970-0137

Prota, A., Cicco, F., Cosenza, E.(2009). Cyclic behavior of smooth steel reinforcing bars: experimental analysis and modeling issues, *Journal of Earthquake Engineering*, Vol. 13, No. 4, pp.( 500–519), ISSN: 1363-2469

A. Ghobarah (1999). Response-based Damage Assessment of Structures, *Earthquake Engineering and Structural Dynamics*, Vol. 28, No. 1, pp.(79–104), ISSN: 1096-9845

Casarotti, C; Pinho, R & Calvi, GM.(2005). *Adaptive pushover-based methods for seismic assessment and design of bridge structures, ROSE Research Report 2005/06*. Pavia, Italy: IUSS Press, ISBN: 0-646-00004-7

Ohsaki, Y.(1978). On the significance of Phase Content in Earthquake Ground Motions. *Earthquake Engineering and Structural Dynamics*, Vol. 7, No.5, ISSN: 0098-8847

Amin, M. & Ang, A. S. (1968). Non-stationary Stochastic Model of Earthquake Motions. *Journal of American Society of Civil Engineers*, Vol.94, No.EM2, ISSN: 1090-0241

A.Moustafa (2011). Damage-Based Design Earthquake Loads for Single-Degree-Of -Freedom Inelastic Structures. *Journal of structural engineering (ASCE)*, Vol.137, No.3, pp. (456-467), ISSN: 0733-9445

# Permissions

The contributors of this book come from diverse backgrounds, making this book a truly international effort. This book will bring forth new frontiers with its revolutionizing research information and detailed analysis of the nascent developments around the world.

We would like to thank Prof. Abbas Moustafa, for lending his expertise to make the book truly unique. He has played a crucial role in the development of this book. Without his invaluable contribution this book wouldn't have been possible. He has made vital efforts to compile up to date information on the varied aspects of this subject to make this book a valuable addition to the collection of many professionals and students.

This book was conceptualized with the vision of imparting up-to-date information and advanced data in this field. To ensure the same, a matchless editorial board was set up. Every individual on the board went through rigorous rounds of assessment to prove their worth. After which they invested a large part of their time researching and compiling the most relevant data for our readers. Conferences and sessions were held from time to time between the editorial board and the contributing authors to present the data in the most comprehensible form. The editorial team has worked tirelessly to provide valuable and valid information to help people across the globe.

Every chapter published in this book has been scrutinized by our experts. Their significance has been extensively debated. The topics covered herein carry significant findings which will fuel the growth of the discipline. They may even be implemented as practical applications or may be referred to as a beginning point for another development. Chapters in this book were first published by InTech; hereby published with permission under the Creative Commons Attribution License or equivalent.

The editorial board has been involved in producing this book since its inception. They have spent rigorous hours researching and exploring the diverse topics which have resulted in the successful publishing of this book. They have passed on their knowledge of decades through this book. To expedite this challenging task, the publisher supported the team at every step. A small team of assistant editors was also appointed to further simplify the editing procedure and attain best results for the readers.

Our editorial team has been hand-picked from every corner of the world. Their multi-ethnicity adds dynamic inputs to the discussions which result in innovative outcomes. These outcomes are then further discussed with the researchers and contributors who give their valuable feedback and opinion regarding the same. The feedback is then collaborated with the researches and they are edited in a comprehensive manner to aid the understanding of the subject.

Apart from the editorial board, the designing team has also invested a significant amount of their time in understanding the subject and creating the most relevant covers. They scrutinized every image to scout for the most suitable representation of the subject and create an appropriate cover for the book.

The publishing team has been involved in this book since its early stages. They were actively engaged in every process, be it collecting the data, connecting with the contributors or procuring relevant information. The team has been an ardent support to the editorial, designing and production team. Their endless efforts to recruit the best for this project, has resulted in the accomplishment of this book. They are a veteran in the field of academics and their pool of knowledge is as vast as their experience in printing. Their expertise and guidance has proved useful at every step. Their uncompromising quality standards have made this book an exceptional effort. Their encouragement from time to time has been an inspiration for everyone.

The publisher and the editorial board hope that this book will prove to be a valuable piece of knowledge for researchers, students, practitioners and scholars across the globe.

# List of Contributors

George C. Lee and Zach Liang
Multidisciplinary Center for Earthquake Engineering Research, University at Buffalo, State University of New York, USA

Junichi Abe, Hiroyuki Sugimoto and Tadatomo Watanabe
Hokubu Consultant Corporation, Hokkai Gakuen University, Japan

Francesco Castelli
Kore University of Enna, Faculty of Engineering and Architecture, Italy

Ernesto Motta
University of Catania, Department of Civil and Environmental Engineering, Italy

Chong-Shien Tsai
Department of Civil Engineering, Feng Chia University, Taichung, Taiwan

Omar A. Pineda-Porras
Energy and Infrastructure Analysis (D-4), Los Alamos National Laboratory, Los Alamos, USA

Mario Ordaz
Instituto de Ingeniería, Ciudad Universitaria, Universidad Nacional Autónoma de México (UNAM), Mexico

Thomas Zimmermann and Alfred Strauss
University of Natural Resources and Life Sciences Vienna, Institute for Structural Engineering, Austria

Triantafyllos K. Makarios
Hellenic Institute of Engineering Seismology & Earthquake Engineering, Greece

Xiaosong Ren, Pang Li, Chuang Liu and Bin Zhou
Institute of Structural Engineer and Disaster Reduction, Tongji University, China

Halil Sezen and Muhammad S. Lodhi
Department of Civil and Environmental Engineering and Geodetic Science, The Ohio State University, Columbus, Ohio, USA

Mohammad Reza Tabeshpour and Amir Azad
Mechanical Engineering Department, Iran

**Ali Akbar Golafshani**
Civil Engineering Department, Sharif University of Technology, Tehran, Iran

**Zhu Jian**
Ning Xia University, China